Virtual Experiments in Mechanical Vibrations

Virtual Experiments in Mechanical Vibrations

Structural Dynamics and Signal Processing

Michael J. Brennan
Department of Mechanical Engineering
São Paulo State University (UNESP)
Ilha Solteira
Brazil

Bin Tang
School of Energy and Power Engineering
Dalian University of Technology
China

This edition first published 2023
© 2023 John Wiley & Sons Ltd

The right of Michael J. Brennan and Bin Tang to be identified as the authors of this work has been asserted in accordance with law.

Registered Office
John Wiley & Sons Ltd, The Atrium, Southern Gate, Chichester, West Sussex, PO19 8SQ, UK

Editorial Office
9600 Garsington Road, Oxford, OX4 2DQ, UK

For details of our global editorial offices, customer services, and more information about Wiley products visit us at www.wiley.com.

Wiley also publishes its books in a variety of electronic formats and by print-on-demand. Some content that appears in standard print versions of this book may not be available in other formats.

Library of Congress Cataloging-in-Publication Data Applied for:

Hardback ISBN: 9781118307977

Cover Design: Wiley
Cover Image: © teekid/Getty Images

Set in 9.5/12.5pt STIXTwoText by Straive, Chennai, India
Printed and bound by CPI Group (UK) Ltd, Croydon, CR0 4YY

C9781118307977_200922

To Our Wives Laura and Xiudan, and Our Children Emma, Jingde, and Jinghui

"All models are wrong, but some are useful"

George Box

Contents

Preface

The idea to write this book came about from many years of interacting with students, both under-graduate and postgraduate. There seemed to be a disconnect between the theoretical treatment of mechanical vibrations and the signal processing procedures needed to measure vibration in the laboratory. They are often treated as separate subjects, sometimes taught in different departments by different lecturers. When the first author of the book came to UNESP Ilha Solteira in Brazil at the end of 2010, he decided to teach a course that combined the two approaches. The notes developed for that course form the basis of this book.

At the beginning of 2010 Bin Tang came as an academic visitor, supported from the China Scholarship Council (Grant No. 2009821053), to the Institute of Sound and Vibration Research (ISVR) in Southampton, UK, where Mike Brennan had a position as professor of engineering dynamics. They worked together for about one year on research related to nonlinear vibrations. Bin Tang then returned to his position as an assistant professor at Dalian University of Technology (DUT), and Mike departed for Brazil. The following year Mike visited Bin Tang in DUT, and about two years later, Bin Tang came to Brazil as an academic visitor, supported by the Brazilian National Council for Scientific and Technological Development (CNPq). He stayed for two years, and during this time they had many discussions about the topics in this book, honing the ideas and concepts. A decision was made to write the book, but this never really began in earnest until the COVID 19 pandemic struck in 2020. This curtailed the much-enjoyed academic activity of travelling and meeting colleagues around the world, and freed up some time to work on the book.

The authors are extremely grateful for the many discussions with both colleagues and students over the years that have helped to form the perspective from which the book is written. The authors would like to acknowledge the financial support of the Brazilian National Council for Scientific and Technological Development (CNPq), (Grant No. 401360/2012-1) and the National Natural Science Foundation of China (Grant No. 11672058). It is hoped that students who are new to the topic, or those who are more experienced in some areas of either vibration or signal processing will find the book useful.

Michael J. Brennan
São Paulo State University (UNESP)
Ilha Solteira
Brazil
Bin Tang
Dalian University of Technology
China
January 2022

List of Abbreviations

CPSD	cross power spectral density
DFT	discrete Fourier transform
DOF	degrees-of-freedom
DTFT	discrete time Fourier transform
Env	envelope
ESD	energy spectral density
FEM	finite element method
FFT	fast Fourier transform
FRF	frequency response function
FS	Fourier series
FT	Fourier transform
IDFT	inverse discrete Fourier transform
IDTFT	inverse discrete time Fourier transform
IFT	inverse Fourier transform
Im	imaginary part
IRF	impulse response function
ln	natural logarithm
LTI	linear time-invariant
MDOF	multi-degree-of-freedom
ODE	ordinary differential equation
PDE	partial differential equation
PSD	power spectral density
Re	real part
rms	root mean square
SDOF	single-degree-of-freedom
SIMO	single input multiple outputs
SISO	single input single output
SNR	signal-to-noise ratio

List of Symbols

Symbol	Description	Units
$a(t)$	Analytic signal (displacement)	[m]
$(A_{n,l})_p$	Modal constant for the p-th mode	[1/kg]
c	Viscous damping coefficient	[Ns/m]
c_B	Phase speed in a beam	[m/s]
c_R	Phase speed in a rod	[m/s]
$\mathbf{C}$	Damping matrix	[Ns/m]
$\tilde{\mathbf{C}}$	Modal damping matrix	[Ns/m]
E	Young's modulus	[N/m^2]
E	Expectation operator	
f	Frequency	[Hz]
$f_c(t)$	Damping force	[N]
$f_e(t)$	Excitation force	[N]
$\hat{f}_e$	Force impulse	[Ns]
$f_i(t)$	Force applied at point i	[N]
$f_k(t)$	Stiffness force	[N]
$f_m(t)$	Inertia force	[N]
f_n	Natural frequency	[Hz]
f_s	Sampling frequency	[Hz]
$\mathbf{f}(t)$	Vector of forces	[N]
$\bar{\mathbf{f}}(\mathrm{j}\omega)$	Vector of complex force amplitudes	[N]
$\mathcal{F}$	Fourier transform operator	
$\mathcal{F}^{-1}$	Inverse Fourier transform operator	
$\bar{F}$	Complex force amplitude	[N]
$\lvert\bar{F}\rvert$	Force amplitude	[N]
$\bar{F}_c$	Complex damping force	[N]
$\bar{F}_k$	Complex stiffness force	[N]
$\bar{F}_m$	Complex inertia force	[N]

Symbol	Description	Units
$F(j\omega)$	FT of $f_e(t)$	[N/Hz]
$\mathbf{g}$	Modal force vector	[N]
$g_p(t)$	Modal force for the p-th mode	[N]
$\tilde{G}_{xx}(f)$	Estimate of the single-sided PSD of $x(t)$	[m^2/Hz]
$h(t)$	Displacement impulse response function	[m/Ns]
$\dot{h}(t)$	Velocity impulse response function	[m/Ns^2]
$\ddot{h}(t)$	Acceleration impulse response function	[m/Ns^3]
$H(j\omega), H(f)$	Receptance FRF	[m/N]
$\mathbf{H}(j\omega)$	Receptance matrix	[m/N]
$H_{\text{vel}}(j\omega)$	Mobility FRF	[m/Ns]
$H_{\text{acc}}(j\omega)$	Accelerance FRF	[m/Ns^2]
$H_1(j\omega)$	H_1 estimator	[m/N]
$H_2(j\omega)$	H_2 estimator	[m/N]
$i(t)$	Train of delta functions	[1/s]
$i_s(t)$	Current supplied to the shaker	[A]
I	Second moment of area for the cross-section of a beam	[m^4]
j	$\sqrt{-1}$	
k	Stiffness	[N/m]
$\tilde{k}_p$	Modal stiffness of the p-th mode	[N/m]
$K(j\omega)$	Dynamic stiffness	[N/m]
$\mathbf{K}$	Stiffness matrix	[N/m]
$\tilde{\mathbf{K}}$	Modal stiffness matrix	[N/m]
m	Mass	[kg]
$\tilde{m}_p$	Modal mass of the p-th mode	[kg]
$M(j\omega)$	Apparent mass	[kg]
$\overline{M}(j\omega)$	Complex moment amplitude	[Nm]
$\mathbf{M}$	Mass matrix	[kg]
$\tilde{\mathbf{M}}$	Modal mass matrix	[kg]
$\mathbf{q}$	Vector of modal displacements	[m]
$q_p(t)$	Modal participation factor of the p-th mode	[m]
R_{ij}	Residual for the modal model	[m/N]
S	Cross-sectional area of a rod or a beam	[m^2]
$S_{ff}(\omega)$	PSD of $f_e(t)$	[N^2/Hz]
$\tilde{S}_{ff}(\omega)$	Estimate of the PSD of $f_e(t)$	[N^2/Hz]
$S_{fx}(j\omega)$	CPSD between $f_e(t)$ and $x(t)$	[Nm/Hz]
$\tilde{S}_{fx}(j\omega)$	Estimate of the CPSD between $f_e(t)$ and $x(t)$	[Nm/Hz]
$S_{xx}(f)$	PSD of $x(t)$	[m^2/Hz]
$\tilde{S}_{xx}(f)$	Estimate of the PSD of $x(t)$	[m^2/Hz]
$S_{xf}(j\omega)$	CPSD between $x(t)$ and $f_e(t)$	[Nm/Hz]
$\tilde{S}_{xf}(j\omega)$	Estimate of the CPSD between $x(t)$ and $f_e(t)$	[Nm/Hz]

Symbol	Description	Units
t	Time	[s]
T	Time duration	[s]
T_d	Damped natural period	[s]
T_n	Undamped natural period	[s]
T_p	Fundamental period of a periodic signal	[s]
$u(t)$	Heaviside function	
$u(x, t)$	Axial displacement of a rod	[m]
$\overline{U}\,(j\omega)$	Complex axial displacement amplitudes for a rod	[m]
$w(x, t)$	Lateral displacement of a beam	
$w(t), W(f)$	windows in the time domain and its FT	[m]
$\overline{W}(x)$	Complex displacement amplitude of a beam	[m]
$\overline{W}\,(j\omega)$	Complex lateral displacement amplitude for a beam	[m]
$x(t)$	Displacement	[m]
$\mathbf{x}(t)$	Vector of displacements	[m]
$\dot{x}(t)$	Velocity	[m/s]
$\dot{\mathbf{x}}(t)$	Vector of velocities	[m/s]
$\ddot{x}(t)$	Acceleration	[m/s^2]
$\ddot{\mathbf{x}}(t)$	Vector of accelerations	[m/s^2]
$\overline{\mathbf{x}}\,(j\omega)$	Vector of complex displacement amplitudes	[m]
$\overline{X}$	Complex displacement amplitude	[m]
$\left|\overline{X}\right|$	Displacement amplitude	[m]
$\left|\overline{X}_n\right|$	Amplitude of the n-th harmonic of the Fourier series	[m]
$\tilde{X}_n$	Complex amplitude of the n-th harmonic of the complex	[m/Hz]
	Fourier series	[m]
$X(j\omega), X(f)$	FT of $x(t)$	
$X_s(f)$	DTFT of $x(t)$	[m]
$X(k\Delta f)$	DFT of $x(n\Delta t)$	[m]
$Z(j\omega)$	Impedance	[Ns/m]

Greek Symbols

α	Time delay	[s]
β_{R}	Wavenumber for a rod	[1/m]
β_{B}	Wavenumber for a beam	[1/m]
$\delta(t)$	Delta function	[1/s]
$\delta(x)$	Delta function	[1/m]
Δf	Frequency resolution	[Hz]
Δt	Time resolution	[s]
ε	Time duration	[s]
ϕ, θ, ψ	Phase angle	[rad]

$\phi_p(x)$	Mode shape for the p-th mode of a rod or a beam	
$\boldsymbol{\phi}_p$	Mode shape vector for the p-th mode of a lumped parameter system	
$\boldsymbol{\Phi}$	Matrix of mode-shape vectors for a lumped parameter system	
$\gamma_{fx}^2(\omega)$	Coherence function between f and x	
γ	Ratio of absorber nat. freq. to host structure nat. freq.	
η	Structural loss factor	
μ	Mass ratio	
ρ	Density	$[\text{kg/m}^3]$
σ	Standard deviation	
ζ	Damping ratio	
ζ_p	Damping ratio of the p-th mode	
ω	Circular excitation frequency	$[\text{rad/s}]$
ω_a	Undamped natural frequency of a vibration absorber	$[\text{rad/s}]$
ω_d	Damped natural frequency	$[\text{rad/s}]$
ω_n	Undamped natural frequency or the n-th harmonic	$[\text{rad/s}]$
ω_p	Undamped natural frequency of the p-th mode	$[\text{rad/s}]$
Ω	Non-dimensional frequency	

About the Companion Website

This book is accompanied by a companion website which has MATLAB files.

www.wiley.com/go/brennan/virtualexperimentsinmechanicalvibrations

1

Introduction

1.1 Introduction

Knowledge of the dynamic behaviour of systems and structures becomes increasingly important as organisations and companies strive to produce devices and products that outperform the competition. This means that engineers from a wide range of disciplines covering, for example, automotive, acoustical, aeronautical, aerospace, civil, mechanical, and marine engineering, are required to have knowledge of vibration engineering. Of course, some will need to be experts in this discipline, but others will simply need to be aware of some basic issues. This means that university engineering programmes for all the disciplines mentioned above generally have a course in mechanical vibrations. These courses tackle the subject in different ways, depending on the particular discipline. For example, civil engineers start from the study of the static behaviour of structures. Once this has been mastered, they move to the dynamic behaviour of structures, i.e. they start at a frequency of 0 Hz, and then investigate the behaviour as frequency increases. This sequence of study is similar for many disciplines, with the exception, perhaps, of physicists and acoustical engineers, who may tackle the subject using a wave description of the structural dynamics. Acoustical engineers generally restrict their frequency range of interest to that of human hearing, which is from about 20 Hz to 20 kHz. Thus, the way in which mechanical vibration is taught may vary enormously from course to course. To illustrate the diversity of the topic, Michael Brennan, the first author of this book, started his career in vibration engineering by investigating high-frequency (>500 Hz) structure-borne noise through a helicopter gear box support strut, whereas Bin Tang, the second author of this book, started his career by investigating the relatively low-frequency torsional vibration (<30 Hz) of a ship's propellor shaft.

The terms 'mechanical vibration' and 'engineering vibration' are used interchangeably in this book. To master this topic from a theoretical and a practical point of view, the student is required to have some knowledge of physics, mathematics, and engineering. This is illustrated schematically in the Venn diagram shown in Figure 1.1. It is acknowledged that not all vibration engineers have the same profile. For example, some have a much more mathematical bias, focusing on theoretical aspects of the subject, perhaps working as researchers in universities, and others follow a much more practical career, working on the implementation of vibration control strategies in consulting or engineering companies. Notwithstanding this, it is the firm belief of the authors, that engineers/researchers will only gain mastery of the topic, if their knowledge base is in the area of the overlapping circles shown in Figure 1.1.

It is not the aim of this book to provide basic knowledge in mechanical vibrations, although it is expected that the reader will gain some insight into the dynamical behaviour of a simple vibrating

Virtual Experiments in Mechanical Vibrations: Structural Dynamics and Signal Processing,
First Edition. Michael J. Brennan and Bin Tang.
© 2023 John Wiley & Sons Ltd. Published 2023 by John Wiley & Sons Ltd.
Companion website: www.wiley.com/go/brennan/virtualexperimentsinmechanicalvibrations

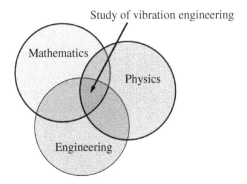

Study of vibration engineering

Figure 1.1 The subject of vibration engineering.

system. There are several textbooks devoted to basic vibration theory, for example, Tse et al. (1978), Thompson (2002), Clough and Penzien (2003), de Silva (2006), Inman (2007), and Rao (2016). There is also the classic book (Den Hartog, 1956) that offers some excellent physical descriptions of vibrating systems. The aim of this book is to provide a text that will help to bridge the gap between vibration theory and laboratory-based experimental work. Many students study vibration from a purely theoretical point of view. In some institutions, the lecturers may not even be experts in vibration engineering, and so they teach by closely following a textbook. Inevitably, this is often a mathematical exposition, with the underlying physics being frequently masked by mathematical complexity. Accordingly, many students do not gain the necessary physical insight, which would be helpful in their future careers. One way to overcome this problem is to formulate vibration problems in a more physical way in terms of variables that are measurable in a laboratory setting. In many situations, these are forces applied to the system or structure and the resulting accelerations/velocities/displacements. This means that before the theory is taught, some thought should be given to an accompanying experiment, to ensure that the output from the theoretical model involves measurable variables. It is, of course, desirable that any course has a practical component to complement and support the theory.

Much of the physical insight gained in vibration engineering, whether it be theoretical or experimental, occurs by viewing data in the frequency domain. However, all vibration signals are measured in the time domain, so these signals must be transformed to the frequency domain using signal processing techniques. This, of course, means that the vibration engineer should have some knowledge of the way in which this is done, and the mathematical basis behind the techniques. The way data are processed in practice is to first sample the data and then to work on them in digitised form using a computer. Processing sampled data brings further complications, which are discussed in Chapter 4. Many students of vibration engineering may have studied some signal processing techniques, such as Fourier analysis, but often this is done in a mathematics department, and therefore is often not related directly to the vibration theory taught in the engineering departments. There can thus be a chasm between the taught vibration theory and the way in which corresponding experimental data are captured and processed to enable comparisons between predictions and reality. It is the intention of this book to bring together these two disciplines and to give the reader some experience in applying the required signal processing techniques on simulated vibration data. There is one book on signal processing, which is specifically tailored for sound and vibration engineers (Shin and Hammond 2008), and there are other more general textbooks on the subject, which may help the reader with some of the more theoretical aspects, for example Papoulis (1962, 1977), Oppenheim and Schafer (1975), Oppenheim et al. (1997), and Bendat and Piersol (1980, 2000).

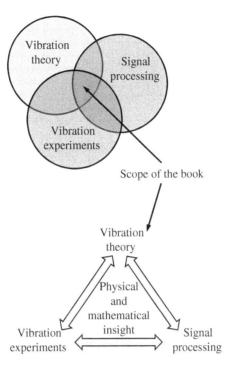

Figure 1.2 Schematic diagram showing the scope of the book.

The scope of the book is encapsulated in the Venn diagram shown in the top part of Figure 1.2. It can be seen that it contains three elements, vibration theory, vibration experiments, and signal processing. At the end of the book the reader will have been exposed to elements of these three topics and will have carried out some 'virtual' experiments using simulated data. Through the theoretical development and exercises in the book, some proficiency should be gained, which hopefully will result in improved physical insight into both vibration theory and the rationale between the choices to be made in the signal processing procedures. At the end of the book, the reader should be in a position to carry out an experiment in the laboratory and process the measured signals, provided that the experimenter has been given some additional tuition on the practical aspects of how to set up an experiment and how to handle the transducers correctly.

1.2 Typical Laboratory-Based Vibration Tests

Two typical vibration tests are shown in Figure 1.3. In the top part of the figure, an electrodynamic shaker is used to excite the structure under test, and in the lower part of the figure an instrumented impact hammer is used to excite the structure. In both cases, the resulting vibration response is measured using an accelerometer, as shown in the figure. Details of some typical signals, which are used to drive the shaker and the type of force signal generated by the impact hammer, are discussed in Chapter 5. For the shaker excitation, a signal is provided by a signal generator, which is then passed through a power amplifier, before supplying the shaker. The signal then has enough power to drive the shaker. In many cases the signal generator forms part of a software package in a computer. The force is measured using a force gauge attached to the structure, and this signal together

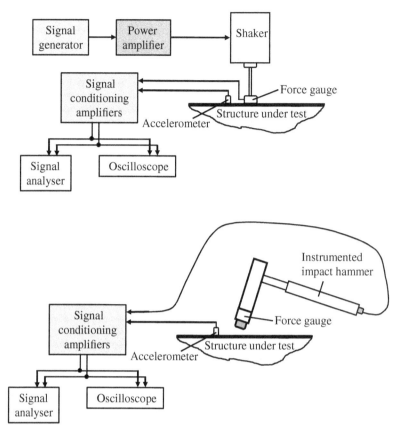

Figure 1.3 Typical experimental set-ups to measure a frequency response function (FRF). Source: Modified from Waters (2013) / Taylor & Francis.

with the signal from the accelerometer are passed through conditioning amplifiers before entering the signal analyser, and being viewed in analogue form using the oscilloscope. For hammer excitation, the force gauge is in the tip of the hammer and measures the force applied to the structure during the impact. The signals from the force gauge and the accelerometer are processed in a similar way for both shaker and force excitation. Further details on how to set up a vibration experiment similar to that shown in Figure 1.3 are given in Waters (2013). General textbooks on vibration testing have been written by Ewins (2000), McConnell and Varoto (2008), Brandt (2011), and Avitabile (2017).

The test set-ups shown in Figure 1.3. are designed to measure a single input and single output (SISO). More accelerometers can be added at different points on the structure to form a single input multi-output (SIMO) system, and an example of this type of measurement is described in Chapter 9. As mentioned above, more insight is gained by examining the data in the frequency domain – specifically the output for a given input at each frequency of excitation. This is achieved by studying this relationship which is called the *frequency response function* (FRF). The FRF is the backbone of this book, both theoretically and experimentally. It is derived analytically for a simple vibrating system in Chapter 2, and the way in which it is estimated from measurements or simulations using time domain force and acceleration data is described in Chapter 8.

1.3   Relationship Between the Input and Output for a SISO System

The relationships between the signals from a vibration measurement are shown in Figure 1.4. However, note that in this figure, displacement rather than acceleration is the response variable. This has been chosen for convenience, but also note that acceleration signals can easily be converted to velocity or displacement, by time-domain integration as discussed in Appendix A. The engineering units are shown for all the variables in Figure 1.4, as this is considered to be important in the context of this book and is rarely provided in books on signal processing. The input to the system is a force $f_e(t)$ which has the SI unit of N, and the displacement response $x(t)$ which has the unit of m. The vibrating system connecting the input to the output has a time domain description $h(t)$, which is the *impulse response function* (IRF) and has units of m/Ns. The displacement output can be determined by convolving $f_e(t)$ with $h(t)$, which is discussed further in Chapter 2, and is used extensively throughout the book.

As mentioned above, it is necessary to transform the data to the frequency domain. This is achieved by using the Fourier transform. The Fourier transform of the force time history is given by $\mathcal{F}\{f_e(t)\}$ and results in $F(j\omega)$, where $j = \sqrt{-1}$ and ω is angular frequency, which has units of rad/s; $F(j\omega)$ has units of N/Hz. Note that in this book time domain quantities are denoted by lower-case italic symbols and frequency domain quantities are denoted by upper-case italic symbols. The Fourier transform of the displacement time history is given by $X(j\omega) = \mathcal{F}\{x(t)\}$, which has units of m/Hz. Chapter 3 is devoted to the Fourier transform applied to continuous and sampled time histories. Note that frequency domain data can be transformed to the time domain, and this is achieved using the inverse Fourier transform, which is also discussed in Chapter 3. The frequency domain description of the system is the FRF, denoted by $H(j\omega)$. This is related to the IRF by the Fourier transform, i.e. $H(j\omega) = \mathcal{F}\{h(t)\}$ and has units of m/N. The output in the frequency domain $X(j\omega)$ can be determined by multiplying $F(j\omega)$ with $H(j\omega)$, and this is discussed in Chapter 2.

You will become aware as you read this book that most of the analysis is conducted using FRFs. The theoretical FRFs shown are analytical because the systems discussed are relatively simple. However, if modelling is carried out using numerical tools such as finite element analysis (FEA), Petyt (2010), which is used extensively in industry, it is also important that structures are modelled so that FRFs can be easily extracted for analysis and comparison with measurements.

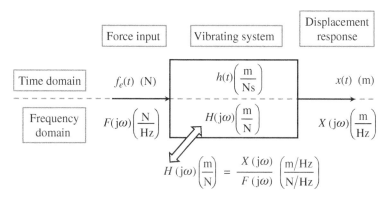

Figure 1.4   Block diagram representing of a simple single-input, single-output vibration test. Note that the response in this case is displacement for convenience, which can be obtained by integrating acceleration twice as described in Appendix A.

1.4 A Virtual Vibration Test

As mentioned previously, the aim of this book is to bridge the gap between vibration theory and vibration experiments. The book can also be used by students who do not have access to a laboratory to conduct experiments. They can carry out 'virtual' experiments. In a real experiment both force input and displacement output are measured, but in a virtual experiment the output data are generated using a model of the system. The concept is shown in Figure 1.5. The virtual experiment has a major advantage as a learning tool, in that the processed data in terms of an IRF or FRF, can be compared with the original model, which was used to generate the output time series. In this way, any artefacts in the data due to the processing can be clearly identified, which is not always possible in a real experiment.

Several methods can be used to determine the displacement output data, three of which are used in this book, and are described in Chapter 6. They are:

1. If the differential equation(s) of the vibrating system are known, then the response can be calculated by numerical integration of the equation(s) of motion. Generally, this is a straightforward procedure using a computer and is described in Appendix D.
2. If the IRF of the vibrating system is known, the response can be determined using convolution. Again, this is a relatively straightforward procedure and is described in Appendix G.
3. If the FRF of the vibrating system is known, the input force time history can be transformed to the frequency domain using the Fourier transform. The frequency domain response can then be calculated by multiplying this by the FRF, which can then be transformed to the time domain using the inverse Fourier transform to give the time history of the response. Alternatively, the FRF can be transformed to give the IRF and then the method in 2 can be used.

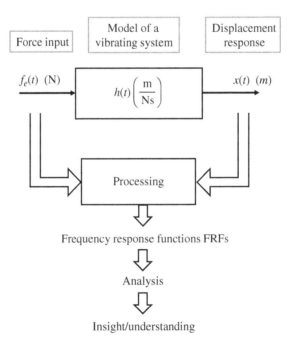

Figure 1.5 The process and rationale for a virtual vibration experiment.

1.5   Some Notes on the Book

The first thing that should be highlighted, is that this book is written in a heuristic way. Some mathematical proofs and details are omitted for ease of understanding. The more mathematically minded reader can readily find these details in the literature cited in this chapter. Secondly, the book is designed to be followed chapter by chapter to cover all the basic topics needed to conduct a virtual experiment. However, if the reader has detailed knowledge of certain topics, for example vibration theory, then Chapter 2 can be skipped, or if the reader is proficient in Fourier analysis, then Chapter 3 can be skipped.

The book is written with a novice in mind, so that very little previous knowledge is assumed. Accordingly, the book could be used as an undergraduate or a postgraduate text. The treatment of most topics, however, is brief, even though it is self-contained, so many readers may need to consult other basic texts, for example Rao (2016) for vibration or Shin and Hammond (2008) for signal processing, for more detailed information.

Each chapter contains some computer programs written using MATLAB, which are provided to illustrate some of the concepts and to give the reader some practice in applying the techniques presented to consolidate their understanding. The programs can be found on the accompanying website. Although MATLAB is used for convenience to illustrate the computational procedures, the code can be readily modified to run in other software packages such as GNU Octave[1] or python[2], which are open source.

References

Avitabile, P. (2017). *Modal Testing: A Practitioner's Guide*. Wiley.

Bendat, J.S. and Piersol, A.G. (1980). *Engineering Applications of Correlation and Spectral Analysis*. Wiley.

Bendat, J.S. and Piersol, A.G. (2000). *Random Data: Analysis and Measurement Procedures*, 3rd Edition. Wiley-Interscience.

Brandt, A. (2011). *Noise and Vibration Analysis: Signal Analysis and Experimental Procedures*. Wiley.

Clough, R.W. and Penzien, J. (2003). *Dynamics of Structures*, 3rd Edition. Computers & Structures, Inc.

Ewins, D.J. (2000). *Modal Testing: Theory, Practice and Application*, 2nd Edition. Research Studies Press.

Den Hartog, J.P. (1956). *Mechanical Vibrations*, 4th Edition. McGraw-Hill.

Inman, D.J. (2007). *Engineering Vibration*, 3rd Edition. Pearson.

McConnell, K.G. and Varoto, P.S. (2008). *Vibration Testing: Theory and Practice*, 2nd Edition. Wiley.

Oppenheim, A.V. and Schafer, R.W. (1975). *Digital Signal Processing*. Prentice Hall International.

Oppenheim, A.V., Willsky, A.S., and Hamid Nawab, S. (1997). *Signals and Systems*, 2nd Edition. Prentice Hall International.

Papoulis, A. (1962). *The Fourier Integral and Its Applications*, McGraw-Hill.

Papoulis, A. (1977). *Signal Analysis*, McGraw-Hill.

Petyt, M. (2010). *Introduction to Finite Element Vibration Analysis*, 2nd Edition. Cambridge University Press.

Rao, S.S. (2016). *Mechanical Vibrations*, 6th Edition. Pearson.

---

1 https://www.gnu.org/software/octave/index (accessed 27 December 2021)
2 https://www.python.org/ (accessed 27 December 2021)

Shin, K. and Hammond, J.K. (2008). *Fundamentals of Signal Processing for Sound and Vibration Engineers*. Wiley.

de Silva, C.W. (2006). *Vibration: Fundamentals and Practice*, 2nd Edition. CRC Press.

Thompson, W.T. (2002). *Theory of Vibration with Applications*, 3rd Edition. CBS Publishers & Distributors.

Tse, F.S., Morse, I.E., and Hinkle, R.T. (1978). *Mechanical Vibrations – Theory and Applications*, 2nd Edition. Ally and Bacon, Inc.

Waters, T.P. (2013). *Vibration Testing, Chapter 9 in Fundamentals of Sound and Vibration*, (eds. F.J. Fahy and D.J. Thompson). CRC Press.

2

Fundamentals of Vibration

2.1 Introduction

A vibrating system can be characterised in both the time and frequency domain. The quantities used to characterise the system can be obtained theoretically or experimentally, and are used extensively in this book. This chapter is devoted to deriving these quantities for a simple mechanical system. Thorough knowledge of such a system is essential for the deeper understanding of mechanical vibrations in general. Further, an understanding of the dynamics of a vibrating system in terms of its physical properties is extremely helpful in the interpretation of experimental data. No previous knowledge of vibrations is assumed in this chapter, as all the results are derived from first principles, requiring only a basic understanding of mechanics.

2.2 Basic Concepts – Mass, Stiffness, and Damping

There are three fundamental physical properties of a vibrating system. They are mass, stiffness, and damping. Although they tend to exist in a distributed form in the real world, for an initial study of vibration it is convenient to represent them in lumped parameter form using idealised elements as shown in Figure 2.1. Note that only translational linear elements are considered for simplicity, rather than rotational and/or nonlinear elements, which also exist in the real world. The interested reader is referred to more-in-depth texts on linear and nonlinear vibration, such as Tse et al. (1978), Inman (2007), Worden and Tomlinson (2001), Thomsen (2003), Kovacic and Brennan (2011), and Rao (2016).

The stiffness element is represented by a linear, massless spring with stiffness k, which has units of N/m. It is shown in Figure 2.1ai. The equations relating the forces at each end of the spring to the corresponding displacements are given by

$$f_1(t) = k(x_1(t) - x_2(t)) \tag{2.1a}$$

and

$$f_2(t) = k(x_2(t) - x_1(t)). \tag{2.1b}$$

Summing Eqs. (2.1a) and (2.1b) results in $f_2(t) = -f_1(t)$, which shows that a force passes unattenuated through the stiffness element. If the right-hand end of the spring is blocked, i.e. it is connected to a rigid foundation as shown in Figure 2.1aii, then it is described simply by $f_k(t) = kx(t)$, where $f_k(t) = f_1(t)$ and $x(t) = x_1(t)$.

Virtual Experiments in Mechanical Vibrations: Structural Dynamics and Signal Processing,
First Edition. Michael J. Brennan and Bin Tang.
© 2023 John Wiley & Sons Ltd. Published 2023 by John Wiley & Sons Ltd.
Companion website: www.wiley.com/go/brennan/virtualexperimentsinmechanicalvibrations

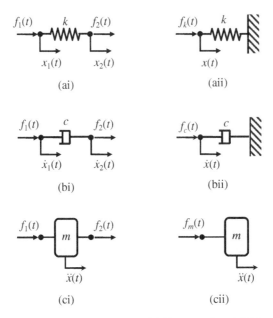

Figure 2.1 Fundamental lumped parameter elements. (ai) Linear spring. (aii) Linear spring with one end attached to ground. (bi) Linear viscous damper. (bii) Linear viscous damper with one end attached to ground. (ci) Lumped mass. (cii) Lumped mass with one free end.

For convenience, the damping element is represented by a linear viscous damper, with damping coefficient c, which has units of Ns/m. The damper has infinitely stiff (rigid), massless links that connect to the damping element, which is shown in Figure 2.1bi. The equations relating the forces at each end of the damper to the corresponding velocities are given by

$$f_1(t) = c(\dot{x}_1(t) - \dot{x}_2(t)) \tag{2.2a}$$

and

$$f_2(t) = c(\dot{x}_2(t) - \dot{x}_1(t)), \tag{2.2b}$$

where the overdot denotes differentiation with respect to time. Summing Eqs. (2.2a) and (2.2b) results in $f_2(t) = -f_1(t)$, which is the same result as for the stiffness element. Thus, a force also passes through the damping element unattenuated. If the right-hand end of the damper is blocked, i.e. it is connected to a rigid foundation as shown in Figure 2.1bii, then it is described simply by $f_c(t) = c\dot{x}(t)$, where again $f_c(t) = f_1(t)$ and $\dot{x}(t) = \dot{x}_1(t)$.

A mass element is represented by a point mass, with mass m, which has units of Ns^2/m or kg. The mass is a point in space, i.e. it has no dimension. It also has no damping, is infinitely stiff, and is shown in Figure 2.1ci. The equation relating the forces to the acceleration of the mass is given by

$$f_1(t) + f_2(t) = m\ddot{x}(t). \tag{2.3}$$

Note that the equation describing the behaviour of a mass is fundamentally different from those that describe the stiffness and damping elements. In this case, a force does not pass through the mass unattenuated, because $f_2(t) = m\ddot{x}(t) - f_1(t)$. It makes no sense to block the mass, so in this case $f_2(t)$ is set to zero as shown in Figure 2.1cii, so that $f_m(t) = m\ddot{x}(t)$.

Note that the mass and stiffness elements store energy in the form of kinetic energy and potential energy, respectively, and the damping element dissipates energy.

2.3 Single Degree-of-Freedom System

As mechanical systems generally consist of a combination of mass, stiffness, and damping, the elements described in Section 2.1 are assembled as shown in Figure 2.2a. This is called a single degree-of-freedom (SDOF) system as only one coordinate is required to describe the motion. The drawing in Figure 2.2a is a conventional diagram of an SDOF system, but an alternative way of drawing the same system is shown in Figure 2.2b, Hixson (1976) and Gardonio and Brennan (2002). Note the convention used to represent a mass in Figure 2.2b, which is the same as that used in (Hixson 1976), which means that the mass is 'connected' to an inertial system of reference. The advantage of using the alternative representation of an SDOF system is that it is clear in this case that the three elements are connected in parallel, so that the excitation force $f_e(t)$ is split three ways, such that

$$f_m(t) + f_c(t) + f_k(t) = f_e(t). \tag{2.4a}$$

Substituting for the forces of the individual elements results in

$$\underbrace{m\ddot{x}(t)}_{\substack{\text{inertia} \\ \text{force}}} + \underbrace{c\dot{x}(t)}_{\substack{\text{damping} \\ \text{force}}} + \underbrace{kx(t)}_{\substack{\text{stiffness} \\ \text{force}}} = f_e(t), \tag{2.4b}$$

which is the equation of motion for an SDOF system that is found in most elementary texts on vibration.

2.4 Free Vibration

The SDOF system vibrates freely when it is set in motion, and the excitation force is removed. The subsequent motion of the mass is then dependent only on the inertia, damping, and stiffness forces. In this case, Eq. (2.4b) becomes

$$m\ddot{x}(t) + c\dot{x}(t) + kx(t) = 0, \tag{2.5a}$$

which means that the inertia force, the damping force, and the stiffness force sum to zero at each instant in time. Because these forces are proportional to the acceleration, velocity, and displacement of the mass, respectively, the time histories corresponding to these quantities must all have the same shape. For this to occur, the displacement should be described as an exponential function (which maintains its shape when differentiated). This can have a real, an imaginary, or a complex exponent depending upon the amount of damping in the system.

Before determining the solution to Eq. (2.5a) it is convenient to divide each term by m to give

$$\ddot{x}(t) + 2\zeta\omega_n\dot{x}(t) + \omega_n^2 x(t) = 0, \tag{2.5b}$$

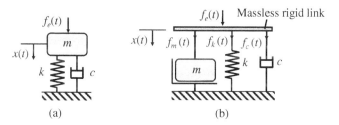

(a)   (b)

Figure 2.2 SDOF mass-spring-damper system. (a) Conventional diagram. (b) Alternative representation. Source: (b) Based on Hixson (1976) and Gardonio and Brennan (2002).

where $\omega_n = \sqrt{k/m}$ is the undamped natural frequency of the system, which is the frequency at which the system oscillates freely in the absence of damping. It has units of rad/s; $\zeta = c/2\sqrt{mk} = c/2m\omega_n$ is the damping ratio, which is the damping in the system c divided by the critical damping given by $2\sqrt{mk}$. Consider first the case when the system is undamped, i.e. when $\zeta = 0$. If the mass is now perturbed from its original position (which is called the static equilibrium position) and then released, it oscillates harmonically. The displacement can thus be described by a sine or a cosine function with an amplitude corresponding to the initial displacement. Because there is no damping in the system it continues to oscillate with this amplitude and frequency indefinitely, with the mass storing kinetic energy and the spring storing potential energy. When the mass is at its extreme position, its velocity is zero, and the mass has no kinetic energy. Thus, at this point in the cycle all the energy in the system is stored as potential energy in the spring. When the mass passes through the static equilibrium position it has maximum velocity, and all the energy in the system is stored as kinetic energy in the mass at this point in the cycle. As the system oscillates through each cycle, the stored energy is continuously passed from the spring to the mass and back again twice within each cycle, because of the reasons described above. When $\zeta = 1$, the system is critically damped and does not oscillate freely. There are also solutions for $\zeta > 1$, which is an overdamped condition, and for $\zeta < 0$, which is an unstable oscillator, but these cases are not considered in this book. The range of damping is restricted to the range $0 \leq \zeta \leq 1$, as this covers most practical vibration cases.

The solution to Eq. (2.5b) has the form

$$x(t) = Ae^{st}, \tag{2.6}$$

where A and s can either be real or complex numbers. Noting that $\dot{x}(t) = Ase^{st}$ and $\ddot{x}(t) = As^2e^{st}$, substituting Eq. (2.6) and its derivatives into Eq. (2.5b) and dividing by Ae^{st} results in

$$s^2 + 2\zeta\omega_n s + \omega_n^2 = 0. \tag{2.7}$$

This is called the characteristic equation of the system. Because the damping ratio is considered to be within the range $0 \leq \zeta \leq 1$, the solution to Eq. (2.7) is given by

$$s_{1,2} = -\zeta\omega_n \pm j\omega_d, \tag{2.8a,b}$$

where $\omega_d = \omega_n\sqrt{1-\zeta^2}$ is the damped natural frequency, which is the frequency of free vibration of an SDOF mass-spring-damper system, and $j = \sqrt{-1}$. Note that when $\zeta = 0$, $\omega_d = \omega_n$, and when $\zeta = 1$, $\omega_d = 0$. Because there are two solutions for s, which are complex conjugates, i.e. $s_2 = s_1^*$, where * denotes the complex conjugate, and the fact that $x(t)$ must be real, means that the solution to Eq. (2.6) should be of the form

$$x(t) = \frac{Ae^{s_1 t} + A^*e^{s_1^* t}}{2} \tag{2.9a}$$

Substituting Eq. (2.8a,b) into Eq. (2.9a), noting that $A = |A|e^{j\psi}$ and rearranging, results in

$$x(t) = \frac{|A|}{2}e^{-\zeta\omega_n t}\left(e^{j(\omega_d t+\psi)} + e^{-j(\omega_d t+\psi)}\right) \tag{2.9b}$$

Using Euler's formula $e^{\pm j\theta} = \cos\theta \pm j\sin\theta$, Eq. (2.9b) can be written as

$$x(t) = |A|e^{-\zeta\omega_n t}\cos(\omega_d t + \psi) \tag{2.9c}$$

or

$$x(t) = \underbrace{|A|e^{-\zeta\omega_n t}}_{\text{envelope}}\underbrace{\sin(\omega_d t + \phi)}_{\text{oscillation}}, \tag{2.9d}$$

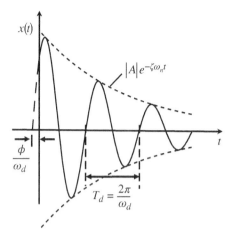

Figure 2.3 Free vibration of an SDOF system.

where $\phi = \psi + \pi/2$. Equations (2.9c) and (2.9d) describe the free vibration displacement of an underdamped mass-spring-damper SDOF system. An illustration of Eq. (2.9d) for a lightly damped system is shown in Figure 2.3. The solid line in this figure gives the complete response and the dashed line gives the envelope of the response. The mass oscillates at the damped natural frequency of ω_d, which corresponds to a damped natural period of $T_d = 2\pi/\omega_d$. It is clear that damping has two effects. In most practical situations, the main effect is to influence the rate at which the vibration decays, and a secondary effect is to reduce the frequency at which the system oscillates.

2.5 Impulse Response Function (IRF)

As mentioned in Chapter 1, the impulse response function (IRF) of a vibrating system is a cornerstone of dynamic analysis, because it relates the time domain input of a system to its output (response) by way of convolution. To determine the theoretical IRF of a system, it is impacted by a force that has the form of a delta function, and the response is calculated. The IRF of the vibrating system is its response to the delta function. Some details concerning the delta function are given in Appendix E.

An impulse is a force which acts on a system for a very short time period, providing finite momentum to the system. Such an impulse is shown in Figure 2.4. The applied force $\hat{f}_e/\varepsilon$, acts for a short

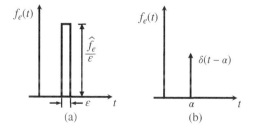

Figure 2.4 An impulsive force and its idealised representation as a delta function in the case when $\varepsilon \to 0$ and $\hat{f}_e = 1$. (a) An impulsive force. (b) A delta function applied at time $t = \alpha$.

time duration ε, so that the impulse, which is the area under the force curve, is given by

$$\hat{f}_e = \text{Area} = \frac{\hat{f}_e}{\varepsilon}\varepsilon, \tag{2.10}$$

which has units of Ns. If $\varepsilon \to 0$ such that the force acts over an infinitesimally small period of time, and $\hat{f}_e = 1$, the result is a unit impulse or the delta function. As shown in Appendix E, the delta function $\delta(t)$ is defined as

$$\delta(t) = 0 \quad \text{for} \quad t \neq 0 \quad \text{and} \quad \int_{-\infty}^{\infty} \delta(t)dt = 1, \tag{2.11}$$

If the delta function is applied at time $t = \alpha$ as shown in Figure 2.4b, rather than at time $t = 0$, the delta function is given by $\delta(t - \alpha)$. In practice it is not possible to apply a delta function because any applied force will act over a finite time period, so an IRF cannot be measured directly, it must be determined by manipulating measured data, and the way this is achieved is discussed in Chapters 3 and 4.

To determine the displacement IRF for an underdamped SDOF system, it is first assumed that before the impulse $\hat{f}_e$ is applied to the mass at time $t = 0$, both the displacement and velocity of the mass are zero. At the instant the force is applied $x(0) = 0$, and there is an initial velocity due to the change in momentum of the mass so that $\hat{f}_e = m\dot{x}(0)$. The problem can thus be thought of as a free vibration problem with initial conditions of $x(0) = 0$ and $\dot{x}(0) = \hat{f}_e/m$. The equation of motion for this system is Eq. (2.5a), which has the solution given by Eq. (2.9d). The initial conditions are applied to determine the amplitude $|A|$ and phase angle ϕ. Considering the initial displacement $x(0) = 0$ results in $0 = |A| \sin \phi$. The nontrivial solution means that $\phi = 0$, (the trivial solution simply corresponds to the solution where the mass remains stationary and so is of no interest), such that Eq. (2.9a) becomes

$$x(t) = |A|e^{-\zeta\omega_n t}\sin(\omega_d t). \tag{2.12}$$

Differentiating Eq. (2.12) with respect to time gives

$$\dot{x}(t) = -\zeta\omega_n|A|e^{-\zeta\omega_n t}\sin(\omega_d t) + \omega_d|A|e^{-\zeta\omega_n t}\cos(\omega_d t). \tag{2.13}$$

Considering the initial velocity $\dot{x}(0) = \hat{f}_e/m$, results in $|A| = \hat{f}_e/m\omega_d$. So, the displacement response of the system is given by

$$x(t) = \frac{\hat{f}_e}{m\omega_d}e^{-\zeta\omega_n t}\sin(\omega_d t) \quad \text{for} \quad t \geq 0. \tag{2.14}$$

Note, the causality constraint ($t \geq 0$), which means that the expression is only valid once the impulsive force has been applied. If $\hat{f}_e$ is a unit impulse then $x(t)$ is the response per unit impulse, so that $x(t) = h(t)$, where $h(t)$ is the displacement IRF, then

$$h(t) = \frac{1}{m\omega_d}e^{-\zeta\omega_n t}\sin(\omega_d t) \quad \text{for} \quad t \geq 0, \tag{2.15a}$$

which has units of m/(Ns). To determine the velocity and acceleration IRFs, the constraint of causality has to be applied mathematically. This can be done by using the Heaviside function, which is zero for $t < 0$ and unity for $t \geq 0$. Thus, Eq. (2.15a) can be written as

$$h(t) = u(t)\overline{h}(t), \tag{2.15b}$$

where $\overline{h}(t) = \frac{1}{m\omega_d}e^{-\zeta\omega_n t}\sin(\omega_d t)$ without the constraint of $t \geq 0$ and $u(t)$ is the Heaviside function. Differentiating Eq. (2.15b) with respect to time gives

$$\dot{h}(t) = \dot{u}(t)\overline{h}(t) + u(t)\dot{\overline{h}}(t), \tag{2.16a}$$

Note that $\dot{u}(t) = \delta(t)$ and $\overline{h}(0) = 0$, so that $\dot{u}(t)\overline{h}(t) = 0$, and Eq. (2.16a) becomes

$$\dot{h}(t) = u(t)\dot{\overline{h}}(t), \tag{2.16b}$$

where $\dot{\bar{h}}(t) = \frac{1}{m\omega_d}e^{-\zeta\omega_n t}(\omega_d\cos(\omega_d t) - \zeta\omega_n\sin(\omega_d t))$. Equation (2.16b) can be written as

$$\dot{h}(t) = \frac{\omega_n}{m\omega_d}e^{-\zeta\omega_n t}\cos(\omega_d t + \theta) \quad \text{for} \quad t \geq 0, \tag{2.16c}$$

where $\theta = \tan^{-1}(\zeta/\sqrt{1-\zeta^2})$. To determine the acceleration IRF, Eq. (2.16b) is differentiated with respect to time to give

$$\ddot{h}(t) = \dot{u}(t)\bar{h}(t) + u(t)\dot{\bar{h}}(t). \tag{2.17a}$$

Now, as previously mentioned $\dot{u}(t) = \delta(t)$ and $\bar{h}(0) = 1/m$, so that $\dot{u}(t)\bar{h}(t) = \delta(t)/m$, and Eq. (2.17a) becomes

$$\ddot{h}(t) = \frac{\delta(t)}{m} + u(t)\ddot{\bar{h}}(t), \tag{2.17b}$$

where $\ddot{\bar{h}}(t) = \frac{-\omega_n^2}{m\omega_d}e^{-\zeta\omega_n t}((1-2\zeta^2)\sin(\omega_d t) + 2\zeta\sqrt{1-\zeta^2}\cos(\omega_d t))$. Equation (2.17b) can be written as

$$\ddot{h}(t) = \frac{\delta(t)}{m} - \frac{\omega_n^2}{m\omega_d}e^{-\zeta\omega_n t}\sin(\omega_d t + \phi) \quad \text{for} \quad t \geq 0, \tag{2.17c}$$

where $\phi = \sin^{-1}(2\zeta\sqrt{1-\zeta^2})$. Note that in the derivation of the acceleration IRF, it is essential to include the Heaviside function; if it had not been included, then the term $\delta(t)/m$ would have been omitted. This term is a scaled delta function and occurs mathematically because the delta function is the derivative of the Heaviside function, i.e. $\delta(t) = \dot{u}(t)$. Physically, it occurs because an impulse imparts finite momentum to the system resulting in an instantaneous change in velocity of the mass, which means that its acceleration at $t = 0$ must be of the form of a delta function. Further details concerning the acceleration IRF are given in Iwanaga et al. (2021).

Time-delayed displacement, velocity, and acceleration IRFs are plotted in Figure 2.5a–c respectively. The IRFs are as result of a delta function being applied as the impulse on the mass at $t = \alpha$. This was done so that the initial parts of the IRFs can be clearly seen. Note that the y axis for the accelerance IRF in Figure 2.5c is not continuous. It is shown this way because the initial response is generally very large compared to the oscillatory part of the IRF.

MATLAB Example 2.1

In this example, the displacement IRF of a single degree-of-freedom system with four different values of damping is plotted.

```
clear all

%% Parameters
m=1;                      % [kg]           % mass
k=10000;                  % [N/m]          % stiffness
wn=sqrt(k/m);             % [rad/s]        % undamped natural frequency
z=0.01;                                    % damping ratio, 0.001, 0.01, 0.1, 1
c=2*z*wn*m;               % [Ns/m]         % damping coefficent
wd=sqrt(1-z^2)*wn;        % [rad/s]        % damped natural frequency

%% Time vector
dt=0.001;                 % [s]            % time resolution in seconds
T=100;                    % [s]            % duration of time signal
t=0:dt:T;                 % [s]            % time vector

%% Impulse response
h=1/(m*wd)*exp(-z*wn*t).*sin(wd*t); % [m/Ns]          % displacement IRF
```

(Continued)

MATLAB Example 2.1   (Continued)

```
%% Plot results
plot(t,h);
grid;axis square
xlabel('Time (s)');
ylabel('Displacement (m)');
```

Results

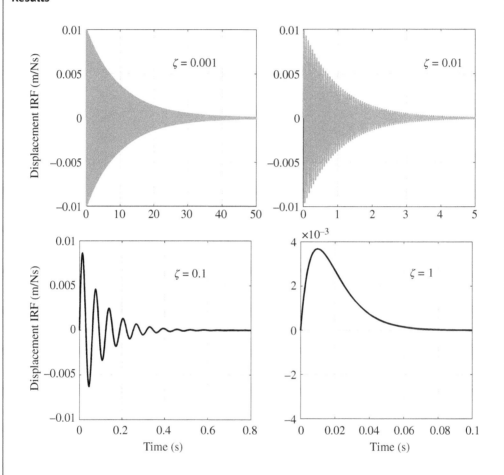

Comments

1. For different values of damping, the time axis must be set to the appropriate length.
2. When $\zeta = 1$, Eq. (2.15a) cannot be used to plot $h(t)$, because $\omega_d = 0$. In this case, ζ can be set to 0.9999 to plot $h(t)$, or the approximation $\sin(\omega_d t) \approx \omega_d t$, so that for $\zeta = 1$, $h(t) \approx \frac{1}{m} t e^{-\omega_n t}$ for $t \geq 0$.
3. An exercise for the reader is to plot $\dot{h}(t)$ and $\ddot{h}(t)$ for the four values of damping used to plot $h(t)$. Also, check the results for $\dot{h}(t)$ and $\ddot{h}(t)$ by numerically differentiating $h(t)$ and $\dot{h}(t)$. See Appendix A, which discusses how to carry out differentiation numerically in MATLAB.

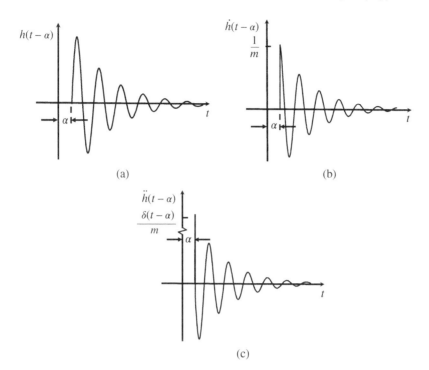

(a)

(b)

(c)

Figure 2.5 Delayed IRFs for an SDOF system when a delta function force impulse is applied at time $t = \alpha$. (a) Displacement IRF. (b) Velocity IRF. (c) Acceleration IRF.

2.6 Determination of Damping from Free Vibration

It was shown in Eq. (2.9d) and Figure 2.3 that the free vibration response of an SDOF system is given by an expression that consists of the product of an oscillatory part and an envelope. The envelope (Env) is given by

$$\text{Env} = |A|e^{-\zeta\omega_n t}. \tag{2.18}$$

Taking the natural logarithms of both sides of Eq. (2.18) results in

$$\ln(\text{Env}) = \ln|A| - \zeta\omega_n t, \tag{2.19}$$

which is the equation for a straight line, where $\ln|A|$ is the intercept with the y axis and $\zeta\omega_n$ is the modulus of the gradient (Grad). It is known that $\omega_d = \omega_n\sqrt{1 - \zeta^2}$ and $\omega_d = 2\pi/T_d$ so that

$$\text{Grad} = \frac{2\pi\zeta}{T_d\sqrt{1 - \zeta^2}}, \tag{2.20}$$

which can be rearranged to give

$$\zeta = \frac{1}{\sqrt{1 + \chi^2}}, \tag{2.21}$$

where $\chi = 2\pi/(T_d\text{Grad})$. To determine the envelope from the measured free vibration decay of the SDOF system, the analytic signal is first determined. This is a complex signal in the time domain. Its real part is the original signal $x(t)$, and the imaginary part $\hat{x}(t)$ is the original signal, but phase shifted by $-90°$. To obtain the analytic signal, the Hilbert transform (Feldman, 2011) is used, a summary of which is given in Appendix B. The analytic signal is given by

$$a(t) = x(t) + j\hat{x}(t). \tag{2.22}$$

The envelope is then simply given by $|a(t)|$.

MATLAB Example 2.2

In this example the damping is determined from the impulse response of an SDOF system, which has a damping ratio of $\zeta = 0.01$.

```
clear all

%% Parameters
m=1;                        % [kg]        % mass
k=10000;                    % [N/m]       % stiffness
wn=sqrt(k/m);               % [rad/s]     % undamped natural frequency
z=0.01;                                   % damping ratio
c=2*z*wn*m;                 % [Ns/m]      % damping coefficient
wd=sqrt(1-z^2)*wn;          % [rad/s]     % damped natural frequency
Td=2*pi/wd;                 % [s]         % damped natural period

%% Time vector
dt=0.001;                   % [s]         % time resolution in seconds
T=5;                        % [s]         % duration of time signal
t=0:dt:T;                   % [s]         % time vector

%% Normalised displacement IRF
h=exp(-z*wn*t).*sin(wd*t);               % normalized impulse response function

%% Calculations
a=hilbert(h);                            % create analytic signal
env=log(abs(a));                         % calculate the log of the envelope
t1=t(500:3500);env1=env(500:3500);       % set a specific time range
p=polyfit(t1,env1,1);                    % fit straight line in time range t1
grad=-p(1);                              % calculate the gradient of the line
gamma=2*pi/(grad*Td);                    % calculate the constant
Est_z=1/sqrt(1+gamma^2)                  % damping ratio estimate

%% Plot results
figure(1);
plot(t,h);                               % normalized IRF
hold on;
plot(t,abs(a),t,-abs(a),'linewidth',2)
grid;axis square
xlabel('time (s)');
ylabel('normalised displacement');

figure(2);
plot(t,env,'linewidth',2);               % envelope
hold on
plot(t1,env1,'linewidth',10,'Color',[.7 .7 .7])
grid;axis square
xlabel('time (s)');
ylabel('ln(envelope)');
```

(Continued)

MATLAB Example 2.2   (Continued)

Results

From the data, the damping ratio is estimated to be 0.0099. The relative error between the actual and the estimated damping ratio is 1.1%.

Comments

1. It can be seen in the example that the envelope is not estimated correctly at the beginning and the end of the time history. This is because the Hilbert transform is calculated in MATLAB using a frequency domain approach, and there is a windowing effect (which is discussed in Chapter 4) due to the truncation of data. There is also an effect due to the non-causality of the Hilbert transform, which is evident at the end of the time history (this effect is also discussed in Chapter 4).
2. Because the envelope is poorly estimated at the beginning and end of the time history, the time period chosen for the estimation of the damping ratio must avoid these regions. In the example this time period is between 0.5 and 3.5 seconds.
3. An exercise for the reader is to repeat the example, but for damping values of $\zeta = 0.001$ and $\zeta = 0.1$. Note that the time period used to estimate the damping ratio must be adjusted in each case.

2.7   Harmonic Excitation

In Section 2.2, the IRF was derived for the SDOF mass-spring-damper system. This is the time domain description of the system. There is a counterpart to this – the frequency response function (FRF), which is the frequency domain description of the system. Before considering this, the response to a harmonic force is first studied. In this case Eq. (2.4b) becomes

$$m\ddot{x}(t) + c\dot{x}(t) + kx(t) = |\overline{F}| \sin(\omega t), \tag{2.23a}$$

where $|\overline{F}|$ is the amplitude of the excitation force acting on the mass, and ω is the angular excitation frequency, which has units of rad/s. Note that $\omega = 2\pi f$, where f is the frequency in

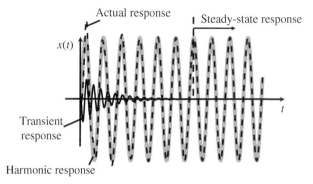

Figure 2.6 Response of an SDOF mass-spring-damper system to harmonic excitation.

cycles per second, or Hz. If the system is initially at rest and is then excited by the force $|\overline{F}| \sin(\omega t)$ there are two components to the response. One is a transient response that decays with increasing time according to the amount of damping in the system, and the second is a harmonic response at the same frequency as the excitation force, but with a phase shift at each frequency due to the time taken for the system to respond to the force. The actual displacement response is plotted in Figure 2.6 together with the transient and the harmonic response. It is clear from Figure 2.6 that after some time, once the transient response has decayed to a negligible level, then the actual response is the same as the harmonic response, which is given by $x(t) = |\overline{X}| \sin(\omega t + \phi)$, where $|\overline{X}|$ is the amplitude of the displacement. This is called the steady-state response and is a particularly important quantity in vibration analysis. The displacement can be differentiated with respect to time to give the velocity, $\dot{x}(t) = \omega |\overline{X}| \cos(\omega t + \phi)$, which in turn can be differentiated to give the acceleration, $\ddot{x}(t) = -\omega^2 |\overline{X}| \sin(\omega t + \phi)$. The expressions for the displacement, velocity, and acceleration can be substituted into Eq. (2.23a) to give

$$\underbrace{-\omega^2 m |\overline{X}| \sin(\omega t + \phi)}_{\text{inertia force}} + \underbrace{\omega c |\overline{X}| \cos(\omega t + \phi)}_{\text{damping force}} + \underbrace{k |\overline{X}| \sin(\omega t + \phi)}_{\text{stiffness force}} = \underbrace{|\overline{F}| \sin(\omega t)}_{\text{excitation force}}. \tag{2.23b}$$

This equation is represented graphically in terms of the force vectors in Figure 2.7. The amplitude of the stiffness force is constant with frequency, the amplitude of the damping force is proportional to frequency, and the amplitude of the inertia force is proportional to the square of frequency. The

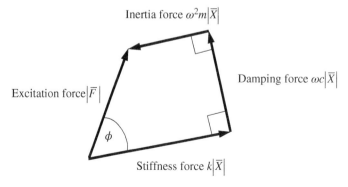

Figure 2.7 Schematic diagram showing the amplitudes of the stiffness, damping, and inertia forces compared to the excitation force, and the phase angles between them.

phase angle between the damping and the stiffness force is 90°, as is the phase angle between the inertia force and the damping force. This is because the stiffness, damping, and inertia forces are proportional to displacement, velocity, and acceleration, respectively, which each differ in phase by 90° as the motion is harmonic. The phase between the displacement response and the excitation force ϕ is clearly shown in Figure 2.7. It should be noted that the amplitudes of the damping and acceleration forces change with frequency, so the shape of the force diagram is a function of frequency, resulting in a different phase shift ϕ at each frequency. If the system is undamped, however, the damping force is zero. Because the stiffness and inertia forces oppose each other, the response will therefore be in-phase (0°) with the excitation force if the stiffness force is greater than the inertia force, and will be in anti-phase (−180°) if the inertia force is greater than the stiffness force. This is further discussed in Section 2.8.

For a given harmonic excitation force of amplitude $|\overline{F}|$ at frequency ω, the object is to determine the steady-state amplitude of the displacement response $|\overline{X}|$ and its phase with respect to the force ϕ. To achieve this, Eq. (2.23b) needs to be solved. However, to do this requires splitting the equation into two parts, which can be solved to give $|\overline{X}|$ and ϕ. It is preferable to use complex numbers by noting that $|\overline{F}|\sin(\omega t) = |\overline{F}|(e^{j\omega t} - e^{-j\omega t})/j2$, which means that Eq. (2.23b) can be split into two equations, one of which has an excitation force $|\overline{F}|e^{j\omega t}$ and the other $|\overline{F}|e^{-j\omega t}$. These two equations can be summed and then divided by j2 to give the original equation. The displacement response can also be written in complex form. In response to the excitation force $|\overline{F}|e^{j\omega t}$, it is $|\overline{X}|e^{j(\omega t+\phi)}$ or $\overline{X}e^{j\omega t}$, where $\overline{X} = |\overline{X}|e^{j\phi}$, which is called the *complex amplitude*. Note that it is written in upper-case italics to differentiate from time domain quantities, which are written in lower-case italics. For convenience, the excitation force can be written as $\overline{F} = |\overline{F}|e^{j\theta}$, in which $\theta = 0$ in the case considered here. The complex displacement amplitude and excitation forces are sometimes written as $\overline{X}(j\omega)$ and $\overline{F}(j\omega)$ to emphasise that they are complex quantities and are a function of frequency. In general, they contain both amplitude and phase information.

Consider Eq. (2.23a), but with a complex excitation force (note that this is a mathematical not a physical quantity, but it serves the purpose of simplifying the analysis), so that

$$m\ddot{x}(t) + c\dot{x}(t) + kx(t) = \overline{F}e^{j\omega t}. \tag{2.24}$$

The steady-state displacement response $x(t) = \overline{X}e^{j\omega t}$ is assumed, which means that the response is at the same frequency as the excitation force, but with a phase difference. This is differentiated to give the velocity $\dot{x}(t) = j\omega\overline{X}e^{j\omega t}$, which in turn is differentiated to give $\ddot{x}(t) = -\omega^2\overline{X}e^{j\omega t}$. Substituting for the displacement, velocity, and acceleration in Eq. (2.24) results in

$$\left(\underbrace{-\omega^2 m\overline{X}}_{\overline{F}_m} + \underbrace{j\omega c\overline{X}}_{\overline{F}_c} + \underbrace{k\overline{X}}_{\overline{F}_k} \right) e^{j\omega t} = \overline{F}e^{j\omega t}, \tag{2.25}$$

where $\overline{F}_m$, $\overline{F}_c$, and $\overline{F}_k$ are the complex inertia, damping, and stiffness forces, respectively. A representation of the system in the steady state is shown in Figure 2.8a (compare this with Figure 2.2b), and the relationship between the forces for three different frequencies is shown in Figure 2.8b. Note that the $e^{j\omega t}$ time dependence has been omitted for clarity in the figure. It is instructive to further examine the forces as frequency increases. First consider zero frequency $\omega = 0$, sometimes called DC (from electrical system theory – direct current). In this case both the inertia and the damping forces are zero, so the stiffness force is equal to the excitation force, i.e. $k\overline{X} = \overline{F}$, which means that the displacement is given by $\overline{X} = \overline{F}/k$.

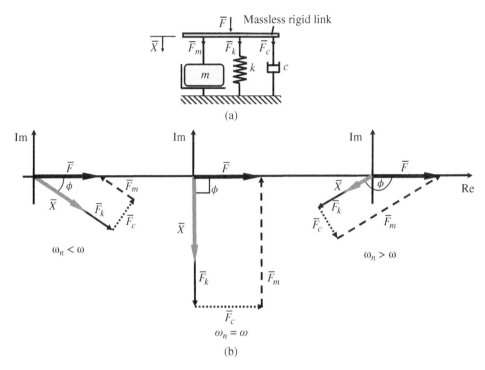

Figure 2.8 Frequency domain representation of an SDOF mass-spring-damper system. (a) Frequency domain representation. (b) Relationship between the forces at different frequencies. Note that ϕ is negative in each case, i.e. the displacement lags the force by ϕ radians.

As frequency increases, the magnitude of the inertia force increases. When the excitation frequency is equal to the natural frequency, i.e. when $\omega = \omega_n = \sqrt{k/m}$, which is called the resonance frequency, then the amplitude of the inertia force is equal to the amplitude of the stiffness force. However, because these forces are in anti-phase they sum to zero, which means that the damping force is equal to the excitation force. For this to occur, the damping force must be in-phase with the excitation force, so that $j\omega_n c\overline{X} = \overline{F}$, which means that the displacement is given by $\overline{X} = \overline{F}/j\omega_n c$. Thus, the displacement lags the excitation force by 90°, which in turn means that the velocity is in phase with the excitation force. Because the inertia and stiffness have effectively disappeared from the point of view of the excitation force, the source only 'sees' a damper at this frequency. Above the resonance frequency, as the excitation frequency increases, the inertia force also increases. When $\omega \gg \omega_n$ the inertia force is much greater than the damping and the stiffness forces so that $-\omega^2 m\overline{X} \approx \overline{F}$; therefore, $\overline{X} \approx \overline{F}/(-\omega^2 m)$. This means that the displacement is in anti-phase with the excitation force, or the acceleration is in-phase with the excitation force. This is because both the damping and stiffness have a negligible effect on the motion of the mass, so from the point of view of the excitation force, it only 'sees' a mass at high frequencies.

2.8 Frequency Response Function (FRF)

The FRF describes the way in which the system responds in the steady-state at each excitation frequency. It is the most fundamental and widely used quantity in vibration engineering (Ewins, 2000), and has a particular relationship with IRF, which is discussed further in Chapter 3. The FRF is a complex quantity, having both an amplitude and a phase, which in general, are frequency dependent. The displacement FRF can be easily determined from Eq. (2.25), by dividing both sides by $e^{j\omega t}$ and rearranging to give

$$H(j\omega) = \frac{\overline{X}(j\omega)}{\overline{F}(j\omega)} = \frac{1}{k - \omega^2 m + j\omega c}. \tag{2.26}$$

Note that the argument is shown to emphasise complex frequency domain quantities. The FRF shown in Eq. (2.26) is called *receptance* and is related to displacement. It has units of m/N. There is also an FRF related to velocity, which is given by $H_{\text{vel}}(j\omega) = j\omega H(j\omega)$, and is called *mobility*, so that

$$H_{\text{vel}}(j\omega) = j\omega \frac{\overline{X}(j\omega)}{\overline{F}(j\omega)} = \frac{j\omega}{k - \omega^2 m + j\omega c}, \tag{2.27}$$

which has units of m/Ns. There is an FRF related to acceleration, which is given by $H_{\text{acc}}(j\omega) = j\omega H_{\text{vel}}(j\omega) = -\omega^2 H(j\omega)$, and is called *accelerance*, so that

$$H_{\text{acc}}(j\omega) = -\omega^2 \frac{\overline{X}(j\omega)}{\overline{F}(j\omega)} = \frac{-\omega^2}{k - \omega^2 m + j\omega c}, \tag{2.28}$$

which has units of m/Ns². The receptance can be written as $H(j\omega) = |H(j\omega)|e^{j\phi(j\omega)}$ where $|H(j\omega)| = 1/\sqrt{(k - \omega^2 m)^2 + (\omega c)^2}$ is the modulus, and $\phi(j\omega) = \tan^{-1}(-\omega c/(k - \omega^2 m))$ is the phase. The modulus and phase of the receptance are plotted in Figure 2.9 for three values of damping ratio, $\zeta = 0.001$, 0.01, and 0.1. There are several features about the receptance that should be noted:

1. The modulus is plotted on log–log axes. This is done because, in general, the modulus has a large range of values (large dynamic range), and details of both the small and large values can be clearly seen over a wide frequency range. Moreover, the high-frequency response becomes

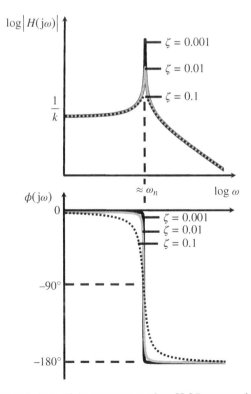

Figure 2.9 Modulus and phase of the receptance of an SDOF mass-spring-damper system.

a straight line rather than a curved line. Instead of plotting the modulus on a log axis it is frequently plotted in dB. The way in which this is done is discussed in Appendix C.

2. It can be seen that the damping only has an effect on the receptance at frequencies close to the resonance frequency. Indeed, at the resonance frequency Eq. (2.26) becomes $|H(j\omega)|_{\omega=\omega_n} = 1/\omega_n c$. At low frequencies the stiffness is the controlling factor, which can be seen from Eq. (2.26) which reduces to $|H(j\omega)|_{\omega \ll \omega_n} \approx 1/k$. In a similar way, it can be seen that at high frequencies the mass is the controlling parameter as $|H(j\omega)|_{\omega \gg \omega_n} \approx 1/\omega^2 m$.

3. The phase is plotted as a linear quantity. However, so that it can be compared with the behaviour of the modulus, which is plotted on a log frequency axis, it is plotted on log-linear axes. It can be seen that the main effect of the damping is to control the rate of change of phase close to the resonance frequency; the larger the damping, the greater the rate of change of phase. In the undamped case, the phase is 0° below the resonance frequency and −180° above the resonance frequency. At the resonance frequency the phase is −90° for all values of ζ.

To show the different frequency regimes, and the low- and high-frequency asymptotes, the modulus of the receptance is plotted in Figure 2.10. The three frequency regimes in which the stiffness, damping, and mass control the dynamic behaviour are evident.

The FRFs in Eqs. (2.26–2.28) are complex quantities. Accordingly, they can be written in terms of their real and imaginary components. As an example, consider the receptance given in Eq. (2.26). The numerator and denominator are each multiplied by the complex conjugate of the denominator, to give

$$H(j\omega) = \frac{1}{k - \omega^2 m + j\omega c} \times \frac{k - \omega^2 m - j\omega c}{k - \omega^2 m - j\omega c} \tag{2.29a}$$

which simplifies to

$$H(j\omega) = \text{Re}\{H(j\omega)\} + j\text{Im}\{H(j\omega)\} \tag{2.29b}$$

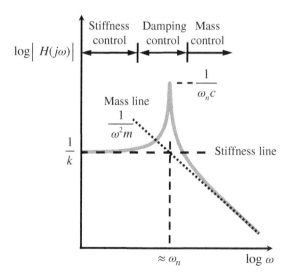

Figure 2.10 Modulus of the receptance of an SDOF mass-spring-damper system showing the parameters that control the response in various frequency regions.

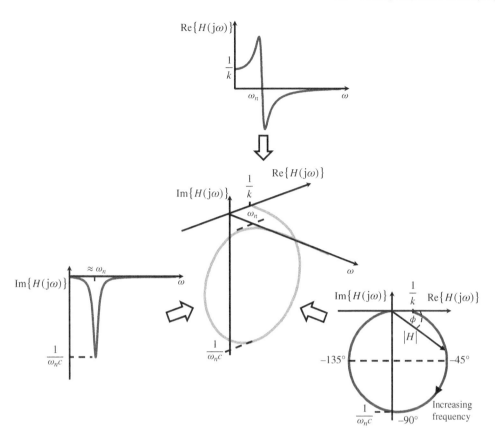

Figure 2.11 Complex representation of the receptance for an SDOF mass-spring-damper system.

where $\mathrm{Re}\{H(\mathrm{j}\omega)\} = \frac{k-\omega^2 m}{(k-\omega^2 m)^2 + (\omega c)^2}$ and $\mathrm{Im}\{H(\mathrm{j}\omega)\} = \frac{-\omega c}{(k-\omega^2 m)^2 + (\omega c)^2}$. Note that the denominators of the real and the imaginary parts are the same and are positive. Thus, the signs of the numerators of the real and imaginary parts govern whether they are positive or negative. The real part is positive when $k > \omega^2 m$, i.e. when $\omega < \omega_n$, it is zero when $\omega = \omega_n$, and is negative when $\omega > \omega_n$. The imaginary part is negative for all frequencies, which is a consequence of energy dissipation in the system (the damping coefficient is positive). To illustrate the behaviour, the real and imaginary parts are plotted as a function of frequency in Figure 2.11, together with a 3-dimensional plot, which captures the complete behaviour of the FRF. Also plotted is the Nyquist or Argand diagram, which is a plot of the imaginary part against the real part. The relationship between the modulus and phase and the Nyquist diagram is clearly marked. Note that the phase angles of −45°, −90°, and −135° are marked, as is the arrow indicating an increase in frequency.

The choice of which FRF plot to use, is made according to the purpose and preference of the analyst. For example, if the energy dissipation characteristics of the system are of interest, then the imaginary part is appropriate. If the frequency range needs to be determined where mass or stiffness is the controlling parameter, then the sign of the real part should be consulted. If damping estimation is of interest, then a variety of plots may be useful, and this is discussed in the next sub-section.

MATLAB Example 2.3

In this example the receptance FRF of an SDOF system is plotted, which has a damping ratio of $\zeta = 0.01$.

```
clear all

%% parameters
m=1;                                    % [kg]          % mass
k=10000;                                % [N/m]         % stiffness
wn=sqrt(k/m);                           % [rad/s]       % undamped natural frequency
z=0.01;                                                 % damping ratio
c=2*z*wn*m;                             % [Ns/m]        % damping coefficient

%% Frequency vector
df=0.001;                               % [Hz]          % frequency resolution in Hz
F=100;                                  % [Hz]          % maximum frequency
f=0:df:F;                               % [Hz]          % frequency vector
w=2*pi*f;                               % [rad/s]

%% Receptance FRF
H=1./(k-w.^2+j*w*c);                    % [m/N]         % receptance FRF

%% Plot results
figure (1)
semilogx(f,20*log10(abs(H)),'linewidth',6)      % modulus
hold on
semilogx(f,20*log10(1./w.^2),'-')
hold on
semilogx(f,20*log10(f./f*1/k),':')
hold on
semilogx(wn/(2*pi),20*log10(1/(wn*c)),'xk')
grid;axis square
xlabel('frequency (Hz)');
ylabel('|receptance| (dB ref 1 m/N)');

figure (2)
semilogx(f,180/pi*angle(H))                     % phase
grid;axis square
xlabel('frequency (Hz)');
ylabel('phase angle (degrees)');

figure (3)
plot(f,real(H))                                 % real part
grid;axis square
xlabel('frequency (Hz)');
ylabel('real(receptance) (m/N)');

figure (4)
plot(f,imag(H))                                 % imaginary part
grid;axis square
xlabel('frequency (Hz)');
ylabel('imag(receptance) (m/N)');

figure (5)
plot(real(H),imag(H),'linewidth',4)             % Nyquist
grid;axis square
xlabel('real (receptance) (m/N)');
ylabel('imag(receptance) (m/N)');
```

(Continued)

MATLAB Example 2.3   (Continued)

Results

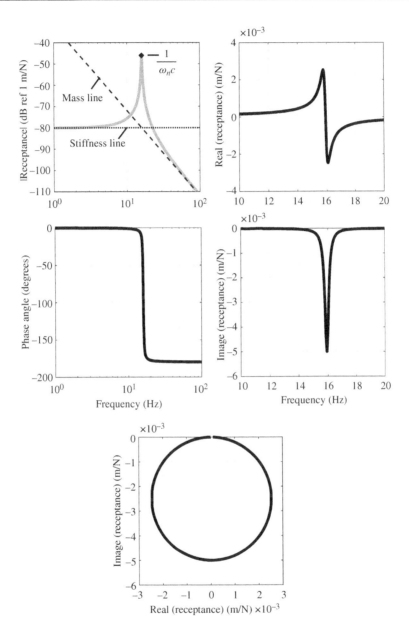

Comments

1. An exercise for the reader is to repeat the example, but for damping values of $\zeta = 0.001$ and $\zeta = 0.1$. Note that the amplitude range will need to be adjusted in each case.
2. An exercise for the reader is to repeat the example, but for mobility and acceleration FRFs.

2.9   Other Features of the Receptance FRF

Apart from the general features of the FRF described in the previous section there are two other factors concerning the damping-controlled frequency range that are worthy of discussion. One of these is the frequency at which the maximum response occurs, and the other is the influence of damping on the FRF at frequencies close to the resonance frequency. To investigate these, it is useful to write Eq. (2.26) as

$$H(j\omega) = \frac{1}{k} \times \frac{1}{1 - \Omega^2 + j2\zeta\Omega}, \tag{2.30}$$

where $\Omega = \omega/\omega_n$. The modulus of Eq. (2.30) can be written as

$$|H(j\omega)| = \frac{1}{k} \times ((1 - \Omega^2)^2 + (2\zeta\Omega)^2)^{-\frac{1}{2}}. \tag{2.31}$$

The maximum of the FRF modulus occurs when the derivative of Eq. (2.31) with respect to frequency is equal to zero, i.e. when

$$-\frac{1}{2}((1 - \Omega^2)^2 + (2\zeta\Omega)^2)^{-\frac{3}{2}} \times \frac{d}{d\Omega}((1 - \Omega^2)^2 + (2\zeta\Omega)^2) = 0, \tag{2.32}$$

which results in $-4\Omega(1 - \Omega^2) + 8\zeta^2\Omega = 0$, so that

$$-4\Omega(1 - \Omega^2 - 2\zeta^2) = 0. \tag{2.33}$$

The positive solution to Eq. (2.33) results in the frequency at which the peak of the FRF modulus occurs, and is given by

$$\Omega_{\text{peak}} = \sqrt{1 - 2\zeta^2} \quad \text{or} \quad \omega_{\text{peak}} = \omega_n\sqrt{1 - 2\zeta^2}. \tag{2.34}$$

Note that if the damping is light, such that $\zeta < 0.1$ then $\omega_{\text{peak}} \approx \omega_n$. For the mobility FRF, $\omega_{\text{peak}} = \omega_n$, and for the accelerance FRF (for light damping), $\omega_{\text{peak}} = \omega_n\sqrt{1 + 2\zeta^2}$. It is left as an exercise for the reader to verify these relationships.

To investigate the effect of damping on the FRF modulus in the damping-controlled frequency range, consider the displacement FRF shown in Figure 2.12. Note that this is plotted in the frequency range close to the resonance frequency, the modulus is plotted in dB, and the frequency axis is linear rather than log. The half power points shown on the modulus plot are 3 dB below the maximum and occur at frequencies ω_1 and ω_2. The phase angles at these frequencies are approximately $-45°$ and $-135°$, respectively. The maximum value of the modulus is $|H(j\omega)|_{\text{max}} = 1/2\zeta k$, so the value of the modulus at the half-power points is $1/2\sqrt{2}\zeta k$ (because $-3\,\text{dB} \approx 20\log_{10}(1/\sqrt{2})$). Therefore, the frequencies at the half-power points can be determined by setting Eq. (2.31) to this value, so that

$$\frac{1}{k((1 - \Omega^2)^2 + (2\zeta\Omega)^2)^{\frac{1}{2}}} = \frac{1}{2\sqrt{2}\zeta k}. \tag{2.35}$$

Squaring both sides of Eq. (2.35) and rearranging gives

$$\Omega^{4^*} - 2(1 - 2\zeta^2)\Omega^2 + (1 - 8\zeta^2) = 0, \tag{2.36}$$

which can be solved to give

$$\Omega_{1,2}^2 = 1 - 2\zeta^2 \pm 2\zeta\sqrt{1 + \zeta^2}. \tag{2.37a,b}$$

Neglecting terms with ζ^2 results in $\Omega_1^2 = 1 - 2\zeta$ and $\Omega_2^2 = 1 + 2\zeta$, which can be combined to give

$$\Omega_2^2 - \Omega_1^2 = (\Omega_2 + \Omega_1)(\Omega_2 - \Omega_1) = 4\zeta. \tag{2.38}$$

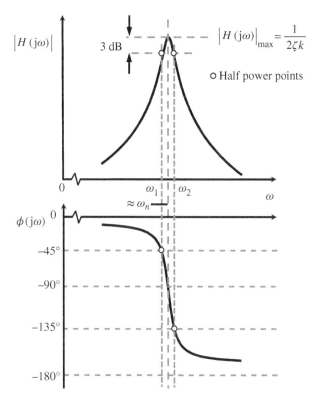

Figure 2.12 Amplitude and phase of the displacement FRF at frequencies close to the resonance frequency.

Because damping is assumed to be light, then both Ω_1 and Ω_2 are close to unity so that $\Omega_2 + \Omega_1 \approx 2$. Therefore $\Omega_2 - \Omega_1 = 2\zeta$, or

$$\zeta = \frac{\omega_2 - \omega_1}{2\omega_n}, \tag{2.39}$$

which relates the damping ratio to the half-power point frequencies, and the undamped natural frequency. Equation (2.39) can be used to determine the damping from experimental data, which is discussed in the next section.

2.10 Determination of Damping from an FRF

The damping ratio can be calculated from an FRF by simply plotting either the amplitude or the phase and determining the three frequency points given in Eq. (2.39). If the damping is very light, special care must be taken to obtain an accurate estimate. In this case, the difference between ω_1 and ω_2, which correspond to phase angles $-45°$ and $-135°$, respectively, is very small. In practice, the FRF is in discrete form, as in MATLAB Example 2.3, so to obtain a good estimate of damping, a fine frequency resolution is required, so that there are enough points within this frequency range. To illustrate this problem, two cases are shown in Figure 2.13. The modulus, phase, and

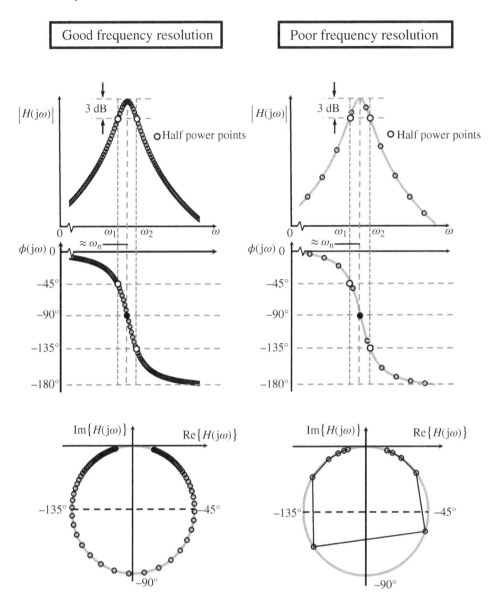

Figure 2.13 Example of the displacement FRF used for damping estimation, showing cases when the frequency resolution is good and when it is poor.

Nyquist plot are depicted in each case. The thick grey line is the actual FRF, and the small circles are data points, from a measurement, for example. On the left is the case where the frequency resolution is adequate for the amount of damping in the system, and on the right is the case is where the frequency resolution is poor, which would lead to a very poor estimate of damping. It is clear that in the case on the left there are many data points between the half-power points, and this is evident in the modulus, phase, and Nyquist plots. However, in the case on the right, there are only two points within this frequency range. This is most clear in the Nyquist plot, which illustrates the value of examining the FRF in this way. Note that, when the frequency resolution is poor, if the measured points are connected with distinct straight lines, then the result is a discontinuous rather than a smooth curve.

To determine the damping using the half-power points from an FRF the following procedure is advised.

1. Examine the Nyquist plot to see if the frequency resolution is adequate. If it is, the Nyquist plot will be smooth, but if it is not the Nyquist plot will have jagged edges as in Figure 2.13. If the Nyquist plot is not smooth then greater frequency resolution is required from the measured data. The way this is achieved is discussed in several of the later chapters. Note that for a lightly damped system, very fine frequency resolution is required to obtain an accurate estimate of damping, and this is illustrated in MATLAB Example 2.4.

2. Determine the half power point frequencies and the resonance frequency. This can be done by examining the modulus, but it can be easier to use the phase plot, by determining the frequencies at which the phase is $-45°$, $-90°$, and $-135°$. Once these frequencies have been determined Eq. (2.39) can be used to estimate the damping ratio. Note that there are alternative ways to determine damping using the Nyquist plot, as discussed by Ewins (2000).

MATLAB Example 2.4

In this example the damping is estimated from the FRF of an SDOF system, which has a damping ratio of $\zeta = 0.01$, using the half-power point method.

```
clear all

%% parameters
m=1;                        % [kg]              % mass
k=10000;                    % [N/m]             % stiffness
wn=sqrt(k/m);               % [rad/s]           % natural frequency
z=0.01;                                         % damping ratio
c=2*z*wn*m;                 % [Ns/m]            % damping coefficient

%% Frequency vector
df=0.05;                    % [Hz]              % frequency resolution in Hz
F=17;                       % [Hz]              % maximum frequency
f=0:df:F;                   % [Hz]              % frequency vector
w=2*pi*f;                   % [rad/s]

%% Receptance FRF
H = 1./(k-w.^2+j*w*c);      % [N/m]             % receptance FRF

%% Plot results
figure (1)
plot(f,20*log10(abs(H)),'o')                    % modulus
grid;axis square
xlabel('frequency (Hz)');
ylabel('|receptance| (dB ref 1 m/N)');

figure (2)
plot(f,180/pi*angle(H),'o')                     % phase
grid;axis square
xlabel('frequency (Hz)');
ylabel('phase angle (degrees)');

figure (3)
plot(real(H),imag(H),'o')                       % Nyquist
grid;axis square
xlabel('real (receptance) (m/N)');
ylabel('imag(receptance) (m/N)');
```

(Continued)

MATLAB Example 2.4   (Continued)

Results

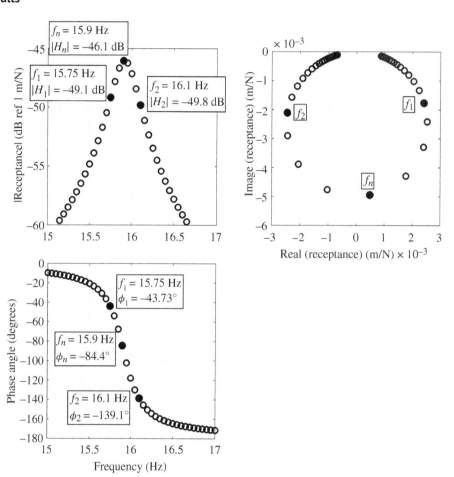

Estimated damping ratio is 0.011, giving a relative error of 10%.

Comments

1. Note the error in the estimate is quite large. The frequency resolution is 0.05 Hz, which is quite a fine frequency resolution. This shows the difficulty in obtaining an accurate damping estimate if the damping is light.
2. An exercise for the reader is to repeat the calculation but use a smaller frequency increment and determine the accuracy of the estimate.
3. An exercise for the reader is to repeat the example, but for damping values of $\zeta = 0.001$ and $\zeta = 0.1$. Note that the amplitude range will need to be adjusted in each case. Compare your results with the damping values estimated using the time domain method in MATLAB Example 2.2.

2.11   Reciprocal FRF

Rather than plot the FRFs of *receptance*, *mobility*, and *accelerance*, sometimes it is preferable to plot the reciprocals of these quantities, which are *dynamic stiffness*, *impedance*, and *apparent mass*, respectively. This can be advantageous if elements are connected in parallel such as the SDOF mass-spring damper system shown in Figure 2.8a. The dynamic stiffnesses of these elements are, respectively, given by

$$K_m(j\omega) = \overset{\text{mass}}{\frac{\overline{F}_m(j\omega)}{\overline{X}(j\omega)}} = -\omega^2 m, \tag{2.40a}$$

$$K_k(j\omega) = \overset{\text{stiffness}}{\frac{\overline{F}_k(j\omega)}{\overline{X}(j\omega)}} = k, \tag{2.40b}$$

$$K_c(j\omega) = \overset{\text{damping}}{\frac{\overline{F}_c(j\omega)}{\overline{X}(j\omega)}} = j\omega c. \tag{2.40c}$$

These sum to give

$$K(j\omega) = \frac{\overline{F}(j\omega)}{\overline{X}(j\omega)} = k - \omega^2 m + j\omega c, \tag{2.41}$$

which is the reciprocal of $H(j\omega)$ given in Eq. (2.26). Note that the impedance and apparent mass can be determined by dividing Eq. (2.41) by $j\omega$ and $-\omega^2$, respectively. Further, it can be seen that the dynamic stiffness has the desirable property that the reactive properties of the system, which store energy, are contained solely in the real part, and the damping properties are contained solely in the imaginary part. If the real part is plotted as a function of the square of frequency and the imaginary part is plotted as a function of frequency, then the graphs are straight lines, and simple curve fitting procedures can be used to estimate the system properties. The real and imaginary parts of the SDOF system are plotted in Figure 2.14 to illustrate plots of the dynamic stiffness.

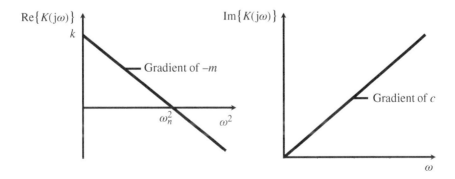

Figure 2.14   Dynamic stiffness of an SDOF system. The real part is plotted as a function of the square of frequency, and the imaginary part is plotted as a function of frequency.

MATLAB Example 2.5

In this example the mass stiffness and damping properties are estimated from the dynamic stiffness FRF of an SDOF system, which has a damping ratio of $\zeta = 0.01$.

```
clear all

%% Parameters
m=1;                        % [kg]                % mass
k=10000;                    % [N/m]               % stiffness
z=0.01;                                           % damping ratio = 0.01
c=2*z*sqrt(m*k);            % [Ns/m]              % damping coefficient

%% Frequency vector
df=1;                       % [Hz]                % frequency resolution in Hz
F=20;                       % [Hz]                % maximum frequency
f=0:df:F;                   % [Hz]                % frequency vector
w=2*pi*f;                   % [rad/s]

%% Dynamic stiffness FRF
K = k-w.^2+j*w*c;           % [N/m]               % receptance FRF

%% Calculations
ff=f.^2;                                          % square of frequency
p = polyfit(ff,real(K),1);                        % least square fit
stiffness=p(2);                                   % estimate of stiffness
mass=-p(1)/(2*pi)^2;                              % estimate of mass

q = polyfit(f,imag(K),1);                         % least square fit
damping=q(1)/(2*pi)/(2*sqrt(mass*stiffness));     % estimate of damping

%% Plot results
figure (1)
plot(ff,real(K),'o')                              % real part vs frequency^2
grid;axis square
xlabel('frequency^2 (Hz^2)');
ylabel('real(dynamic stiffness) (N/m)');

figure (2)
plot(f,imag(K),'o')                               % imaginary part vs frequency
grid;axis square
xlabel('frequency (Hz)');
ylabel('imag(dynamic stiffness) (N/m)');
```

Results

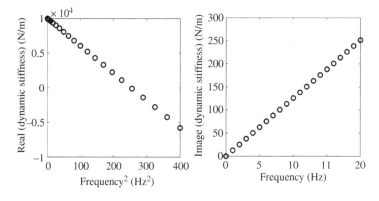

The mass, stiffness, and damping ratio are estimated exactly.

(Continued)

MATLAB Example 2.5   (Continued)

Comments

1. Note that this method can be used to estimate all the properties with relative ease by fitting straight lines to measured data. However, the method relies on good low-frequency measurements, which may not be possible in some cases using accelerometers, so laser displacement sensors may need to be used.
2. An exercise for the reader is to repeat the example, but for damping values of $\zeta = 0.001$ and $\zeta = 0.1$. Compare your results with the damping values estimated using the time domain method in MATLAB Examples 2.2 and 2.4.
3. The method can be modified to estimate the parameters using acceleration data to form the apparent mass, or using velocity data to form impedance. This is left as an exercise for the reader.

2.12   Summary

This chapter has introduced the fundamental idealised components of a vibrating system, namely mass, stiffness, and damping. These are connected is parallel to form an SDOF, which has been studied in detail. Time and frequency domain descriptors, namely the impulse and frequency response functions (IRFs and FRFs), have been introduced and derived for the SDOF system. These are given in Table 2.1.

Table 2.1 Relationships between IRFs and FRFs for an SDOF system.

IRF, ($t \geq 0$)		FRF	
$h(t) = \dfrac{1}{m\omega_d} e^{-\zeta\omega_n t} \sin(\omega_d t)$	$\dfrac{\mathrm{m}}{\mathrm{Ns}}$	Receptance, $H(\mathrm{j}\omega)$ $\dfrac{1}{k - \omega^2 m + \mathrm{j}\omega c}$	$\dfrac{\mathrm{m}}{\mathrm{N}}$
$\dot{h}(t) = \dfrac{\omega_n}{m\omega_d} e^{-\zeta\omega_n t} \cos(\omega_d t + \theta)$	$\dfrac{\mathrm{m}}{\mathrm{Ns}^2}$	Mobility, $H_{\mathrm{vel}}(\mathrm{j}\omega)$ $\dfrac{\mathrm{j}\omega}{k - \omega^2 m + \mathrm{j}\omega c}$	$\dfrac{\mathrm{m}}{\mathrm{Ns}}$
$\ddot{h}(t) = \dfrac{\delta(t)}{m} - \dfrac{\omega_n^2}{m\omega_d} e^{-\zeta\omega_n t} \sin(\omega_d t + \phi)$	$\dfrac{\mathrm{m}}{\mathrm{Ns}^3}$	Accelerance, $H_{\mathrm{acc}}(\mathrm{j}\omega)$ $\dfrac{-\omega^2}{k - \omega^2 m + \mathrm{j}\omega c}$	$\dfrac{\mathrm{m}}{\mathrm{Ns}^2}$

where,

Undamped natural frequency $\omega_n = \sqrt{\dfrac{k}{m}}$

Viscous damping ratio $\zeta = \dfrac{c}{2\sqrt{km}} = \dfrac{c}{2m\omega_n}$

Damped natural frequency $\omega_d = \omega_n \sqrt{1 - \zeta^2}$

$\theta = \tan^{-1}\left(\dfrac{\zeta}{\sqrt{1 - \zeta^2}}\right)$

$\phi = \tan^{-1}\left(\dfrac{2\zeta\sqrt{1 - \zeta^2}}{1 - 2\zeta^2}\right)$

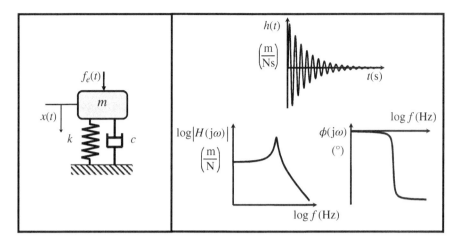

Figure 2.15 Displacement IRF and FRF for an SDOF mass-spring-damper system.

The IRF and FRF are fundamental quantities that are important for all dynamical systems and a thorough understanding of these for an SDOF system is essential prior to studying more complex systems and interpreting experimental data. The displacement IRF and FRF are summarised in Figure 2.15.

For a lightly damped SDOF system, the damping can be estimated in the time domain by

$$\zeta \approx \frac{\text{Grad}}{\omega_n}$$

where ω_n is the undamped natural frequency given in Table 2.1, and Grad is the modulus of the slope of the natural log of the envelope of the decay of free vibration. If the damping is light, it can also be determined in the frequency domain from the resonance frequency ω_n and the half power point frequencies ω_1 and ω_2 by

$$\zeta = \frac{\omega_2 - \omega_1}{2\omega_n}$$

in which the phase angles at ω_1, ω_n, and ω_2 are $-45°$, $-90°$, and $-135°$, respectively. The definition of the FRFs and their reciprocals is given in Table 2.2. Note that if the model or set of measurements involves several force and response positions, such that there is a matrix of FRFs, then reciprocal FRFs are calculated by inverting the matrix rather than by simply inverting an individual FRF.

Table 2.2 Definitions of the FRFs and their reciprocals.

Receptance	$H(j\omega) = \dfrac{\overline{X}(j\omega)}{\overline{F}(j\omega)}$	Dynamic stiffness	$K(j\omega) = \dfrac{1}{H(j\omega)} = \dfrac{\overline{F}(j\omega)}{\overline{X}(j\omega)}$
Mobility	$H_{\text{vel}}(j\omega) = j\omega H(j\omega)$	Impedance	$Z(j\omega) = \dfrac{1}{H_{\text{vel}}(j\omega)}$
Accelerance	$H_{\text{acc}}(j\omega) = -\omega^2 H(j\omega)$	Apparent mass	$M(j\omega) = \dfrac{1}{H_{\text{acc}}(j\omega)}$

References

Ewins, D.J. (2000). *Modal Testing: Theory, Practice and Application*, 2nd Edition. Research Studies Press.

Feldman, M. (2011). *Hilbert Transform Applications in Mechanical Vibration*. Wiley.

Inman, D.J. (2007). *Engineering Vibration*, 3rd Edition. Pearson.

Gardonio, P. and Brennan, M.J. (2002). On the origins and development of mobility and impedance methods in structural dynamics. *Journal of Sound and Vibration*, 249(3), 557–573. https://doi.org/10.1006/jsvi.2001.3879.

Hixson, E.L. (1976). *Mechanical Impedance, Chapter 10 in Shock and Vibration Handbook*, 2nd Edition. (eds. C.M. Harris and C.E. Crede). McGraw-Hill.

Iwanaga, M.K., Brennan, M.J., Tang, B., et al. (2021). Some features of the acceleration impulse response function, *Meccanica*, 56, 169–177. https://doi.org/10.1007/s11012-020-01265-4.

Kovacic, I. and Brennan, M.J. eds. (2011). *The Duffing Equation: Nonlinear Oscillators and Their Behaviour*. Wiley.

Rao, S.S. (2016). *Mechanical Vibrations*, 6th Edition. Pearson.

Thomsen, J.J. (2003). *Vibrations and Stability*. Springer.

Tse, F.S., Morse, I.E., and Hinkle, R.T. (1978). *Mechanical Vibrations – Theory and Applications*, 2nd Edition. Ally and Bacon, Inc.

Worden, K., and Tomlinson, G.R. (2001). *Nonlinearity in Structural Dynamics: Detection, Identification and Modelling*. IoP Publishing.

3

Fourier Analysis

3.1 Introduction

In Chapter 2, it was shown that the way in which a vibrating system is generally characterised, is either by its impulse response function (IRF) in the time domain, or by its frequency response function (FRF) in the frequency domain. Most analysis is conducted in the frequency domain because it is much easier to determine system properties such as natural frequencies and damping, and to relate the system behaviour to its physical properties such as mass, stiffness, and damping. However, data captured during vibration measurements are in the time domain, and many numerical simulations often involve time domain operations. The data can be displacement, velocity, or acceleration depending upon the specific situation and the sensor used, or the quantity of interest. Thus, it is important to be able to transform time domain data to frequency domain data (and vice versa). The way in which this is achieved is by using either the Fourier series (FS) or the Fourier transform (FT), which is named after Jean-Baptiste Joseph Fourier, who developed the techniques in the early part of the nineteenth century (Fourier, 1822). The Fourier transform is arguably the most important mathematical operation in the field of vibration engineering, and as such it is imperative for the vibration engineer to be thoroughly acquainted with Fourier analysis. Nowadays, most data are sampled (or digitised) before being processed, so the Fourier transform that operates on such data – the so-called discrete Fourier transform (DFT) – is of particular interest. In this chapter, this transform and its inverse are derived, and their key features are discussed. The derivations are compact, and include various assumptions that are not necessarily stated explicitly. For further detailed information the reader can consult (Shin and Hammond, 2008), and there are other more general textbooks on the subject, which may be of help to the reader such as Papoulis (1962, 1977), Oppenheim and Schafer (1975), Oppenheim et al. (1997), and Bendat and Piersol (1980, 2000).

3.2 The Fourier Transform (FT)

As mentioned above, signals are generally transformed from the time domain to the frequency domain using the FT, which is sometimes called the Fourier integral. The starting point for the derivation of this transform is the FS, as engineering students are familiar with this from their undergraduate studies. The FS is used to describe a periodic signal, such as that shown in Figure 3.1. The signal has a fundamental period of T_p seconds. It can be decomposed into a DC signal and the sum of harmonics of the fundamental frequency $\omega_1 = 2\pi/T_p$ as shown in the right part of

Virtual Experiments in Mechanical Vibrations: Structural Dynamics and Signal Processing,
First Edition. Michael J. Brennan and Bin Tang.
© 2023 John Wiley & Sons Ltd. Published 2023 by John Wiley & Sons Ltd.
Companion website: www.wiley.com/go/brennan/virtualexperimentsinmechanicalvibrations

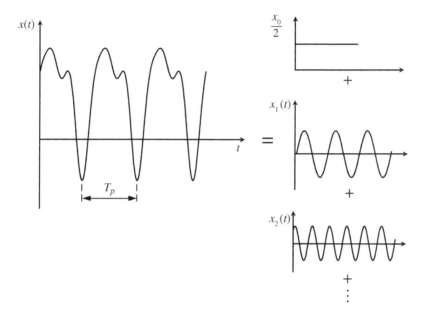

Figure 3.1 A periodic displacement time history and its Fourier components.

Figure 3.1. Each harmonic is defined by an amplitude and phase with respect to the fundamental component. In principle, the sum can involve an infinite number of harmonics, but a finite sum of a few harmonics may give a good approximation to many signals. This is because the components of a signal tend to have smaller amplitudes at higher frequencies, especially for signals corresponding to displacements. A periodic displacement time history can be written as

$$x(t) = \frac{x_0}{2} + \sum_{n=1}^{\infty} x_n(t), \tag{3.1}$$

where $x_n(t) = |\overline{X}|_n \cos(\omega_n t + \phi_n)$, in which $|\overline{X}|_n$ and ϕ_n are the amplitude and phase of the n-th harmonic at frequency ω_n, respectively, where $\omega_n = n\omega_1$, $n = 1, 2, \ldots$. Thus, provided that the amplitude and phase of each harmonic are known then $x(t)$ can be written as a Fourier series. To determine the unknown quantities, it is preferable to write down Eq. (3.1) as

$$x(t) = \frac{a_0}{2} + \sum_{n=1}^{\infty} [a_n \cos(\omega_n t) + b_n \sin(\omega_n t)], \tag{3.2}$$

in which the amplitude $|\overline{X}|_n = \sqrt{a_n^2 + b_n^2}$ and the phase $\phi_n = \tan^{-1}(-b_n/a_n)$. The coefficients in Eq. (3.2) are given by

$$\frac{a_0}{2} = \frac{1}{T_p} \int_0^{T_p} x(t) \mathrm{d}t, \tag{3.3a}$$

which is the mean or DC value,

$$a_n = \frac{2}{T_p} \int_0^{T_p} x(t) \cos(\omega_n t) \mathrm{d}t \quad n = 1, 2, \ldots, \tag{3.3b}$$

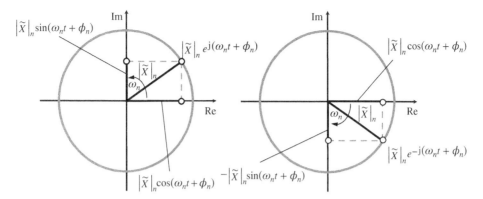

Figure 3.2 Basis functions for the complex Fourier series: two contra rotating vectors, $|\tilde{X}|_n e^{\mathrm{j}(\omega_n t+\phi_n)}$ and $|\tilde{X}|_n e^{-\mathrm{j}(\omega_n t+\phi_n)}$.

and

$$b_n = \frac{2}{T_p} \int_0^{T_p} x(t)\sin(\omega_n t)\mathrm{d}t \quad n = 1,2,\ldots,$$ (3.3c)

are the harmonic components.

The next stage in the derivation of the Fourier transform is to express the Fourier series in terms of complex exponential functions, i.e. phasors or vectors that rotate in the clockwise and anti-clockwise directions as shown in Figure 3.2. Two contra-rotating vectors are shown; $|\tilde{X}|_n e^{\mathrm{j}(\omega_n t+\phi_n)}$, which corresponds to a vector with amplitude $|\tilde{X}|_n$ and phase ϕ_n rotating with an angular velocity of ω_n in the anti-clockwise direction, and $|\tilde{X}|_n e^{-\mathrm{j}(\omega_n t+\phi_n)}$, rotating in the clockwise direction. Note that $\tilde{X}_n = \overline{X}_n/2$ and that these vectors can be written as $\tilde{X}_n e^{\mathrm{j}\omega_n t}$ and $\tilde{X}_n^* e^{-\mathrm{j}\omega_n t}$, where * denotes the complex conjugate, because $\tilde{X}_n = |\tilde{X}|_n e^{\mathrm{j}\phi_n}$ and $\tilde{X}_n^* = |\tilde{X}|_n e^{-\mathrm{j}\phi_n}$. As can be seen in Figure 3.2, the cosine function is related to the projection of the rotating vector onto the real axis because $\cos(\omega_n t) = \mathrm{Re}\left\{e^{\mathrm{j}\omega_n t}\right\}$, and the sine function is related to the projection of the rotating vector onto the imaginary axis because $\sin(\omega_n t) = \mathrm{Im}\left\{e^{\mathrm{j}\omega_n t}\right\}$. Alternatively, sine and cosine functions can be written as a combination of the rotating vectors using Euler's formula $e^{\pm\mathrm{j}\theta} = \cos\theta \pm \mathrm{j}\sin\theta$, so that

$$\cos(\omega_n t) = \frac{e^{\mathrm{j}\omega_n t} + e^{-\mathrm{j}\omega_n t}}{2}$$ (3.4a)

and

$$\sin(\omega_n t) = \frac{e^{\mathrm{j}\omega_n t} - e^{-\mathrm{j}\omega_n t}}{\mathrm{j}2}.$$ (3.4b)

Equations (3.4a) and (3.4b) can be substituted into Eq. (3.2) to give

$$x(t) = \frac{a_0}{2} + \sum_{n=1}^{\infty}\left[\frac{a_n}{2}\left(e^{\mathrm{j}\omega_n t} + e^{-\mathrm{j}\omega_n t}\right) + \frac{b_n}{\mathrm{j}2}\left(e^{\mathrm{j}\omega_n t} - e^{-\mathrm{j}\omega_n t}\right)\right],$$ (3.5)

which can be rearranged to give

$$x(t) = \frac{a_0}{2} + \sum_{n=1}^{\infty}\frac{a_n - \mathrm{j}b_n}{2}e^{\mathrm{j}\omega_n t} + \sum_{n=1}^{\infty}\frac{a_n + \mathrm{j}b_n}{2}e^{-\mathrm{j}\omega_n t}.$$ (3.6)

Letting $\overline{X}_n = a_n - jb_n$, so that $\overline{X}_n^* = a_n + jb_n$ and noting that $\tilde{X}_n = \overline{X}_n/2$ results in

$$x(t) = \tilde{X}_0 + \sum_{n=1}^{\infty} \tilde{X}_n e^{j\omega_n t} + \sum_{n=1}^{\infty} \tilde{X}_n^* e^{-j\omega_n t}. \tag{3.7}$$

Note that this is simply the sum of a DC term and an infinite sum of contra-rotating vectors. Because $\tilde{X}_n = |\tilde{X}|_n e^{j\phi_n}$, each vector has an amplitude $|\tilde{X}|_n$ and phase ϕ_n and rotates at angular velocity ω_n. Thus, the basis functions for the complex Fourier series are rotating vectors rather than the sine and cosine functions used in the real Fourier series. Noting that $\tilde{X}_n = (a_n - jb_n)/2$, and $\tilde{X}_n^* = (a_n + jb_n)/2$, Eqs. (3.3a, 3.3b, 3.3c) can be combined to give

$$\tilde{X}_0 = \frac{1}{T_p} \int_0^{T_p} x(t) dt, \tag{3.8a}$$

$$\tilde{X}_n = \frac{1}{T_p} \int_0^{T_p} x(t) e^{-j\omega_n t} dt, \tag{3.8b}$$

$$\tilde{X}_n^* = \frac{1}{T_p} \int_0^{T_p} x(t) e^{j\omega_n t} dt. \tag{3.8c}$$

Thus, $\tilde{X}_n^* = \tilde{X}_{-n}$, so that $\sum_{n=1}^{\infty} \tilde{X}_n^* e^{-j\omega_n t}$ can be written as $\sum_{n=-\infty}^{-1} \tilde{X}_n e^{j\omega_n t}$, which means that Eq. (3.7) can be written as

$$x(t) = \sum_{n=-\infty}^{\infty} \tilde{X}_n e^{j\omega_n t}, \tag{3.9}$$

where $\tilde{X}_n = \frac{1}{T_p} \int_0^{T_p} x(t) e^{-j\omega_n t} dt$.

The complex Fourier series contains twice as many Fourier components as the real Fourier series, so each component of the complex Fourier series is half the size of that in the real Fourier series. For each component at a positive frequency, there is a corresponding component at a negative frequency that has the same modulus but opposite phase. This means that the modulus is an even function, and the phase is an odd function. Similarly, the real part of the complex Fourier series is an even function and the imaginary part is an odd function. It is also a discrete spectrum as shown in Figure 3.3. Note that if the periodic time history is symmetric about $t = 0$, then it is an even function. In this case the Fourier series is also an even function, which means that the phase of each Fourier coefficient is either 0° or 180°. Note also that if the time series is symmetric about $x(t) = 0$, then there is no DC component and the Fourier series consists of only odd harmonics. This latter situation occurs for forced vibrations of a linear system if the system vibrates about its static equilibrium position.

The next part of the derivation for the Fourier transform starts from the complex Fourier series. Consider the periodic signal shown in Figure 3.4. For convenience it is taken to be symmetric about $t = 0$ and has a fundamental period of T_p seconds. Following on from Eq. (3.9), the Fourier coefficients for the periodic signal in Figure 3.4 are then given by

$$\tilde{X}_n = \frac{1}{T_p} \int_{-T_p/2}^{T_p/2} x(t) e^{-j\omega_n t} dt. \tag{3.10}$$

Note that $\tilde{X}_n$ is a coefficient for the complex Fourier series, which has a double-sided spectrum (it has both positive and negative frequencies), and thus it is half of the equivalent coefficient for the corresponding real Fourier series, which has a single-sided spectrum (it has only positive frequencies). As shown in Figure 3.4, T_p is extended so that in the limit $T_p \to \infty$. In this case the frequency

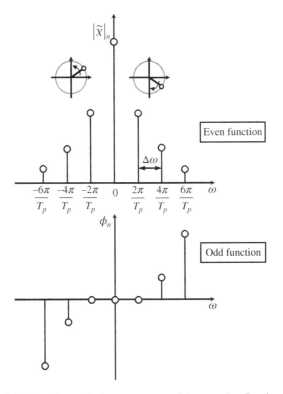

Figure 3.3   Modulus and phase spectrum of the complex Fourier series.

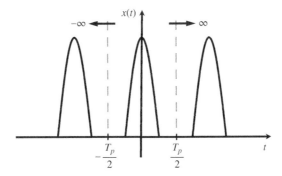

Figure 3.4   A periodic signal with period T_p that is extended to infinity.

difference between the harmonics $\Delta f = \Delta\omega/2\pi = 1/T_p \to 0$, so that Eq. (3.10) becomes

$$\tilde{X}_n \big|_{\substack{T_p \to \infty \\ \Delta f \to 0}} = \Delta f \int_{-T_p/2}^{T_p/2} x(t)e^{-j\omega_n t}\,dt. \tag{3.11}$$

Thus, as T_p increases the number of frequency components in the spectrum increases. However, to maintain the same level of energy in the spectrum there must be a reduction in the amplitude of the Fourier components, so that as $\Delta f \to 0$ then $\tilde{X}_n \to 0$. To overcome this situation, both sides of

Eq. (3.11) are divided by Δf and the limit taken, which results in

$$X(f) = \int_{-\infty}^{\infty} x(t)e^{-j2\pi ft}dt, \tag{3.12}$$

where, because the spectrum is now a continuum, $2\pi f = \omega$ instead of ω_n is written, and $X(f) = \tilde{X}_n/\Delta f|_{T_p \to \infty \atop \Delta f \to 0}$, which is a complex amplitude divided by a frequency bandwidth (sometimes called an amplitude density). Equation (3.12) is the Fourier transform (FT) of $x(t)$. Note that this is fundamentally different to the Fourier series, in which a Fourier coefficient has the same units as the time domain quantity. For example, if $x(t)$ is a displacement and has units of metres (m), the Fourier coefficients also have the units of m. However, with the Fourier transform, the units are different. If the Fourier transform of $x(t)$ is calculated according to Eq. (3.12), then the units are m/Hz. It is important to note that whereas the Fourier series is used to determine the spectrum of a periodic time series, the Fourier transform can be used to determine the spectrum (with different units of course) of any time series.

3.2.1 Example – SDOF system

Consider the SDOF mass-spring-damper system in which m is the mass, k is the stiffness, and c is the damping, described in Chapter 2. It is represented in block diagram form in Figure 3.5, in which displacement is the measured response. The units are shown to make it clear why the force input and displacement output spectra are in terms of amplitude densities, i.e. N/Hz and m/Hz respectively, whereas the unit for the displacement FRF, (called receptance), is m/N, which is not an amplitude density. The relationship between the receptance FRF $H(j\omega)$ and the IRF $h(t)$ is given by

$$H(j\omega) = \int_{-\infty}^{\infty} h(t)e^{-j\omega t}dt, \tag{3.13}$$

where $h(t) = \frac{1}{m\omega_d}e^{-\zeta\omega_n t}\sin(\omega_d t)$ for $t \geq 0$, which is given by Eq. (2.15a), where ω_n is the undamped natural frequency of the system, ζ is the damping ratio, and $\omega_d = \omega_n\sqrt{1-\zeta^2}$ is the

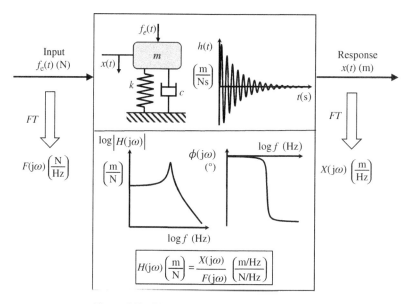

Figure 3.5 Block diagram of an SDOF system.

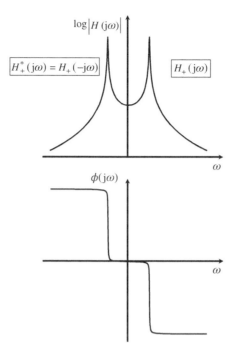

Figure 3.6 Double-sided receptance FRF.

damped natural frequency. Substituting for $\sin(\omega_d t) = \left(e^{j\omega_d t} - e^{-j\omega_d t}\right)/j2$ in $h(t)$ results in

$$h(t) = \frac{u(t)}{j2m\omega_d}(e^{-st} - e^{-s^*t}), \tag{3.14}$$

where $u(t)$ is the Heaviside function, which ensures that $h(t) = 0$ for $t < 0$, $s = \zeta\omega_n - j\omega_d$, and $*$ denotes the complex conjugate. Substituting Eq. (3.14) into Eq. (3.13) and noting that the Fourier transform of $u(t)e^{-st}$ is $1/(s+j\omega)$ results in

$$H(j\omega) = \frac{1}{m\left(\omega_n^2 - \omega^2 + j2\zeta\omega\omega_n\right)}, \quad \text{for } -\infty < \omega < \infty. \tag{3.15a}$$

As $\omega_n = \sqrt{k/m}$ and $\zeta = c/2\sqrt{mk}$, Eq. (3.15a) can be written as

$$H(j\omega) = \frac{1}{k - \omega^2 m + j\omega c}, \quad \text{for } -\infty < \omega < \infty. \tag{3.15b}$$

Note that Eq. (3.15b) is identical to Eq. (2.26), except that it is defined for both positive and negative frequencies. It is plotted in Figure 3.6. It can be seen that the part of the receptance at negative frequencies is equal to the complex conjugate of the receptance at positive frequencies, i.e. $H_+^*(j\omega) = H(-j\omega)$.

3.3 The Discrete Time Fourier Transform (DTFT)

In the previous section it was shown that the Fourier series of a continuous time history results in a discrete spectrum, and the Fourier transform of the same time history results in a continuous spectrum. In many cases, however, a measured time history is not continuous, but is a sequence of sampled data points, such as that illustrated in Figure 3.7. This is the case in numerical simulations,

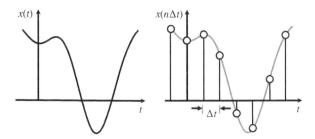

Figure 3.7 A continuous time history $x(t)$ and the same time history sampled every Δt seconds.

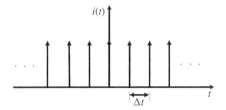

Figure 3.8 An impulse train of delta functions.

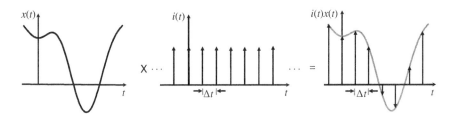

Figure 3.9 Impulse train representation of a sampled time history.

and when measured data are sampled prior to processing in a computer. The time period between each data point is Δt, and the value of the time series $x(t)$ at the n-th data point is $x(n\Delta t)$. The question is, what operation needs to be performed on the sampled data that will give a similar result to the Fourier transform of the continuous time series? This is not tackled directly, but is studied using an impulse train of delta functions $i(t)$, such as that shown in Figure 3.8. The delta function and its properties are described in Appendix E. If the continuous time history is multiplied by a train of delta functions, the result is an impulse train of delta functions modulated by the time history as shown in Figure 3.9. Note that $i(t)x(t) \neq x(n\Delta t)$, (Shin and Hammond, 2008), because the delta function does not have the same units as the sampled signal $x(n\Delta t)$. However, using the train of delta functions in this way is mathematically convenient in the development of the discrete time Fourier transform (DTFT).

The train of delta functions, spaced at time Δt apart, is described by

$$i(t) = \sum_{n=-\infty}^{\infty} \delta(t - n\Delta t), \tag{3.16}$$

and the Fourier transform of $i(t)x(t)$ is given by

$$X_s(f) = \int_{-\infty}^{\infty} i(t)x(t)e^{-j2\pi ft}dt, \tag{3.17}$$

where the subscript s is used to show that $X_s(f)$ is different from $X(f)$. Substituting Eq. (3.16) into Eq. (3.17) results in

$$X_s(f) = \int_{-\infty}^{\infty} x(t) \sum_{n=-\infty}^{\infty} \delta(t - n\Delta t)e^{-j2\pi ft}\mathrm{d}t. \tag{3.18}$$

Changing the order of the summation and integration, and noting from the sifting property of the delta function described in Appendix E, that $\int_{-\infty}^{\infty} x(t)e^{-j2\pi ft}\delta(t - n\Delta t)\mathrm{d}t = x(n\Delta t)e^{-j2\pi fn\Delta t}$, Eq. (3.18) becomes

$$X_s(f) = \sum_{n=-\infty}^{\infty} x(n\Delta t)e^{-j2\pi fn\Delta t}, \tag{3.19}$$

where $x(n\Delta t)$ is the sampled time history as shown in Figure 3.7. The operation described in Eq. (3.19) is known as the discrete time Fourier transform (DTFT). The question remains as to how $X_s(f)$ is related to $X(f)$. To show this, an alternative representation of the impulse train of delta functions is used. In Appendix E it is shown that $i(t)$ can be written in terms of its Fourier series as

$$i(t) = \frac{1}{\Delta t} \sum_{n=-\infty}^{\infty} e^{j2\pi nt/\Delta t}. \tag{3.20}$$

Substituting Eq. (3.20) into Eq. (3.17) results in

$$X_s(f) = \frac{1}{\Delta t} \int_{-\infty}^{\infty} x(t) \sum_{n=-\infty}^{\infty} e^{j2\pi nt/\Delta t}e^{-j2\pi ft}\mathrm{d}t. \tag{3.21}$$

Changing the order of the summation and integration, and noting that $f_s = 1/\Delta t$, which is the frequency at which the data are sampled (sampling frequency), Eq. (3.21) can be written as

$$X_s(f) = f_s \sum_{n=-\infty}^{\infty} \int_{-\infty}^{\infty} x(t)e^{-j2\pi(f-nf_s)t}\mathrm{d}t. \tag{3.22}$$

Noting that $X(f - nf_s) = \int_{-\infty}^{\infty} x(t)e^{-j2\pi(f-nf_s)t}\mathrm{d}t$, Eq. (3.22) can be written as

$$X_s(f) = f_s \sum_{n=-\infty}^{\infty} X(f - nf_s) \tag{3.23}$$

This is an important result, which shows how the DTFT of a sampled time history is related to the FT of the same time history in continuous form. An example of the FT of $x(t)$, which is the displacement of an SDOF system when subject to an impulse force in the form of a delta function, and the DTFT of the sampled version $x(n\Delta t)$ are shown in Figure 3.10. Examining this figure and Eq. (3.23) it can be seen that there are several important features of the DTFT of $x(n\Delta t)$ compared to the FT of $x(t)$:

- The spectrum $X_s(f)$ is a scaled periodic version of $X(f)$, repeating every f_s Hz, and has components at both positive and negative frequencies. Note that the units for $X_s(f)$ are the same as the units for $x(n\Delta t)$, i.e. it is not an amplitude density which is the case for $X(f)$. This can be easily seen by examining Eq. (3.19), and Eq. (3.23), which contains the scaling factor f_s.
- $X_s(f)$ is a continuous function, even though it is derived from a discrete time history.
- The part of the spectrum of $X_s(f)$, which contains all the information related to the original continuous time history is in the frequency range $0 \le f \le f_s/2$.
- The part of the spectrum of $X_s(f)$ in the frequency range $f_s/2 < f \le f_s$ is the complex conjugate of $X_s(f)$ in the frequency range $0 \le f \le f_s/2$.

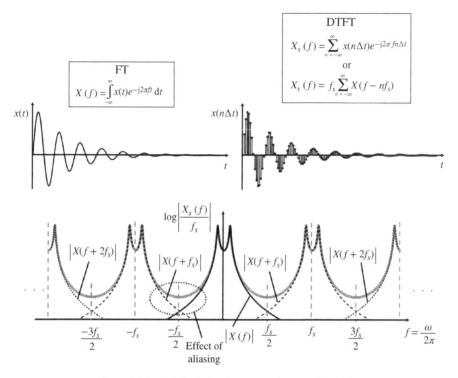

Figure 3.10 DTFT of $x(n\Delta t)$ compared to the FT of $x(t)$.

- When $X_s(f)$ is calculated from $x(n\Delta t)$ using Eq. (3.19), the spectrum $X(f)$ cannot be faithfully determined for all frequencies in the range $0 \leq f \leq f_s/2$. This is due to the phenomenon of aliasing, which can occur when data are sampled. The effect of aliasing can be seen in Figure 3.10 and is discussed in detail in Chapter 4.

3.4 The Discrete Fourier Transform (DFT)

The final step in the derivation of the DFT is to discretise $X_s(f)$, so that $X_s(f)$ is sampled every Δf Hz, and there are a finite number of data points, 1, 2, 3, ..., N in the spectrum. Thus, there are N data points in the time domain, with the first data point at $t = 0$ and the N-th data point coinciding with the time length of the data T, and there are N data points in the frequency domain with the first data point at $f = 0$ and the N-th data point coinciding with the sampling frequency f_s. The frequency spacing (or frequency resolution) is given by $\Delta f = f_s/N$, and the k-th frequency is given by $k\Delta f = kf_s/N$. In this case, by substituting for $f = kf_s/N$ into Eq. (3.19), and taking into account the finite number of data points N in the sampled time history, results in

$$X(k\Delta f) = \sum_{n=0}^{N-1} x(n\Delta t)e^{-j(2\pi/N)kn}, \tag{3.24}$$

which is the DFT of $x(n\Delta t)$. Note $X(k\Delta f)$ is a periodic function, with a period of N. Note also that the sampling in frequency effectively imposes a periodic structure on the sampled time series $x(n\Delta t)$. Thus, when the DFT is used to transform a finite length signal with duration T, into the frequency domain, it is implicit that the signal is a single period of a periodic sequence that repeats every T seconds.

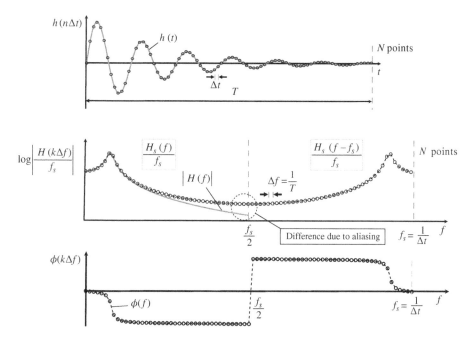

Figure 3.11 DFT of $h(n\Delta t)$ compared to the FT of $h(t)$ of the displacement IRF of an SDOF system.

The DFT is calculated by a computer using an algorithm known as the fast Fourier transform (FFT). The modern generic FFT algorithm is generally attributed to (Cooley and Tukey, 1965), but a similar version had been used by Karl Friedrich Gauss in 1805 to interpolate the orbits of asteroids from sample observations (Heideman et al., 1984). More details about FFT algorithms can be found in (Kumar et al., 2019).

To illustrate the key parameters to be chosen when using the DFT to transform sampled data from the time domain to the frequency domain, consider the sampled displacement IRF $h(n\Delta t)$ shown in Figure 3.11. The two parameters are the sampling frequency of $f_s = 1/\Delta t$ Hz and the time duration T seconds. Overlaid in the figure is the continuous IRF $h(t)$. It is important to have a long enough time history to capture the decay of vibration, which is governed by the product of the damping ratio and the undamped natural frequency in this case, i.e. $\zeta\omega_n$. The samples are denoted by small circles. Note that although the samples are explicitly shown in this figure, in many cases they are not shown. Lines are generally drawn between the samples, giving a plot the appearance of being continuous.

The amplitude and phase of the receptance FRF are also shown in Figure 3.11. The theoretical FRF and the FRF calculated using the DFT of the sampled IRF are overlaid. Note how the parameters chosen when sampling the IRF affect the FRF, which has N points – the same number as the IRF, a frequency range from $0 - f_s$ Hz, and a frequency resolution of $\Delta f = 1/(N\Delta t)$. All the information from $h(n\Delta t)$ is contained in the lower half of the frequency range from $0 \leq f < f_s/2$ Hz. The information in the frequency range from $f_s/2$ to f_s Hz is that shown in Figure 3.6 for negative frequencies. As noted previously, this is simply the complex conjugate of the corresponding component for positive frequencies. As can be seen in Figure 3.11, the spectrum calculated using the DFT is restricted to the frequency range $0 - f_s$ Hz, in which there are N points, and $H(N\Delta f - k\Delta f) = H^*(k\Delta f)$. It can also be seen in Figure 3.11 that, as with the DTFT, the scaling factor between the actual spectrum and that calculated using the DFT, is the sampling frequency f_s.

In Figure 3.11, it is clear that at frequencies close to $f_s/2$ there is a difference between the theoretical FRF and that calculating using the DFT of the sampled IRF. This occurs because of under sampling, which means that high-frequency components of the original time series are not accurately captured and manifest themselves at low frequencies. This effect is called aliasing and is discussed in detail in Chapter 4.

There is a subtle difference between the spectrum calculated using the DFT if there is an odd or an even number of points. To illustrate the difference, consider two simple situations shown in Figure 3.12. Two identical time histories of 1 second duration are sampled at different rates, one at 4 Hz and one at 5 Hz. The frequency resolutions of the two spectra are thus both 1 Hz, but the number of points in each case, given by $N = f_s T + 1$, is 5 and 6, respectively. In the first case (an odd number of points), the values of the spectral components up to and including $f_s/2 = 2$ Hz are the DC component, and X_1 at 1 Hz, and X_2 at 2 Hz. With the exception of the DC term, these components appear in the second half of the spectrum in complex conjugate form, so that X_1^* and X_2^* occur at 3 and 4 Hz, respectively. In the second case (an even number of points), there is no component at $f_s/2 = 2.5$ Hz. As in the previous cases the spectral components in the lower part of the spectrum are the DC component, and X_1 at 1 Hz, and X_2 at 2 Hz. However, now the corresponding complex

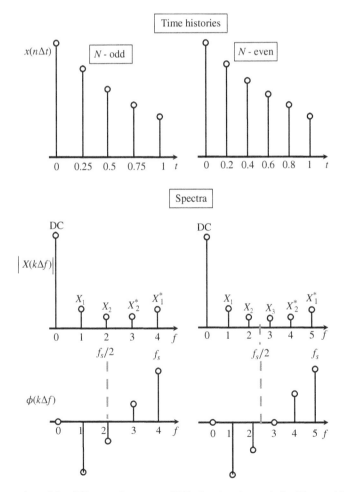

Figure 3.12 Illustration of the difference between a DFT of a signal sampled with an odd number of points compared to a signal sampled with an even number of points.

conjugates occur at frequencies 'mirrored' in the component at 3 Hz, such that X_1^* and X_2^* occur at 4 and 5 Hz, respectively. Note that the component at 3 Hz is purely real.

MATLAB Example 3.1

In this example, the DFT of the displacement IRF of an SDOF system is compared with the theoretical receptance FRF.

```
clear all

%% Parameters
m = 1;                      % [kg]       % mass
k = 10000;                  % [N/m]      % stiffness
z = 0.001;                              % damping ratio
c = 2*z*sqrt(m*k);          % [Ns/m]     % damping factor
wn=sqrt(k/m);               % [rad/s]    % natural frequency
wd=sqrt(1-z^2)*wn;          % [rad/s]    % damped natural frequency

%% Time and frequency parameters
T=100;                      % [s]        % duration of time signal
fs=1000;                    % [Hz]       % sampling frequency
dt=1/fs;                    % [s]        % time resolution
t=0:dt:T;                   % [s]        % time vector
df=1/T;                     % [Hz]       % frequency resolution
f=0:df:fs;                  % [Hz]       % frequency vector
N=fs*T;                                 % number of points - 1

%% Theoretical IRF
h=1/(m*wd)*exp(-z*wn*t).*sin(wd*t); % [m/Ns]    % impulse response

%% Calculation of DFT
H=dt*fft(h);                            % calculation of receptance FRF

%% Theoretical FRF
dff=.001;                   % [Hz]       % frequency resolution
fr=0:dff:fs/2;              % [Hz]       % frequency vector
w=2*pi*fr;                  % [rad/s]    % frequency in rad/s
HH=1./(k-w.^2*m+j*w*c);     % [m/N]      % theoretical receptance FRF

%% Plot the results
figure (1)
plot(t,h,'linewidth',2,'Color',[.6 .6 .6])      % IRF
grid;axis square
xlabel('time (s)');
ylabel('displacement IRF (m/Ns)');

figure (2)
plot(fr,20*log10(abs(HH)))                       % modulus
hold on
plot(f,20*log10(abs(H)))
grid;axis square
xlabel('frequency (Hz)');
ylabel('|receptance| (dB ref 1 m/N)');

figure (3)
plot(fr,180/pi*angle(HH))                        % phase
hold on
plot(f,180/pi*angle(H))
grid;axis square
xlabel('frequency (Hz)');
ylabel('phase (degrees)');
```

(Continued)

MATLAB Example 3.1 (Continued)

```
figure (4)
plot(real(HH),imag(HH))                              % Nyquist
hold on
plot(real(H(1:N/2+1)),imag(H(1:N/2+1)))
grid;axis square
xlabel('real\{receptance\} (m/N)');
ylabel('imag\{receptance\} (m/N)');

figure (5)
semilogx(fr,20*log10(abs(HH)))                       % modulus
hold on
semilogx(f(1:N/2+1),20*log10(abs(H(1:N/2+1))))
grid;axis square
xlabel('frequency (Hz)');
ylabel('|receptance| (dB ref 1 m/N)');

figure (6)
semilogx(fr,180/pi*angle(HH))                        % phase
hold on
semilogx(f(1:N/2+1),180/pi*angle(H(1:N/2+1)))
grid;axis square
xlabel('frequency (Hz)');
ylabel('phase (degrees)');
```

Results

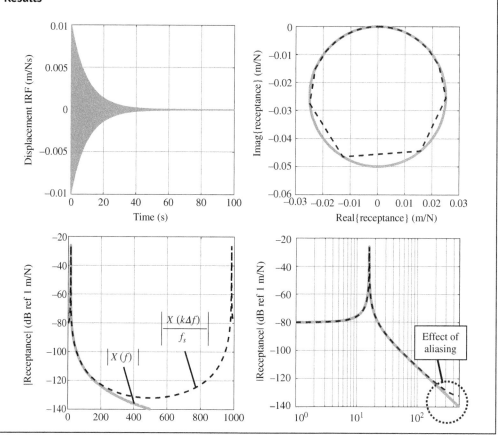

MATLAB Example 3.1   (Continued)

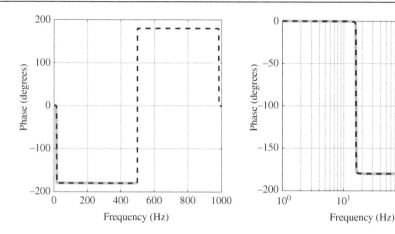

Comments:

1. An exercise for the reader is to explore what happens to the calculated FRF when the sampling frequency/time resolution is changed, and when the length of the time history/frequency resolution is changed.
2. An exercise for the reader is to repeat the exercise for different values of damping.

3.5   Inverse Fourier Transforms

In the same way that the FT transforms time domain data into frequency domain data, the inverse Fourier transform (IFT) transforms frequency domain data into time domain data. The FT, the DTFT, and the DFT all have inverse transforms. These transforms can all be derived in a straightforward manner.

First consider the IFT, in which the continuous amplitude density $X(f)$ is transformed to a continuous time series $x(t)$. Recall Eq. (3.9), which describes the complex Fourier series written in terms of $\tilde{X}_n$ rather than $\overline{X}_n$ to give

$$x(t) = \sum_{n=-\infty}^{\infty} \tilde{X}_n e^{j\omega_n t}, \tag{3.25}$$

where $\tilde{X}_n$ is the displacement amplitude of the n-th harmonic. In the case when the frequency difference between the harmonics $\Delta f \to 0$, then $X(f_n) = \tilde{X}_n / \Delta f$, so that Eq. (3.25) can be written as $x(t) = \sum_{n=-\infty}^{\infty} X(f_n) e^{j2\pi f_n t} \Delta f$, where $2\pi f_n = \omega_n$. This can be further written in continuous form to give the IFT, as

$$x(t) = \int_{-\infty}^{\infty} X(f) e^{j2\pi f t} \mathrm{d}f. \tag{3.26}$$

The inverse discrete time Fourier transform (IDTFT) can be derived from Eq. (3.26), by substituting for $t = n\Delta t$. Noting that within the frequency range $-f_s/2 \leq f \leq f_s/2$, $X(f) = X_s(f)/f_s$, and $X_s(f)$ is periodic in frequency with period of f_s. Eq. (3.26) becomes

$$x(n\Delta t) = \frac{1}{f_s} \int_{-f_s/2}^{f_s/2} X_s(f) e^{j2\pi f n \Delta t} \mathrm{d}f, \tag{3.27}$$

which is the definition of the IDTFT.

The inverse discrete Fourier transform (IDFT) can be derived from Eq. (3.27). First note from Figure 3.10 that $X_s(f)$ for $f_s/2 < f \leq f_s$ is equal to $X_s(f)$ for $-f_s/2 \leq f < 0$ so that Eq. (3.27) becomes

$$x(n\Delta t) = \frac{1}{f_s} \int_0^{f_s} X_s(f) e^{j2\pi f n \Delta t} \mathrm{d}f. \tag{3.28}$$

Next, writing Eq. (3.28) in terms of discrete frequency $k\Delta f$ results in

$$x(n\Delta t) = \frac{\Delta f}{f_s} \sum_{k=0}^{N-1} X_s(k\Delta f) e^{j2\pi k \Delta f n \Delta t}. \tag{3.29}$$

Finally, because $\Delta f = f_s/N$ and $f_s = 1/\Delta t$, Eq. (3.29) becomes

$$x(n\Delta t) = \frac{1}{N} \sum_{k=0}^{N-1} X(k\Delta f) e^{j(2\pi/N)kn}, \tag{3.30}$$

which is the definition of the IDFT. Note that as with $X(k\Delta f)$, $x(n\Delta t)$ is also periodic with period N. Thus, the process of calculating $x(n\Delta t)$ from $X(k\Delta f)$ imposes a periodic structure on $x(n\Delta t)$ in the same way that sampling the time history results in a periodic structure in the frequency domain, as discussed previously. The effect of this is discussed further in Chapter 4.

To illustrate the IDFT, consider again the receptance function shown in Figure 3.11. The starting point is the analytical expression given by $H(j\omega) = 1/(k - \omega^2 m + j\omega c)$. This is sampled in the frequency range with a frequency resolution Δf within the frequency range 0 to $f_s/2$. The 'double-sided' spectrum is then formed by using the complex conjugates of the frequency components of the sampled version of $H(j\omega)$ within this frequency range. Care must be taken to ensure that the part of the spectrum from $f_s/2$ to f_s is formed correctly depending on whether there is an even or an odd number of points, as discussed in Section 3.4. The receptance is shown in Figure 3.13, labelled as $H(f)$ together with the reconstructed complex conjugate mirror of the spectrum. Also shown is $H(k\Delta f)/f_s$, to show the difference between the double-sided spectrum reconstructed from the theoretical FRF compared to that determined from the IRF using the DFT. The difference is because the double-sided spectrum formed from the theoretical FRF only has frequency content up to $f_s/2$, whereas $H(k\Delta f)/f_s$ contains this plus components from an infinite number of aliased frequencies (which is discussed in detail in Chapter 4). Thus, the two spectra are not exactly the same, so when their IDFTs are calculated the IRFs are not exactly the same. These are plotted in the lower part of Figure 3.13. It can be seen that the differences between the IRFs are small. However, this is not always the case, as illustrated in Chapter 4. The difference in the IRF calculated from the theoretical FRF and $h(n\Delta t)$ occurs at the beginning of the time history. This can be seen by first moving the end of the time history labelled A to the beginning as shown in Figure 3.13, which allows a complete picture of the IRF for $t < 0$. Of

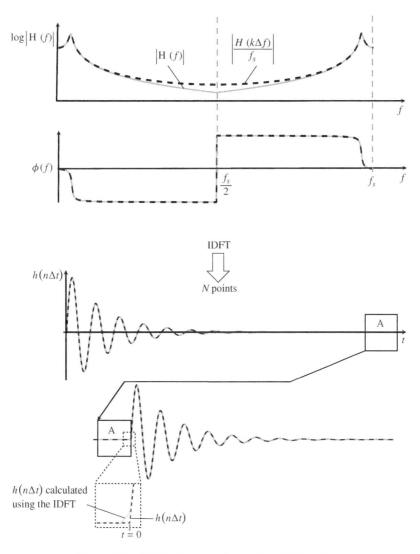

Figure 3.13 IDFT of the receptance of an SDOF system.

course, this should be zero in principle, because the system should remain at rest until impacted (the constraint of causality). Examining the close-up of the beginning of the time history in Figure 3.13, it can be seen that the IRF calculated from the theoretical FRF has a very small non-zero value for $t < 0$. This is due to the implicit low-pass filter applied to the theoretical FRF, by the restriction of the maximum frequency content to $f_s/2$. Issues such as this, concerning the IDFT in transforming data from the frequency to the time domain, are discussed further in Chapter 4.

MATLAB Example 3.2

In this example, the IDFT of the receptance FRF of an SDOF system is compared with the theoretical displacement IRF.

```
clear all

%% Parameters                              % see MATLAB example 3.1
m = 1;                        % [kg]
k = 10000;                    % [N/m]
z = 0.08; c = 2*z*sqrt(m*k);  % [Ns/m]
wn=sqrt(k/m); wd=sqrt(1-z^2)*wn;    % [rad/s]

%% Time and frequency parameters            % see MATLAB example 3.1
T=2;                          % [s]
fs=400;                       % [Hz]
dt=1/fs; t=0:dt:T;            % [s]
df=1/T; f=0:df:fs;            % [Hz]

%% Theoretical IRF
h=1/(m*wd)*exp(-z*wn*t).*sin(wd*t); % [m/Ns]    % IRF

%% Calculation of DFT
H=dt*fft(h);                              % calculation of receptance FRF

%% Theoretical FRF                         % see MATLAB example 3.1
dff=df; fr=0:df:fs/2;         % [Hz]
w=2*pi*fr;                    % [rad/s]
HH=1./(k-w.^2*m+j*w*c);       % [m/N]

%% Calculation of IDFT
Hd=[HH fliplr(conj(HH))];                 % form the double-sided spectrum
Hdd=Hd(1:length(Hd)-1);
hd=fs*ifft(Hdd);              % [m/Ns]    % IDFT

%% Plot results
figure(1)
plot(f,20*log10(abs(Hdd)))                % modulus
hold on
plot(f,20*log10(abs(H)))
xlabel('frequency (Hz)');
ylabel('|receptance| (dB ref 1 m/N)');

figure(2)
plot(f,180/pi*angle(Hdd))                 % phase
hold on
plot(f,180/pi*angle(H))
xlabel('frequency (Hz)');
ylabel('phase (degrees)');

figure(3)
plot(t,hd,t,h)                            % IRF
xlabel('time (s)');
ylabel('IRF (m/Ns)');
```

(Continued)

MATLAB Example 3.2   (Continued)

Results

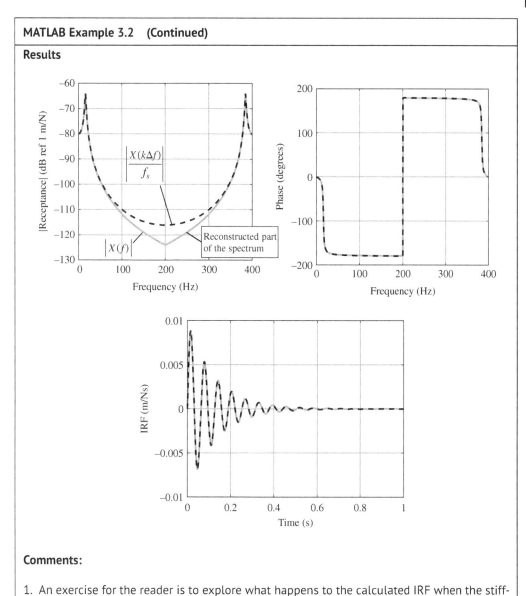

Comments:

1. An exercise for the reader is to explore what happens to the calculated IRF when the stiffness, damping, and mass are changed. Also, investigate what happens to the IRF when the sampling frequency or the frequency resolution is changed.

3.6   Summary

Starting from the real Fourier series (FS), this chapter has derived the complex Fourier series, the Fourier transform (FT), the discrete time Fourier transform (DTFT), and the discrete Fourier transform (DFT). These relationships are summarised in Table 3.1, in which it is assumed that the time series is a displacement that has the unit of metre. The DFT is arguably the most important transform in vibration engineering as it transforms sampled data, which is measured or simulated in the

Table 3.1 Fourier transforms and inverse Fourier transforms for a displacement time history.

	Time → Frequency	Units		Frequency → Time	Units
Fourier series (FS)	$\widetilde{X}(\omega_n) = \dfrac{1}{T_p}\displaystyle\int_0^{T_p} x(t)e^{-j\omega_n t}\,dt$ ↑ discrete ↑ continuous	m		$x(t) = \displaystyle\sum_{n=-\infty}^{\infty} \widetilde{X}(\omega_n)e^{j\omega_n t}$ ↑ continuous ↑ discrete	m
Fourier transform (FT)	$X(\omega) = \dfrac{1}{2\pi}\displaystyle\int_{-\infty}^{\infty} x(t)e^{-j\omega t}\,dt$ ↑ continuous ↑ continuous	$\dfrac{m}{rad/s}$	Inverse Fourier transform (IFT)	$x(t) = \displaystyle\int_{-\infty}^{\infty} X(\omega)e^{j\omega t}\,d\omega$ ↑ continuous ↑ continuous	m
	$X(f) = \displaystyle\int_{-\infty}^{\infty} x(t)e^{-j2\pi ft}\,dt$ ↑ continuous ↑ continuous	$\dfrac{m}{Hz}$		$x(t) = \displaystyle\int_{-\infty}^{\infty} X(f)e^{j2\pi ft}\,df$ ↑ continuous ↑ continuous	m
Discrete time Fourier transform (DTFT)	$X_s(f) = \displaystyle\sum_{n=-\infty}^{\infty} x(n\Delta t)e^{-j2\pi fn\Delta t}$ ↑ continuous ↑ discrete	m	Discrete time inverse Fourier transform (DTIFT)	$x(n\Delta t) = \dfrac{1}{f_s}\displaystyle\int_{-f_s/2}^{f_s/2} X_s(f)e^{j2\pi fn\Delta t}\,df$ ↑ discrete ↑ continuous	m
Discrete Fourier transform (DFT)	$X(k\Delta f) = \displaystyle\sum_{n=0}^{N-1} x(n\Delta t)e^{-j(2\pi/N)kn}$ ↑ discrete ↑ discrete	m	Discrete inverse Fourier transform (DIFT)	$x(n\Delta t) = \dfrac{1}{N}\displaystyle\sum_{k=0}^{N-1} X(k\Delta f)e^{j(2\pi/N)kn}$ ↑ discrete ↑ discrete	m

time domain, to frequency domain data, where most of the analysis is conducted. As shown in this chapter, there are some issues when working with sampled data, such as aliasing and scaling, and because the actual continuous spectrum is important from a physical point view, the relationship between the sampled and continuous spectra is extremely important. This has been studied briefly in this chapter, but it is explored in depth in Chapter 4. Also of importance is the inverse Fourier transform (IFT), which transforms data from the frequency domain to the time domain.

References

Bendat, J.S. and Piersol, A.G. (1980). *Engineering Applications of Correlation and Spectral Analysis*. Wiley.

Bendat, J.S. and Piersol, A.G. (2000). *Random Data: Analysis and Measurement Procedures*, 3rd Edition, Wiley-Interscience.

Cooley, J.W. and Tukey, J.W. (1965). An algorithm for the machine calculation of complex Fourier series. *Mathematics of Computation*, 19(90), 297-301. https://doi.org/10.1090/S0025-5718-1965-0178586-1.

Fourier, J-B. J. (1822). *Théorie analytique de la chaleur*, Gauthier-Villars et Fils, Imprimeurs-Libraires du Bureau, des Longitudes, de L'École Polythechnique.

Heideman, T.M., Johnson, D.H., and Burrus, C.S. (1984). Gauss and the history of the fast Fourier transform. *IEEE ASSP Magazine*, 1(4), 14-21. https://doi.org/10.1109/MASSP.1984.1162257.

Kumar, G.G., Sahoo, A.K. and Meher, P.K. (2019). 50 years of FFT algorithms and applications. *Circuits, Systems, and Signal Processing* 38(12), 5665–5698. https://doi.org/10.1007/s00034-019-01136-8.

Oppenheim, A.V. and Schafer, R.W. (1975). *Digital Signal Processing*. Prentice Hall International.

Oppenheim, A.V., Willsky, A.S., and Hamid Nawab, S. (1997). *Signals and Systems*, 2nd Edition. Prentice Hall International.

Papoulis, A. (1962). *The Fourier Integral and its Applications*. McGraw-Hill.

Papoulis, A. (1977). *Signal Analysis*. McGraw-Hill.

Shin, K. and Hammond, J.K. (2008). *Fundamentals of Signal Processing for Sound and Vibration Engineers*. Wiley.

4

Numerical Computation of the FRFs and IRFs of an SDOF System

4.1 Introduction

In Chapter 2 the impulse response function (IRF), a time domain quantity, and the frequency response function (FRF), a frequency domain quantity, were introduced as important descriptors of vibrating systems. The way in which the FRF can be calculated from the IRF using the Fourier transform (FT), and the calculation of the IRF from the FRF using the inverse Fourier transform (IFT) were discussed in Chapter 3. In practice these operations are carried out using sampled data in the time and the frequency domains using the discrete versions of these transforms, i.e. the DFT and the IDFT, which were also discussed in Chapter 3. The processing of sampled data results in IRFs and FRFs that have some differences to their continuous counterparts. When analysing these quantities, it is important to know which of their features are due to the physical properties of the vibrating system, and which of their features are artefacts due to signal processing. In this chapter, three IRFs and their counterpart FRFs are analysed in detail for an SDOF system. They are the displacement, velocity, and acceleration IRFs, and the receptance, mobility, and acceleration FRFs, as these are quantities most frequently involved in the modelling and measurement of vibrating systems. The theoretical IRFs and FRFs are plotted in Figure 4.1, together with the input–output relationship of an SDOF system shown in block diagram form. The FRFs are plotted on log–log axes so that the low- and high-frequency asymptotes governed by stiffness and mass, respectively, are represented by straight lines as they are proportional to $f^0, f^{\pm 1}$, or $f^{\pm 2}$ depending on the FRF.

When a continuous time series is sampled and its DFT calculated, two problems occur. The first is aliasing, which was seen in Chapter 3, and discussed at length in Appendix F. The other is distortion of the signal if it has a non-zero value at the beginning or end of the time history. Aliasing occurs for all the FRFs considered, but distortion only occurs for velocity and acceleration IRFs, and their corresponding FRFs. Distortion does not occur for the displacement IRF and the receptance FRF, because the IRF is zero at the beginning, and is zero at the end provided that the time window used to capture the data is long enough.

4.2 Effect of Sampling on the FRFs

To gain some insight into the parameters that cause aliasing and distortion, some analytical expressions are derived for the FRFs at zero frequency $f = 0$, and at half the sampling frequency $f = f_s/2$.

Virtual Experiments in Mechanical Vibrations: Structural Dynamics and Signal Processing,
First Edition. Michael J. Brennan and Bin Tang.
© 2023 John Wiley & Sons Ltd. Published 2023 by John Wiley & Sons Ltd.
Companion website: www.wiley.com/go/brennan/virtualexperimentsinmechanicalvibrations

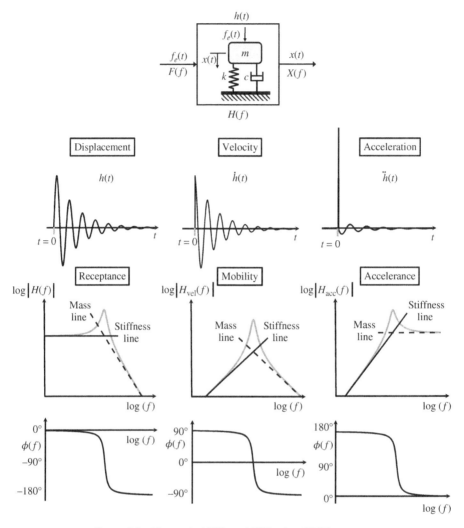

Figure 4.1 Theoretical IRFs and FRFs of an SDOF system.

The effects are different in the receptance, mobility, and accelerance FRFs, so they are considered separately.

4.2.1 Receptance

As discussed in Chapters 2 and 3, the displacement IRF is given by $h(t) = \frac{1}{m\omega_d}e^{-\zeta\omega_n t}\sin(\omega_d t)$ for $t \geq 0$ and the receptance FRF is given by $H(j\omega) = 1/(k - \omega^2 m + j\omega c)$, for $-\infty < \omega < \infty$.

Aliasing at f = 0
The value of the DFT($h(t)$)/f_s at a frequency of zero is the sum of the actual value of $H(0)$ plus all of the aliases, which occur at frequencies of $f = nf_s$ for integer values of n between $-\infty$ and ∞, as shown in Appendix F. It is given by

$$\frac{\text{DFT}(h(t))}{f_s}\bigg|_{f=0} = \sum_{n=-\infty}^{\infty} H(nf_s). \tag{4.1}$$

where $H(nf_s) = 1/(k - (2\pi nf_s)^2 m + j2\pi nf_s c)$. Now, $H(0) = 1/k$, and provided that $f_s \gg f_n$ in which $f_n = (\sqrt{k/m})/2\pi$, then $H(nf_s) \approx -1/[(2\pi nf_s)^2 m]$, so that Eq. (4.1) becomes

$$\left.\frac{\text{DFT}(h(t))}{f_s}\right|_{f=0} \approx \frac{1}{k} - \frac{1}{2(\pi f_s)^2 m}\sum_{n=1}^{\infty}\frac{1}{n^2}. \tag{4.2a}$$

Noting that $\sum_{n=1}^{\infty}\frac{1}{n^2} = \frac{\pi^2}{6}$ (Abramowitz and Stegun, 2014), Eq. (4.2a) becomes

$$\left.\frac{\text{DFT}(h(t))}{f_s}\right|_{f=0} \approx \frac{1}{k} - \frac{1}{12f_s^2 m}. \tag{4.2b}$$

The value of the aliased version of $H(0)$, given by Eq. (4.2b) divided by the approximate true value of $H(0) = 1/k$, is $1 - \frac{\pi^2}{3}\left(\frac{f_n}{f_s}\right)^2$. Note that the aliased version of $H(0)$ is less than the true value of $H(0)$, and that the amount by which the FRF is changed due to aliasing is a function of the ratio of the natural frequency of the system to the sampling frequency.

Aliasing at $f = f_s/2$

Following the procedure for determining the value of $\text{DFT}(h(t))/f_s$ at $f = 0$, the value of the $|\text{DFT}(h(t))|/f_s$ at a frequency of $f_s/2$ is given by

$$\left.\left|\frac{\text{DFT}(h(t))}{f_s}\right|\right|_{f=f_s/2} = \left|\sum_{n=-\infty}^{\infty} H((n-1/2)f_s)\right|. \tag{4.3}$$

If $f_s \gg f_n$, then $H((n-1/2)f_s) \approx -1/\left[(2n-1)^2\pi^2 f_s^2 m\right]$, so that Eq. (4.3) becomes

$$\left.\left|\frac{\text{DFT}(h(t))}{f_s}\right|\right|_{f=f_s/2} \approx \frac{2}{(\pi f_s)^2 m}\sum_{n=1}^{\infty}\frac{1}{(2n-1)^2}. \tag{4.4a}$$

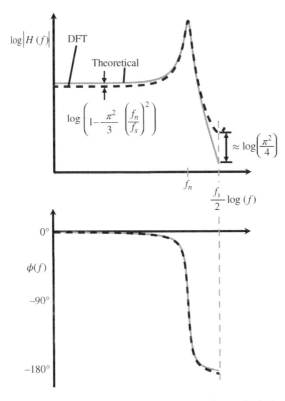

Figure 4.2 Receptance FRF of an SDOF system, showing the effects of aliasing on the $\text{DFT}(h(t))/f_s$.

Noting that $\sum_{n=1}^{\infty} \frac{1}{(2n-1)^2} = \frac{\pi^2}{8}$ (Abramowitz and Stegun, 2014), Eq. (4.4a) becomes

$$\left|\frac{\text{DFT}(h(t))}{f_s}\right|_{f=f_s/2} \approx \frac{1}{4f_s^2 m}. \tag{4.4b}$$

The value of the aliased version of $|H(f_s/2)|$ given by Eq. (4.4b), divided by the approximate true value of $|H(f_s/2)| = 1/\left(\pi^2 f_s^2 m\right)$, is simply $\pi^2/4$. This shows that the difference between the value of the aliased version of $|H(f_s/2)|$ and its actual value is a constant, independent of the sampling frequency, which means that aliasing *always* occurs.

An example of the modulus and phase of the receptance FRF, and its estimate calculated by applying the DFT to the displacement IRF, is shown in Figure 4.2. The differences between the actual FRF and the estimate at $f = 0$ and $f = f_s/2$ due to aliasing are also shown.

MATLAB Example 4.1

In this example, the effects due to aliasing on the receptance FRF of an SDOF system are examined. Two cases are considered, one where aliasing is apparent at very low frequencies and one where aliasing is apparent at high frequencies.

```
clear all

%% Parameters                                  % see MATLAB Example 3.1
m = 1;                      % [kg]             % large k to illustrate aliasing
k = 500000;%k=10000;        % [N/m]            at low frequencies, and small k
z = 0.01; c = 2*z*sqrt(m*k);  % [Ns/m]         to illustrate aliasing at high
wn=sqrt(k/m); wd=sqrt(1-z^2)*wn; % [rad/s]     frequencies

%% Time and frequency parameters               % see MATLAB Example 3.1
T=100;                      % [s]
fs=400;                     % [Hz]
dt=1/fs; t=0:dt:T;          % [s]
df=1/T; f=0:df:fs;          % [Hz]
N=length(t);

%% Theoretical receptance IRF
h=1/(m*wd)*exp(-z*wn*t).*sin(wd*t); % [m/Ns]   % IRF

%% Calculation of DFT
H=dt*fft(h);                                   % calculation of receptance FRF

%% Theoretical receptance FRF
dff=.001; fr=0:dff:fs;      % [Hz]
w=2*pi*fr;                  % [rad/s]
HH=1./(k-w.^2*m+j*w*c);     % [m/N]            % theoretical FRF

%% Calculation of aliased response
for p=1:20;
    f1=(p-1)*fs:dff:p*fs;   % [Hz]             % frequency vector
    w1=2*pi*f1;             % [rad/s]
    HP(p,:)=1./(k-w1.^2*m+j*w1*c);  % [m/N]    % aliased FRF for +ve freq.
    f2=-p*fs:dff:-(p-1)*fs; % [Hz]             % frequency vector
    w2=2*pi*f2;             % [rad/s]
    HM(p,:)=1./(k-w2.^2*m+j*w2*c);  % [m/N]    % aliased FRF for -ve freq.
end
```

(Continued)

MATLAB Example 4.1   (Continued)

```
MP=sum(HP); MS=(sum(HM));                    % summing all aliases
HT1=MP+MS;                                   % total aliased FRF
HT=HT1(1:(N+1)/2);
HA=[HT fliplr(conj(HT))];                    % double sided spectrum
HA1=HA(1:length(f));
H0=1/k-1/(12*fs^2*m);                        % value of aliased FRF at f=0
H1_2=1/(4*fs^2*m);                           % value of aliased FRF at f=fs/2

%% Plot results
figure (1)                                   % figure for small k
semilogx(fr,20*log10(abs(HH)))               % theoretical FRF
hold on
semilogx(f,20*log10(abs(H)))                 % FRF from DFT
hold on
semilogx(fr,20*log10(abs(HT1)))              % theoretical aliased FRF
hold on
plot(1,20*log10(H0),'o')                     % value of HT1 at f=0
plot(1,20*log10(abs(1/k)),'o')               % value of HH at f=0
xlabel('frequency (Hz)');
ylabel('|receptance| (dB ref 1 m/Ns)');

figure (2)                                    % figure for large k
semilogx(fr,20*log10(abs(HH)))               % theoretical FRF
hold on
semilogx(f,20*log10(abs(H)))                 % FRF from DFT
hold on
semilogx(fr,20*log10(abs(HT1)))              % theoretical aliased FRF
hold on
plot(fs/2,20*log10(abs(1/((pi*fs).^2*m))),'o')  % value of abs(HT1) at f=fs/2
plot(fs/2,20*log10(H1_2),'o')                % value of abs(HH) at f=fs/2
xlabel('frequency (Hz)');
ylabel('|receptance| (dB ref 1 m/Ns)');
```

Results

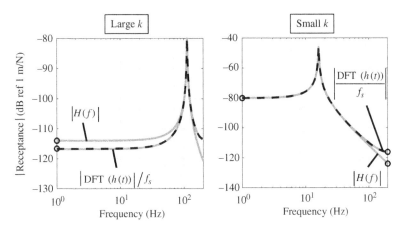

Comments

1. An exercise for the reader is to explore what happens to the FRF calculated using the DFT when the damping and mass are changed. Also, plot the phase as a function of frequency to demonstrate that aliasing does not have a large effect on the phase spectrum.

4.2.2 Mobility

The mobility FRF is the Fourier transform of the velocity IRF discussed in Chapter 2. The velocity IRF is given by $\dot{h}(t) = \frac{\omega_n}{m\omega_d} e^{-\zeta\omega_n t} \cos(\omega_d t + \theta)$ for $t \geq 0$, where $\theta = \tan^{-1}(\zeta/\sqrt{1-\zeta^2})$, and the mobility FRF is determined by calculating the Fourier transform of $\dot{h}(t)$ to give $H_{vel}(j\omega) = j\omega/(k - \omega^2 m + j\omega c)$, for $-\infty < \omega < \infty$. The effects of aliasing can be determined in a similar way to that described for receptance, but there is an additional complication in this case due to sampling, which distorts the velocity IRF. This is shown in Figure 4.3. The velocity IRF changes instantaneously from 0 to $1/m$ at $t = 0$, because of the change in momentum of the mass due to the excitation by the impulsive force described by a delta function. However, this behaviour cannot be captured by sampling. To illustrate what happens, a close-up of $\dot{h}(t)$ is shown in the inset in Figure 4.3. It can be seen that an additional component has to be considered, which is given by

$$\dot{h}_a(t) = \frac{1}{m\Delta t} t + \frac{1}{m} \qquad -\Delta t \leq t \leq 0, \tag{4.5}$$

due to a sample, which has a value of zero at $t = -\Delta t$. This is because the Fourier transform assumes that the time history to be transformed is periodic, and this is the last sample of the previous period. The modified velocity IRF is, therefore, given by

$$\dot{h}_M(t) = \dot{h}(t) + \dot{h}_a(t). \tag{4.6}$$

The Fourier transform of $\dot{h}_M(t)$ is equal to $\mathcal{F}\left\{\dot{h}(t)\right\} + \mathcal{F}\left\{\dot{h}_a(t)\right\}$, where the $\mathcal{F}\left\{\dot{h}_a(t)\right\}$ is given by[1]

$$H_a(j\omega) = \frac{j}{\omega m}(1 - e^{j\omega\Delta t/2}\mathrm{sinc}(f\Delta t)). \tag{4.7}$$

Note that the $\mathrm{DFT}(\dot{h}(t))/f_s$ is the same as the aliased frequency domain sampled version of $\mathcal{F}\{\dot{h}_M(t)\}$.

***The value of the* $\mathrm{DFT}(\dot{h}(t))/f_s$ *at* $f = 0$**
The value of the $\mathrm{DFT}(\dot{h}(t))/f_s$ at zero frequency is the sum of $H_{vel}(0) + H_a(0)$ plus all of their aliases, which occur at frequencies of $f = nf_s$ for integer values of n between $-\infty$ and ∞. It is

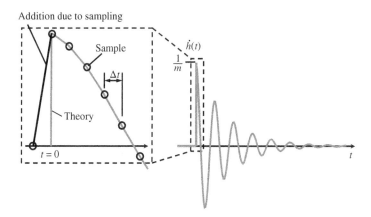

Figure 4.3 Velocity IRF of an SDOF system. The inset shows the details of sampling at the start of the IRF, close to $t = 0$.

---

1 $\mathrm{sinc}(f\Delta t) = \sin(\pi f \Delta t)/(\pi f \Delta t)$, which is the normalized sinc function.

given by

$$\frac{\text{DFT}(\dot{h}(t))}{f_s}\bigg|_{f=0} = \sum_{n=-\infty}^{\infty} (H_{\text{vel}}(nf_s) + H_a(nf_s)). \tag{4.8a}$$

As discussed in Chapter 3, the part of the FRF for negative frequencies is related to the part of the FRF for positive frequencies by its complex conjugate, so that Eq. (4.8a) becomes

$$\frac{\text{DFT}(\dot{h}(t))}{f_s}\bigg|_{f=0} = H_{\text{vel}}(0) + 2\sum_{n=1}^{\infty} \text{Re}\{H_{\text{vel}}(nf_s)\} + H_a(0) + 2\sum_{n=1}^{\infty} \text{Re}\{H_a(nf_s)\}. \tag{4.8b}$$

Now $H_{\text{vel}}(0) = 0$, so the contribution to the DC value of H_{vel} is only due to aliasing. The FRF at frequency nf_s is given by $H_{\text{vel}}(nf_s) = \text{j}2n\pi f_s/(k - (2n\pi f_s)^2 m + \text{j}2n\pi f_s c)$, which can be approximated at high frequencies, when $f_s \gg f_n$, to

$$H_{\text{vel}}(nf_s) \approx \frac{\text{j}}{-2n\pi f_s m + \text{j}c}. \tag{4.9a}$$

This can be further approximated to

$$H_{\text{vel}}(nf_s) \approx \frac{c - \text{j}2n\pi f_s m}{(2n\pi f_s m)^2}, \tag{4.9b}$$

which means that $2\sum_{n=1}^{\infty} \text{Re}\{H_{\text{vel}}(nf_s)\} \approx \frac{2c}{(2\pi f_s m)^2} \sum_{n=1}^{\infty} \frac{1}{n^2}$. Noting that $\sum_{n=1}^{\infty} \frac{1}{n^2} = \frac{\pi^2}{6}$ (Abramowitz and Stegun, 2014), this becomes

$$2\sum_{n=1}^{\infty} \text{Re}\{H_{\text{vel}}(nf_s)\} \approx \frac{c}{12f_s^2 m^2}, \tag{4.10}$$

which is the contribution to response at zero frequency of the mobility FRF due to aliasing. The contribution to the response at zero frequency from the additional component due to the sampling effect contrasts with that for the mobility FRF. There are no aliases of this component because from Eq. (4.7), $\text{sinc}(nf_s\Delta t) = \text{sinc}(n) = 0$ provided that $n \neq 0$; however, $H_a(0) = \frac{1}{2f_s m}$ (which can be verified by writing a series expansion of the terms in Eq. (4.7)). The $\text{DFT}(\dot{h}(t))/f_s\big|_{f=0}$ can be determined by substituting the component parts into Eq. (4.8b) to give

$$\frac{\text{DFT}(\dot{h}(t))}{f_s}\bigg|_{f=0} \approx \frac{1}{2f_s m} + \frac{c}{12f_s^2 m^2}, \tag{4.11a}$$

which can be written as, $\frac{\text{DFT}(\dot{h}(t))}{f_s}\big|_{f=0} \approx \frac{1}{2f_s m}\left(1 + \frac{2\pi}{3}\frac{f_n}{f_s}\zeta\right)$. Provided that $f_n \ll f_s$, this becomes

$$\frac{\text{DFT}(\dot{h}(t))}{f_s}\bigg|_{f=0} \approx \frac{1}{2f_s m}. \tag{4.11b}$$

which means that, provided that the approximation holds, the DFT of the sampled velocity IRF at zero frequency is predominantly due to the additional component added to the IRF by sampling. This is illustrated in an example after the sampling effects at $f = f_s/2$ are considered.

The value of the $\text{DFT}(\dot{h}(t))/f_s$ at $f = f_s/2$

The value of the $\text{DFT}(\dot{h}(t))/f_s$ at $f = f_s/2$ is determined in a similar way to that for $f = 0$. It is given by

$$\frac{\text{DFT}(\dot{h}(t))}{f_s}\bigg|_{f=f_s/2} = \sum_{n=-\infty}^{\infty} (H_{\text{vel}}((n - 1/2)f_s) + H_a((n - 1/2)f_s)), \tag{4.12a}$$

which can also be written as

$$\left.\frac{\text{DFT}(\dot{h}(t))}{f_s}\right|_{f=f_s/2} = 2\sum_{n=1}^{\infty}\text{Re}\{H_{\text{vel}}((n-1/2)f_s)\} + 2\sum_{n=1}^{\infty}\text{Re}\{H_a((n-1/2)f_s)\}. \tag{4.12b}$$

As before, the two components to the aliased FRF are calculated separately, with the mobility calculated first, followed by the additional component due to sampling. The mobility FRF at frequencies of $(n-1/2)f_s$ is given by $H_{\text{vel}}((n-1/2)f_s) = j(2n-1)\pi f_s/(k - ((2n-1)\pi f_s)^2 m + j(2n-1)\pi f_s c)$, which can be approximated at high frequencies, when $f_s \gg f_n$, to

$$H_{\text{vel}}((n-1/2)f_s) \approx \frac{1}{c + j(2n-1)\pi f_s m}. \tag{4.13a}$$

This can be further approximated to

$$H_{\text{vel}}((n-1/2)f_s) \approx \left(\frac{c}{(\pi f_s m)^2} - j\right)\frac{1}{(2n-1)^2}, \tag{4.13b}$$

which means that $2\sum_{n=1}^{\infty}\text{Re}\{H_{\text{vel}}((n-1/2)f_s)\} \approx \frac{2c}{(\pi f_s m)^2}\sum_{n=1}^{\infty}\frac{1}{(2n-1)^2}$. Noting that $\sum_{n=1}^{\infty}\frac{1}{(2n-1)^2} = \frac{\pi^2}{8}$ (Abramowitz and Stegun, 2014), this becomes

$$2\sum_{n=1}^{\infty}\text{Re}\{H_{\text{vel}}((n-1/2)f_s)\} \approx \frac{c}{4f_s^2 m^2}. \tag{4.14}$$

To determine the additional component due to sampling, it is first noted that at frequencies of $(n-1/2)f_s$ Eq. (4.7) becomes

$$H_a((n-1/2)f_s) = \frac{j}{2\pi(n-1/2)f_s m}\left(1 - e^{j\pi(n-1/2)f_s\Delta t}\text{sinc}((n-1/2)f_s\Delta t)\right), \tag{4.15a}$$

which can be written as

$$H_a((n-1/2)f_s) = \frac{1}{\pi f_s m(2n-1)}\left(\frac{2}{\pi(2n-1)} + j\right), \tag{4.15b}$$

so that $2\sum_{n=1}^{\infty}\text{Re}\{H_a((n-1/2)f_s)\} \approx \frac{4}{\pi^2 f_s m}\sum_{n=1}^{\infty}\frac{1}{(2n-1)^2}$. Again, noting that $\sum_{n=1}^{\infty}\frac{1}{(2n-1)^2} = \frac{\pi^2}{8}$, this becomes

$$2\sum_{n=1}^{\infty}\text{Re}\{H_a((n-1/2)f_s)\} \approx \frac{1}{2f_s m}. \tag{4.16}$$

The $\text{DFT}(\dot{h}(t))/f_s|_{f=f_s/2}$ is determined by substituting the component parts into Eq. (4.12b), to give

$$\left.\frac{\text{DFT}(\dot{h}(t))}{f_s}\right|_{f=f_s/2} \approx \frac{c}{4f_s^2 m^2} + \frac{1}{2f_s m}, \tag{4.17a}$$

which can be written as $\left.\frac{\text{DFT}(\dot{h}(t))}{f_s}\right|_{f=f_s/2} \approx \frac{1}{2f_s m}\left(1 + 2\pi\zeta\frac{f_n}{f_s}\right)$. Thus, provided that $f_n \ll f_s$ then

$$\left.\frac{\text{DFT}(\dot{h}(t))}{f_s}\right|_{f=f_s/2} \approx \frac{1}{2f_s m}, \tag{4.17b}$$

which means that, provided the approximation holds, the DFT of the sampled velocity IRF at frequency $f = f_s/2$, is predominantly due to the additional component added to the IRF by sampling and its aliases. The value of the mobility FRF calculated from the DFT of the velocity IRF given by Eq. (4.17b), divided by the approximate value of $H_{\text{vel}}(f_s/2) = 1/(\pi f_s m)$, is simply $\pi/2$.

To illustrate the features discussed above, the modulus and phase of the mobility FRF, and their estimates calculated numerically using the DFT, are shown in Figure 4.4. The differences between

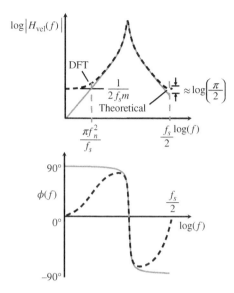

Figure 4.4 Mobility FRF of an SDOF system, showing the effects of sampling and aliasing on the DFT($\dot{h}(t)$)/f_s.

the actual FRF and the estimate at $f = 0$ and $f = f_s/2$ due to the distortion of the IRF caused by sampling and aliasing are also shown. Note, in particular, the profound difference in the phase at $f = 0$ and $f = f_s/2$, which is predominantly because of the distortion at the beginning of the IRF due to sampling, rather than aliasing, as shown in the analysis above. It is possible to determine an approximate value for the lowest frequency at which the DFT will give a reasonably accurate estimate for the modulus of the FRF. This can be calculated by setting the low-frequency asymptote of the mobility FRF given by $2\pi f/k$ to be equal to the mobility FRF calculated using the DFT, which is $1/(2f_s m)$. The result is the minimum frequency of $\pi f_n^2/f_s$, which is shown in Figure 4.4. This is a useful expression as it relates the sampling frequency to the minimum frequency at which the computed mobility FRF is valid. Note also that the frequency resolution is given by the reciprocal of the duration of the time history of the velocity IRF ($1/T$). Thus, care should also be taken in the choice of this parameter if very low-frequency information is required from the FRF.

MATLAB Example 4.2

In this example, the effects due to sampling the velocity IRF and aliasing on the mobility FRF of an SDOF system are examined.

```
clear all

%% Parameters                              % see MATLAB example 3.1
m = 1;                    % [kg]
k = 10000;                % [N/m]
z = 0.01; c = 2*z*sqrt(m*k);   % [Ns/m]
wn=sqrt(k/m); wd=sqrt(1-z^2)*wn;   % [rad/s]

%% Time and frequency parameters           % see MATLAB example 3.1
T=100;                    % [s]
fs=1000;                  % [Hz]
dt=1/fs; t=0:dt:T;        % [s]
df=1/T; f=0:df:fs;        % [Hz]
N=length(t);
```

(Continued)

MATLAB Example 4.2   (Continued)

```
%% Theoretical velocity IRF
hv1=1/(m*wd)*exp(-z*wn*t);
hv2=(wd*cos(wd*t)-z*wn*sin(wd*t));
hv=hv1.*hv2;                        % [m/Ns²]        % velocity IRF

%% Calculation of DFT
Hv=dt*fft(hv);                                       % calculation of mobility FRF

%% Theoretical mobility FRF                          % see Matlab example 3.1
dff=.001; fr=0:dff:fs;         % [Hz]
w=2*pi*fr;                     % [rad/s]
HV=j*w./(k-w.^2*m+j*w*c);      % [m/Ns]               % mobility FRF

%% Calculation of aliased response
for p=1:20;
  f1=(p-1)*fs:dff:p*fs;          % [Hz]               % frequency vector
  w1=2*pi*f1;                    % [rad/s]
  HP(p,:)=j*w1./(k-w1.^2*m+j*w1*c); % [m/Ns]          % aliased FRF for +ve freq.
  EP1=j./(w1*m);                                      % aliased additional component
  EP2= exp(j*w1*dt/2).*sinc(w1/(2*pi)*dt);            for +ve freq.
  EP(p,:)=EP1.*(1-EP2);          % [m/Ns]
  f2=-p*fs:dff:-(p-1)*fs;        % [Hz]               % frequency vector
  w2=2*pi*f2;                    % [rad/s]
  HM(p,:)=j*w2./(k-w2.^2*m+j*w2*c); % [m/N]           % aliased FRF for -ve freq.
  EM1=j./(w2*m);                                      % aliased additional component
  EM2= exp(j*w2*dt/2).*sinc(w2/(2*pi)*dt);            for -ve freq.
  EM(p,:)=EM1.*(1-EM2);          % [m/Ns]
end
MP=sum(HP+EP); MS=(sum(HM+EM));                       % sum of all aliases
HT1=MP+MS;                                            % total aliased FRF
HT=HT1(1:(N+1)/2);
HA=[HT fliplr(conj(HT))];                             % double sided spectrum
HA1=HA(1:length(f));

H0=1/(2*fs*m);                                        % value of aliased FRF at f=0
H1_2=H0;                                              % value of aliased FRF at f=fs/2

%% Plot the results
figure (1)                                           % modulus
semilogx(fr,20*log10(abs(HV)))                        % theoretical FRF
hold on
semilogx(f,20*log10(abs(Hv)))                         % FRF from DFT
hold on
semilogx(fr,20*log10(abs(HT1)))                       % theoretical aliased FRF
hold on
plot(fs/2,20*log10(abs(1/(pi*fs*m))),'o')             % value of HT1 at f=0
plot(fs/2,20*log10(H1_2),'o')                         % value of HH at f=0
plot(0.1,20*log10(H0),'o')
xlabel('frequency (Hz)');
ylabel('|mobility| (dB ref 1 m/Ns)');
grid;axis square

figure (2)                                           % phase
semilogx(fr,180/pi*(angle(HV)))                       % theoretical FRF
hold on
semilogx(f,180/pi*unwrap(angle(Hv)))                  % FRF from DFT
hold on
semilogx(fr,180/pi*(angle(HT1)))                      % theoretical aliased FRF
xlabel('frequency (Hz)');
ylabel('phase (degrees)');
```

(Continued)

MATLAB Example 4.2   (Continued)

Results

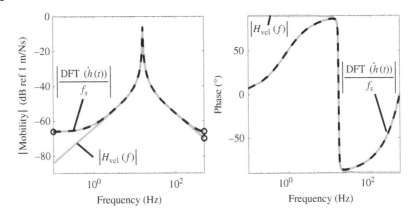

Comments

1. An exercise for the reader is to explore what happens to the calculated FRF when the damping and mass are changed.
2. An exercise for the reader is to explore what happens when the time T is reduced so that the IRF is truncated.

4.2.3 Accelerance

The accelerance FRF is the Fourier transform of the acceleration IRF discussed in Chapter 2. Acceleration is probably the most frequently measured quantity in vibration measurements because the sensors are inertial and are generally small. The acceleration IRF, however, is more complicated than the displacement and velocity IRFs, because it involves a scaled delta function in addition to the oscillatory part, so that $\ddot{h}(t) = \frac{\delta(t)}{m} - \frac{\omega_n^2}{m\omega_d} e^{-\zeta\omega_n t} \sin(\omega_d t + \phi)$ for $t \geq 0$, where $\phi = \sin^{-1}(2\zeta\sqrt{1-\zeta^2})$ (Iwanaga et al., 2021). The accelerance FRF is determined by calculating the Fourier transform of $\ddot{h}(t)$ to give $H_{\text{acc}}(j\omega) = -\omega^2/(k - \omega^2 m + j\omega c)$, for $-\infty < \omega < \infty$, which has a simple form. When the acceleration IRF is sampled there are effectively three components, as shown in Figure 4.5, which are summed to give the effective sampled IRF. The first component is the approximation to the scaled delta function, which occurs due to sampling. The delta function is approximated by a triangle with the three samples, at $t = 0$, and $t = \pm\Delta t$, as shown in the close-up in Figure 4.5. The sample at $t = -\Delta t$ is because of the assumed periodicity of the IRF during Fourier transformation, as discussed for the velocity IRF. The area of the triangle is $1/m$ so that the apex of the triangle is at $1/(m\Delta t)$. The representation of the sampled version of the delta function is given by

$$
\begin{aligned}
\ddot{h}_d(t) &= \frac{1}{m(\Delta t)^2}t + \frac{1}{m\Delta t} && -\Delta t \leq t \leq 0 \\
&= -\frac{1}{m(\Delta t)^2}t + \frac{1}{m\Delta t} && 0 \leq t \leq \Delta t.
\end{aligned}
\tag{4.18a}
$$

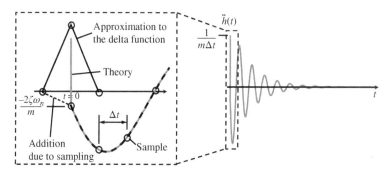

Figure 4.5 Acceleration IRF of an SDOF system. The inset shows the details of sampling at the start of the IRF, close to $t = 0$.

The second component is the oscillatory term given by

$$\ddot{h}_o(t) = -\frac{\omega_n^2}{m\omega_d}e^{-\zeta\omega_n t}\sin(\omega_d t + \phi). \tag{4.18b}$$

Note that $\ddot{h}_o(0) \neq 0$ because of the damping in the system. It has a value of $\ddot{h}_o(0) = -2\zeta\omega_n/m$. As this is not zero, there is an additional component, which is shown in the close-up in Figure 4.5. The reason for this additional component is the same as that for the velocity IRF, and is given by

$$\ddot{h}_a(t) = -\left(\frac{2\zeta\omega_n}{m\Delta t}t + \frac{2\zeta\omega_n}{m}\right) \qquad -\Delta t \le t \le 0. \tag{4.18c}$$

The effective acceleration IRF due to sampling is the sum of Eqs. (4.18a), (4.18b), and (4.18c), i.e.

$$\ddot{h}_M(t) = \ddot{h}_d(t) + \ddot{h}_o(t) + \ddot{h}_a(t). \tag{4.19}$$

The Fourier transform of $\ddot{h}_M(t)$ is equal to $\mathcal{F}\{\ddot{h}_d(t)\} + \mathcal{F}\{\ddot{h}_a(t)\} + \mathcal{F}\{\ddot{h}_o(t)\}$, where the $\mathcal{F}\{\ddot{h}_d(t)\}$ is given by

$$H_d(j\omega) = \frac{1}{m}(\text{sinc}(f\Delta t))^2, \tag{4.20}$$

the $\mathcal{F}\{\ddot{h}_a(t)\}$ is given by

$$H_a(j\omega) = \frac{-j2\zeta\omega_n}{\omega m}(1 - e^{j\omega\Delta t/2}\text{sinc}(f\Delta t)), \tag{4.21}$$

and the $\mathcal{F}\{\ddot{h}_o(t)\}$ is given by

$$H_o(j\omega) = \frac{-\omega^2}{k - \omega^2 m + j\omega c} - \frac{1}{m}. \tag{4.22}$$

Note that the $\text{DFT}(\ddot{h}(t))/f_s$ is the same as the aliased frequency domain sampled version of $\mathcal{F}\{\ddot{h}_M(t)\}$.

The value of the $\text{DFT}(\ddot{h}(t))/f_s$ at $f = 0$

The value of the $\text{DFT}(\ddot{h}(t))/f_s$ at zero frequency is the sum of $H_d(0) + H_a(0) + H_o(0)$ plus all of their aliases, which occur at frequencies of $f = nf_s$ for integer values of n between $-\infty$ and ∞. Following the procedure for the mobility FRF, it is given by

$$\left.\frac{\text{DFT}(\ddot{h}(t))}{f_s}\right|_{f=0} = H_d(0) + 2\sum_{n=1}^{\infty}\text{Re}\{H_d(nf_s)\} + H_a(0) + 2\sum_{n=1}^{\infty}\text{Re}\{H_a(nf_s)\}$$
$$+ H_o(0) + 2\sum_{n=1}^{\infty}\text{Re}\{H_o(nf_s)\}. \tag{4.23}$$

Examining Eq. (4.20), it can be seen that $H_d(0) = 1/m$, and because, $\text{sinc}(nf_s\Delta t) = \text{sinc}(n) = 0$ for $n \neq 0$, there are no aliases of this component. Similarly, there are no aliases for $H_a(nf_s)$, but it has a value of $H_a(0) = \frac{-2\pi\zeta}{m}\frac{f_n}{f_s}$.

Now, the value of $H_o(0) = -1/m$, and the aliased components can be determined in a similar manner as for receptance and mobility. The FRF of the oscillatory component at frequency nf_s is given by

$$H_o(nf_s) = \frac{-(2n\pi f_s)^2}{k - (2n\pi f_s)^2 m + j2n\pi f_s c} - \frac{1}{m}. \tag{4.24}$$

Assuming that $f_s \gg f_n$, and after some algebraic manipulation, Eq. (4.24) becomes

$$H_o(nf_s) \approx \frac{1}{m}\left[\left(\frac{f_n}{f_s}\right)^2(1 - 4\zeta^2)\frac{1}{n^2} + j2\zeta\left(\frac{f_n}{f_s}\right)\left(1 + 2\left(\frac{f_n}{f_s}\right)^2\frac{1}{n^2}\right)\frac{1}{n}\right], \tag{4.25}$$

which results in $2\sum_{n=1}^{\infty}\text{Re}\{H_o(nf_s)\} \approx \frac{2}{m}\left[\left(\frac{f_n}{f_s}\right)^2(1 - 4\zeta^2)\right]\sum_{n=1}^{\infty}\frac{1}{n^2}$. Noting that $\sum_{n=1}^{\infty}\frac{1}{n^2} = \frac{\pi^2}{6}$, and assuming that damping is small so that $4\zeta^2$ can be neglected, then $2\sum_{n=1}^{\infty}\text{Re}\{H_o(nf_s)\} \approx \frac{\pi^2}{3m}\left(\frac{f_n}{f_s}\right)^2$. Substituting for the component parts into Eq. (4.23) results in

$$\left.\frac{|\text{DFT}(\ddot{h}(t))|}{f_s}\right|_{f=0} \approx \frac{\pi}{m}\frac{f_n}{f_s}\left(\frac{\pi}{3}\frac{f_n}{f_s} - 2\zeta\right), \tag{4.26}$$

which means that at frequency $f = 0$, the value of the DFT of the sampled IRF is partly due to the oscillatory component and partly due to the artefact added to oscillatory component because of sampling.

The value of the $\text{DFT}(\ddot{h}(t))/f_s$ at $f = f_s/2$

The value of the $\text{DFT}(\ddot{h}(t))/f_s$ at $f = f_s/2$ is determined in a similar way to that for $f = 0$. It is given by

$$\left.\frac{|\text{DFT}(\ddot{h}(t))|}{f_s}\right|_{f=f_s/2} = 2\sum_{n=1}^{\infty}\text{Re}\{H_d((n - 1/2)f_s)\} + 2\sum_{n=1}^{\infty}\text{Re}\{H_a((n - 1/2)f_s)\}$$

$$+ 2\sum_{n=1}^{\infty}\text{Re}\{H_o((n - 1/2)f_s)\}. \tag{4.27}$$

First, consider the term related to the approximation to the delta function, given by Eq. (4.20). At a frequency $f = (n - 1/2)f_s$, this becomes

$$H_d((n - 1/2)f_s) = \frac{1}{m}(\text{sinc}((n - 1/2)f_s\Delta t))^2, \tag{4.28}$$

which can be written as $H_d((n - 1/2)f_s) = \frac{1}{m}(\text{sinc}(n - 1/2))^2$, so that $2\sum_{n=1}^{\infty}\text{Re}\{H_d((n - 1/2)f_s)\} = \frac{2}{m}\sum_{n=1}^{\infty}(\text{sinc}(n - 1/2))^2$. Now, as $\sum_{n=1}^{\infty}(\text{sinc}(n - 1/2))^2 = 1/2$ (Abramowitz and Stegun, 2014), then

$$2\sum_{n=1}^{\infty}\text{Re}\{H_d((n - 1/2)f_s)\} = \frac{1}{m}. \tag{4.29}$$

Now consider the term related to $H_a(j\omega)$ given by Eq. (4.21). At a frequency of $f = (n - 1/2)f_s$, this becomes

$$H_a((n - 1/2)f_s) = \frac{-j2\zeta\omega_n}{2\pi(n - 1/2)f_s m}\left(1 - e^{j\pi(n-1/2)f_s\Delta t}\text{sinc}((n - 1/2)f_s\Delta t)\right), \tag{4.30}$$

which can be written as $H_a((n-1/2)f_s) = \frac{-4\zeta f_n}{(2n-1)f_s m}\left(\frac{2}{(2n-1)\pi}+j\right)$, so that $2\sum_{n=1}^{\infty}\text{Re}\{H_a((n-1/2)f_s)\} = \frac{-16\zeta f_n}{\pi m f_s}\sum_{n=1}^{\infty}\frac{1}{(2n-1)^2}$. As noted previously, $\sum_{n=1}^{\infty}\frac{1}{(2n-1)^2} = \frac{\pi^2}{8}$, then

$$2\sum_{n=1}^{\infty}\text{Re}\{H_a((n-1/2)f_s)\} = -\frac{2\pi\zeta}{m}\frac{f_n}{f_s}. \tag{4.31}$$

Finally, consider the term related to $H_o(j\omega)$ given by Eq. (4.22). At a frequency $f = (n-1/2)f_s$, this becomes

$$H_o((n-1/2)f_s) = \frac{-(2n-1)^2\pi^2 f_s^2}{k-((2n-1)\pi f_s)^2 m + j(2n-1)\pi f_s c} - \frac{1}{m}. \tag{4.32}$$

Assuming that $f_s \gg f_n$, and after some algebraic manipulation, Eq. (4.32) becomes

$$H_o((n-1/2)f_s) \approx \frac{4}{m}\left[\left(\frac{f_n}{f_s}\right)^2(1-4\zeta^2)\frac{1}{(2n-1)^2} + j\zeta\left(\frac{f_n}{f_s}\right)\left(1+8\left(\frac{f_n}{f_s}\right)^2\frac{1}{(2n-1)^2}\right)\frac{1}{(2n-1)}\right], \tag{4.33}$$

which results in $2\sum_{n=1}^{\infty}\text{Re}\{H_o((n-1/2)f_s)\} \approx \frac{8}{m}\left[\left(\frac{f_n}{f_s}\right)^2(1-4\zeta^2)\right]\sum_{n=1}^{\infty}\frac{1}{(2n-1)^2}$. Again, noting that $\sum_{n=1}^{\infty}\frac{1}{(2n-1)^2} = \frac{\pi^2}{8}$, and assuming that damping is small so that $4\zeta^2$ can be neglected, results in

$$2\sum_{n=1}^{\infty}\text{Re}\{H_o((n-1/2)f_s)\} \approx \frac{\pi^2}{m}\left(\frac{f_n}{f_s}\right)^2. \tag{4.34}$$

Substituting for the component parts into Eq. (4.23) gives

$$\left|\frac{\text{DFT}(\ddot{h}(t))}{f_s}\right|_{f=f_s/2} \approx \frac{1}{m}\left(1+\pi^2\left(\frac{f_n}{f_s}\right)^2 - 2\pi\zeta\frac{f_n}{f_s}\right). \tag{4.35a}$$

Provided that $f_n \ll f_s$ then this reduces to

$$\left|\frac{\text{DFT}(\ddot{h}(t))}{f_s}\right|_{f=f_s/2} \approx \frac{1}{m} \tag{4.35b}$$

which means that at the frequency $f = f_s/2$, the $\text{DFT}(\ddot{h}(t))/f_s\big|_{f=f_s/2}$ is approximately equal to the value of the theoretical accelerance FRF. However, this is serendipitous because there are three effects due to sampling and they combine to cancel each other. One of these is due to the representation of a delta function as a triangle, the second is the addition of a component due to the none zero value of the oscillatory part of the acceleration impulse response at $t = 0$, and the third is due to the aliasing of the complete acceleration impulse response, which is mainly due to the aliasing of the representation of the delta function.

An example of the modulus and phase of the accelerance FRF, and their estimates calculated by applying the DFT to the acceleration IRF, is shown in Figure 4.6. The differences between the actual FRF and the estimate at $f = 0$ due to both sampling and aliasing are also shown in the figure. The additional components at $f = 0$ can be clearly seen, as is the apparent accuracy of the result at $f = f_s/2$ because of the reasons discussed above. However, note that the value of the aliased FRF at zero frequency can change sign depending on the value of damping and the ratio of the natural frequency to the sampling frequency. For the graph shown in Figure 4.6, the damping is such that the DC value of the aliased FRF is negative. If damping is reduced, this can

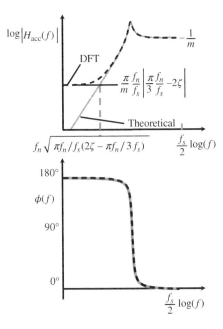

Figure 4.6 Accelerance FRF of an SDOF system, showing the effects of sampling and aliasing on the DFT$(\dot{h}(t))/f_s$.

become positive and the shape of the graph changes. This is illustrated in MATLAB Example 4.3. For the graph shown in Figure 4.6, the frequency above which the DFT gives a reasonably accurate estimate for the modulus of the FRF can be calculated in a similar way to that for the mobility FRF, and is given by $f_n\sqrt{\pi f_n/f_s(2\zeta - \pi f_n/3f_s)}$. Examining the phase in Figure 4.6, it can be seen that distortion due to sampling the IRF and aliasing does not have a profound effect on the phase spectrum. As with any spectrum calculated using the DFT, the minimum frequency component of the accelerance FRF above zero Hz is the reciprocal of the duration of the time history of the acceleration IRF $(1/T)$.

MATLAB Example 4.3

In this example, the effects due to sampling the acceleration IRF and aliasing on the accelerance FRF of an SDOF system are examined.

```
clear all

%% Parameters                              % see MATLAB example 3.1
m = 10;                     % [kg]
k = 10000;                  % [N/m]
z = 0.01; c = 2*z*sqrt(m*k);  % [Ns/m]     % choose z=0.001 for low damping
wn=sqrt(k/m); wd=sqrt(1-z^2)*wn;  % [rad/s]
fn=sqrt(k/m)/(2*pi);        % [Hz]

%% Time and frequency parameters           % see MATLAB example 3.1
T=100;                      % [s]
fs=1000;                    % [Hz]
```

(Continued)

MATLAB Example 4.3   (Continued)

```
dt=1/fs; t=0:dt:T;                  % [s]
df=1/T; f=0:df:fs;                  % [Hz]
N=length(t);

%% Theoretical accelerance IRF
ha1=1/(m*wd)*exp(-z*wn*t);                      % IRF
ha2=-(wd^2*sin(wd*t)+z*wn*wd*cos(wd*t));
ha3=-z*wn*(wd*cos(wd*t)-z*wn*wd*sin(wd*t));
ha=ha1.*(ha2+ha3);
ha(1)=1/(dt*m)+ha(1);               % [m/Ns^3]   % addition of a delta function
                                                  and oscillatory term

%% Calculation of DFT
Ha=dt*fft(ha);                                  % calculation of accelerance FRF

%% Theoretical FRF
dff=.001; fr=0:dff:fs;              % [Hz]       % see Matlab example 3.1
w=2*pi*fr;                          % [rad/s]
HA=-w.^2./(k-w.^2*m+j*w*c);         % [m/Ns^2]

%% Calculation of aliased response
for p=1:20;
  f1=(p-1)*fs:dff:p*fs;                          % frequency vector
  w1=2*pi*f1;
  HP(p,:)=-w1.^2./(k-w1.^2*m+j*w1*c)-1/m;        % aliased FRF for +ve freq.
  AP1=-j*2*z*wn./(m*w1);
  AP2= (1-exp(j*w1*dt/2)).*sinc(w1/(2*pi)*dt);
  AP(p,:)=AP1.*AP2;                              % aliased additional component
  DP(p,:)=1/m*(sinc(w1/(2*pi)*dt)).^2;           % aliased FRF of delta function
  f2=-p*fs:dff:-(p-1)*fs;                        % frequency vector
  w2=2*pi*f2;
  HM(p,:)=-w2.^2./(k-w2.^2*m+j*w2*c)-1/m;        % aliased FRF for -ve freq.
  AM1=-j*2*z*wn./(m*w2);
  AM2= (1-exp(j*w2*dt/2)).*sinc(w2/(2*pi)*dt);
  AM(p,:)=AM1.*AM2;                              % aliased additional component
  DM(p,:)=1/m*(sinc(w2/(2*pi)*dt)).^2;           % aliased FRF of delta function
end
MP=sum(HP+AP+DP);MS=sum(HM+AM+DM);               % sum all aliases
HT1=MP+MS;                                       % total aliased FRF
HT=HT1(1:(N+1)/2);
HAA=[HT fliplr(conj(HT))];                       % double sided spectrum
HA1=HAA(1:length(f));
H0=(1/m*(pi^2/3*fn^2/fs^2 - 2*pi*z*fn/fs));      % value of aliased FRF at f=0

%% Plot the results                              % modulus
semilogx(fr,20*log10(abs(HA)))                   % theoretical FRF
hold on
semilogx(f,20*log10(abs(Ha)))                    % FRF from DFT
hold on
semilogx(fr,20*log10(abs(HT1)))                  % theoretical aliased FRF
plot(df,20*log10(abs((2*pi*df)^2/k)),'o')        % value of HT1 at f=df
plot(df,20*log10(abs(H0)),'o')                   % value of FRF at f=df
axis([df,fs/2,-140,20])
xlabel('frequency (Hz)');
ylabel('|accelerance| (dB ref 1 m/Ns^2)');
grid; axis square;
```

(Continued)

MATLAB Example 4.3   (Continued)

Results

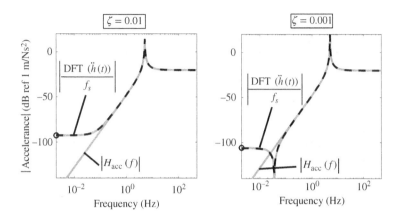

Comments

1. An exercise for the reader is to explore what happens to the calculated FRF when the mass and stiffness are changed.
2. An exercise for the reader is to explore what happens when the time T is reduced so that the IRF is truncated.
3. An exercise for the reader is to plot the non-aliased version of the FT of the sampled IRFs using the sum of Eqs. (4.20)–(4.22) and compare this with the theoretical accelerance FRF.

4.3   Effect of Data Truncation

It was shown in Chapter 2 that when an SDOF system is subject to an impact, the time that it takes for the ensuing vibration to decay to a negligibly small level, depends on the force applied and the system properties. This is the case for displacement, velocity, or acceleration. It is possible that the 'tail' of the vibration response may not be captured correctly from either experimental data or numerical simulations, especially for lightly damped systems. This can have a profound effect when the time domain data are transformed to the frequency domain. As the transformation is generally carried out on sampled data, this effect combines with the effects due to sampling, which were discussed in Section 4.2. For clarity, a continuous time series is first considered in this section to illustrate the effects of data truncation alone.

An IRF is captured over a finite time period, so it can be thought of as a time history multiplied by a rectangular window of unit amplitude, such as that shown at the top of Figure 4.7. If the time duration of the window is sufficiently long so that the whole of the IRF is captured, then there is no data truncation. However, if the time duration T is too short, which is clearly the case in Figure 4.7, the IRF is truncated to give $h_T(t)$ which is different to the actual IRF $h(t)$. They are related by

$$h_T(t) = w(t) \times h(t) \tag{4.36}$$

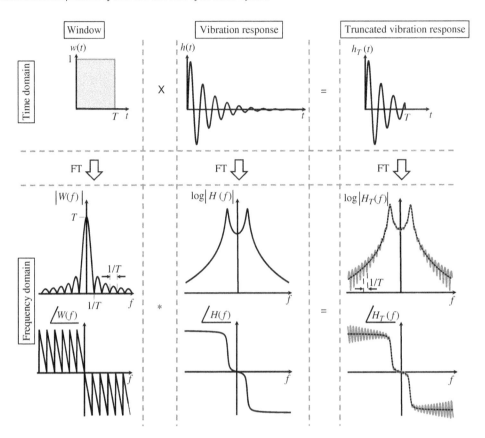

Figure 4.7   Illustration of the effects of data truncation.

To determine the effect on frequency domain data, the FTs of each component of Eq. (4.36) are calculated to give

$$H_T(f) = W(f) * H(f), \tag{4.37a}$$

where $H_T(f) = \mathcal{F}\{h_T(t)\}$, $W(f) = \mathcal{F}\{w(t)\}$, $H(f) = \mathcal{F}\{h(t)\}$, and $*$ denotes convolution, which is defined by

$$H_T(f) = \int_{-\infty}^{\infty} W(f)H(f-g)\mathrm{d}g. \tag{4.37b}$$

The concept of convolution is described in detail in Appendix G. Note that because the window and the IRF are *multiplied* together in the time domain, their Fourier transforms are *convolved* in the frequency domain, as discussed in Appendix G (multiplication in one domain becomes convolution in the other domain). The Fourier transform of the IRF is simply the corresponding FRF, and in the example considered here, it is the receptance given by $H(f)$, and the FT of the rectangular window is given by Shin and Hammond (2008)

$$W(f) = T\frac{\sin(\pi f T)}{\pi f T}e^{-j\pi f T}, \tag{4.38}$$

which is a sinc function that also has phase because the window $w(t)$ is not centred about $t = 0$. Equation (4.38) is plotted in the left-hand column of Figure 4.7. It can be seen that the peak of the main lobe has a value of T, and the modulus is zero when $f = \pm n/T$, $n = 1, 2, 3\ldots$. Note that if

$T \to \infty$, $W(f) \to \delta(f)$. If this is substituted into Eq. (4.37b), then using the sifting properties of the delta function described in Appendix E, it is found that $H_T(f)|_{T \to \infty} = H(f)$. When data truncation occurs, however, as shown in Figure 4.7, then ripples appear in both the amplitude and phase of the FRF, which is evident in the third column of the figure. The ripple has a period of $1/T$, and is clearly caused by the window, as it follows the ripple pattern in $W(f)$.

If sampled data are truncated there are further considerations. There is, of course, the problem of aliasing discussed in Section 4.2, but there is also an issue with frequency resolution Δf, which is related to the duration of the time window T, by $\Delta f = 1/T$. To illustrate this effect, three cases are considered using the displacement IRF of an SDOF system. These cases are shown in the upper left part of Figure 4.8 and are compared in the frequency domain with the actual modulus of the

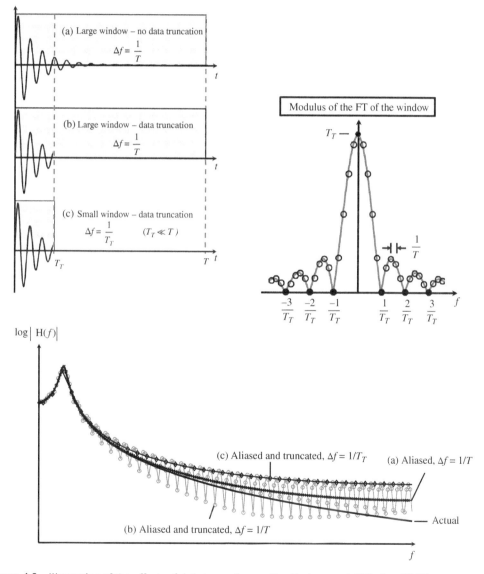

Figure 4.8 Illustration of the effects of data truncation on the displacement FRF of an SDOF system, calculated using the DFT.

theoretical FRF in the lower part of Figure 4.8. The first case, labelled (a), is where the window is long enough to capture the whole of the IRF. In this case the difference between the FRF calculated using the DFT and the actual FRF is only due to aliasing. In case (b) the IRF is truncated at time T_T, but the frequency resolution is maintained by keeping the overall window duration to time T by setting each data point between time T_T and T to zero (this is sometimes called zero-padding). The resulting FRF is the rippled response in the lower part of Figure 4.8. This plot should be compared with the FRF of truncated data in Figure 4.7, in which the ripples are clearly attributed to the rectangular window. However, it should be noted that the plot in Figure 4.8 also contains additional components due to aliasing. In the final case, labelled (c), the IRF is also truncated, but now the frequency resolution $\Delta f = 1/T_T$ is significantly reduced as $T_T \ll T$. This means that there are no ripples in the FRF, which occurs because the frequency resolution is also the period of the ripple, and this is $1/T$ as discussed above. Note that this response also contains aliased components. The modulus of the rectangular window spectrum is shown in the right-hand part of Figure 4.8 to further clarify why there are no ripples in the truncated spectrum of the IRF calculated using the DFT. It is clear from this figure, in which the data points for the window function with time duration T_T are shown as solid circles, that this is due to the limited frequency resolution because of the short duration window.

It is evident from the example given in Figure 4.8 that care must be taken in the interpretation and analysis of sampled data in the frequency domain. This is because the process of capturing the data and transforming it from the time to the frequency domain using the DFT can cause significant distortion, particularly due to aliasing and data truncation.

MATLAB Example 4.4

In this example, the convolution of a sinc function and the FRF of an SDOF system is illustrated by way of an animation.

```
clear all

%% Parameters                          % see MATLAB example 3.1
m = 10;                    % [kg]
k = 10000;                 % [N/m]
z = 0.1; c = 2*z*sqrt(m*k);   % [Ns/m]
wn=sqrt(k/m); wd=sqrt(1-z^2)*wn;   % [rad/s]

%% Time and frequency parameters       % see MATLAB example 3.1
T=0.1;                     % [s]       % time duration of data (window)
fs=400; df=1;              % [Hz]
dt=1/fs; t=0:dt:T;         % [s]

%% Theoretical IRF
h=1/(m*wd)*exp(-z*wn*t).*sin(wd*t);   % [m/Ns]      % IRF

%% plot IRF
plot(t,h,'k','linewidth',4)
set(gca,'fontsize',24)
xlabel('time (s)');
ylabel('displacement IRF (m/Ns)');
```

(Continued)

MATLAB Example 4.4   (Continued)

```
%% Theoretical FRF and sinc function
f=-fs/2:df:fs/2;                    % [Hz]       % frequency vector
w=2*pi*f;                           % [rad/s]
H=1./(k-w.^2*m+j*w*c);              % [m/N]      % FRF
fr=-1.5*fs/2:df:1.5*fs/2;
W=T*sinc(T*fr).*exp(-j*pi*fr*T);    % [s]        % sinc function
delta=sinc(fr).*exp(-j*pi*fr);      % [s]        % approx. delta function

%% Convolve sinc function with FRF
C=conv(W,H)*df; Ca=conv(delta,H)*df;            % with and without truncation

% Plots
fmin=3*min(f);fmax=3*max(f);                    % set freq. range for graph
Wf=fliplr(W);ff=fliplr(-fr);                    % flip the sinc function
ff = ff + ( min(f)-max(ff) );                   % slide range of W
fc = [ff f(2:end)]; fc = fc+max(fr);            % range of convolved function
set(figure,'Position',[40, 40, 1450, 700]);     % set the position of animation
subplot(2,1,1);
HN=abs(H)/max(abs(H)); WN=abs(W)/max(abs(W));    % normalized FRF and W
p=plot(f,HN,'k','linewidth',4);hold on           % plot of normalized FRF
gr=[.6 .6 .6];                                   % define grey colour
q=plot(fr,WN,'k','linewidth',4,'Color',gr);      % plot of normalized W
axis([fmin,fmax,0,1.1])
set(gca,'fontname','arial','fontsize',12)
xlabel('frequency (Hz)');
ylabel('normalised modulus');

sl=line([min(f) min(f)],[1.1 1.1],'color','k');  % vertical line for overlap
hold on; grid on;
sg = rectangle('Position',[min(f) 1 0 0],...     % shaded region
            'FaceColor', [.9 .9 .9]);
subplot(2,1,2);
CdB=20*log10(abs(C)); CadB=20*log10(abs(Ca));    % convolved values in dB
r=plot(fc,CdB,'linewidth',3,'color',gr); hold on; % plot of convolved solution
s = plot(fc,CadB,'k','linewidth',4);             % as above no truncation
grid on; hold on
axis([fmin,fmax,-140,max(CdB)+10]);
set(gca,'fontname', 'arial','fontsize',12);
xlabel('frequency (Hz)');
ylabel('modulus (dB ref 1N/m)');

%% animation block
for n=1:length(fc)
  pause(0);                                      % controls animation speed
  ff=ff+df;
  set(q,'XData',ff,'YData',WN);

  sx=min(max(ff(1),min(f)),max(f));              % left-hand boundary of overlap
  sxa = [sx sx];
  set(sl,'XData',sxa);

  ex=min(ff(end),max(f));                        % right-hand boundary of overlap
  exa=[ex ex];
  set(sl,'XData',exa);
```

(Continued)

MATLAB Example 4.4   (Continued)

```
rpos = [sx 0 max(0.0001,ex-sx) 1.1];          % shading of overlap region
set(sg,'Position',rpos);
uistack(sg,'bottom');

set(r,'XData',fc(1:n),'YData',CdB(1:n));       % plot of convolved function
set(s,'XData',fc(1:n),'YData',CadB(1:n));      % as above no truncation
end
```

Results

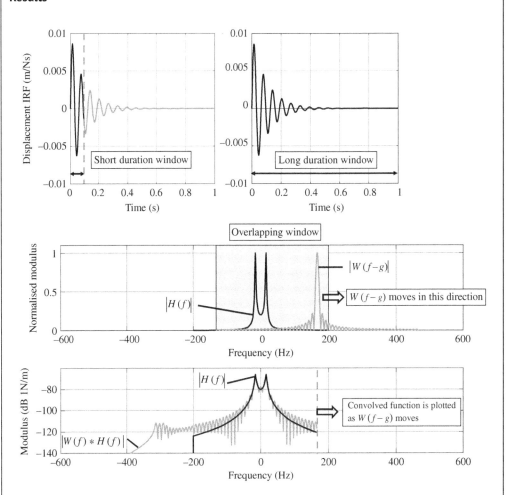

Comments

1. An exercise for the reader is to explore what happens when the duration of the time window is changed.

MATLAB Example 4.5

In this example, the effect of truncating an IRF on the FRF of an SDOF system calculated using the DFT is illustrated.

```
clear all

%% Parameters                                % see MATLAB example 3.1
m = 10;                      % [kg]
k = 10000;                   % [N/m]
z = 0.1; c = 2*z*sqrt(m*k);  % [Ns/m]
wn=sqrt(k/m); wd=sqrt(1-z^2)*wn;   % [rad/s]

%% Time and frequency parameters             % see MATLAB example 3.1
fs=400;                      % [Hz]
Th=1;dt=1/fs; t=0:dt:Th;     % [s]
df=1/Th; f=0:df:fs;          % [Hz]
Tw=.2; tw=0:dt:Tw;           % [s]
dfw=1/Tw; fw=0:dfw:fs;       % [Hz]

%% Theoretical IRFs
h=1/(m*wd)*exp(-z*wn*t).*sin(wd*t);     % [m/Ns]   % IRF long window
ht=1/(m*wd)*exp(-z*wn*tw).*sin(wd*tw);  % [m/Ns]   % IRF short window
w=[ones(1,length(tw)) zeros(1,(length(t)...        % long window
   -(length(tw))))];
hw=h.*w;                                           % truncated IRF

%% Plot IRFs
plot(t,h,'k','linewidth',4)                        % Plot IRF - change the parame-
xlabel('time (s)');                                ters to plot the three figures
ylabel('displacement IRF (m/Ns)');
grid;axis square

%% Calculate DFTs of IRFs
H=dt*fft(h);                                        % DFT - no truncation
HTT=dt*fft(ht);                                     % DFT - truncation, short window
HW=dt*fft(hw);                                      % DFT - truncation, long window

%% Theoretical FRF
dff=0.1; fr=0:dff:fs/2;      % [Hz]
w=2*pi*fr;                   % [rad/s]
HH=1./(k-w.^2*m+j*w*c);      % [m/N]

% Plots of FRFs                                     % modulus
figure
plot(f,20*log10(abs(HW)))
hold on
plot(fr,20*log10(abs(HH)))
hold on
plot(f,20*log10(abs(H)))
hold on
plot(fw,20*log10(abs(HTT)))
xlabel('frequency (Hz)');
ylabel('|receptance| (dB ref 1 m/N)');
axis([0 fs/2 -130 -60])
```

(Continued)

MATLAB Example 4.5   (Continued)

```
figure                                          % phase
plot(f,180/pi*unwrap(angle(HW)))
hold on
plot(fr,180/pi*unwrap(angle(HH)))
hold on
plot(f,180/pi*unwrap(angle(H)))
hold on
plot(fw,180/pi*unwrap(angle(HTT)))
xlabel('frequency (Hz)');
ylabel('phase (^o)');
grid;axis square
axis([0 fs/2 -250 0])
```

Results

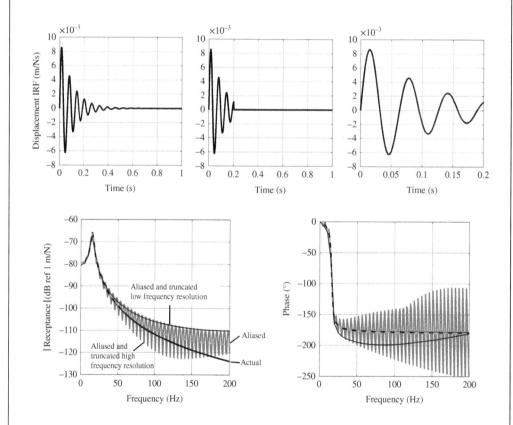

Comments

1. Compare your results with Figure 4.8.
2. An exercise for the reader is to explore what happens when the length of the short duration time window is changed.

4.4 Effects of Sampling on the IRFs Calculated Using the IDFT

In Section 4.2, the effects of sampling on the FRF calculated using the DFT were discussed, and in Section 4.3 the additional effect due to data truncation was discussed. Both of these effects are also important when calculating the IRF from the FRF using the IDFT. This process is often undertaken when using experimental data and is illustrated for a virtual experiment in Chapter 9.

The first case considered in this section is the receptance FRF and the corresponding displacement IRF. This was illustrated at the end of Chapter 3, but is discussed in more detail here. The results are shown in Figure 4.9, in which both the modulus and phase of the FRF and the IRF are plotted. As discussed in Chapter 3, the double-sided spectrum must first be formed from the FRF given by $H(f) = 1/(k - (2\pi f)^2 m + j2\pi fc)$, where $f = k\Delta f$, in which Δf is the frequency resolution. Recall that the FRF in the frequency range from $f_s/2$ to f_s is formed using the complex conjugate of the spectrum in the frequency range from 0 to $f_s/2$ (see MATLAB Example 3.2). The magnitude and phase of the FRF reconstructed double-sided FRF, and that calculated using the DFT of the displacement are shown on the left part of Figure 4.9. The difference due to aliasing is clear. The IRFs calculated using the IDFT are shown in the right-hand plot in Figure 4.9. The IRF calculated from the theoretical FRF is shown as sampled data for clarity. Note that the time resolution $\Delta t = 1/f_s$. Both plots are shifted to the right, and the 'tail' of the IRF is moved to the beginning of the IRF, so that the behaviour of the IRF for $t < 0$ can be visualised (this procedure is discussed in Chapter 3 and illustrated in Figure 3.13). A third plot is shown in the IRF graph, which is the theoretical IRF convolved with the IFT of the rectangular window from 0 to $f_s/2$. The windowing effect was discussed for time domain data in Section 4.3, but when transforming from the frequency domain to the time domain, the rectangular window is in the frequency domain, resulting in a sinc function in the time domain. It is clear from the inset of the IRF, which shows the part close to $t = 0$, that the effect of applying the window in the frequency domain is to cause two artefacts in the IRF. One

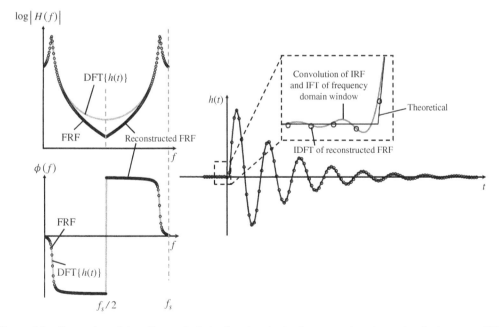

Figure 4.9 Illustration of the effects of windowing data in the frequency domain on the displacement IRF of an SDOF system, calculated using the IDFT.

of the artefacts is the ripples in the IRF which are due to the ripples in the sinc function, and the second is that there is a displacement response before $t = 0$. This means that the system responds before it is impacted resulting in an *acausal* system. Of course, this is not physically possible. It is caused by the sinc function, which is an even function about $t = 0$, i.e. it has components for $t < 0$, which manifest themselves in the IRF when convolved with the sinc function. This acausal behaviour occurs for all IRFs when they are calculated from a sampled simulated or measured FRF. The acausal effect can be minimised by increasing the sampling frequency f_s. Careful examination of the acausal part of the IRF in the inset in Figure 4.9 shows that the ripple which is evident in the IFT of the windowed FRF does not occur in the same way in the IDFT of the windowed sampled FRF. This is because of the limited time resolution, which is a function of the sampling frequency, and the samples occur approximately at times when the ripples pass through zero.

The second case considered is the mobility FRF and the corresponding velocity IRF. The results are shown in Figure 4.10, in which both the modulus and phase of the FRF and the IRF are plotted. Similar features as in Figure 4.9 can be seen. However, the acausal effect is greater than for the displacement IRF. This is because the mobility FRF has higher frequency content than the receptance, and the window removes this in the IRF calculated using the IFT. Hence, windowing the frequency domain data has a greater effect on the velocity IRF. Moreover, it can be seen that the acausal effect is greater on the IRF calculated using the IDFT. This is because the samples in the IRF do not occur at time samples where the ripple passes through zero, as can be seen in the inset in Figure 4.10. They are shifted by a quarter of a cycle (or 90°) because of the phase relationship between velocity and displacement, and therefore occur at the peaks in the ripple, making the acausal effect more prominent.

The final case considered is the accelerance FRF and the corresponding acceleration IRF. The results are shown in Figure 4.11. Similar effects as for the velocity and displacement IRFs can be seen, but the acausal effect is greater. This is because the accelerance contains even higher

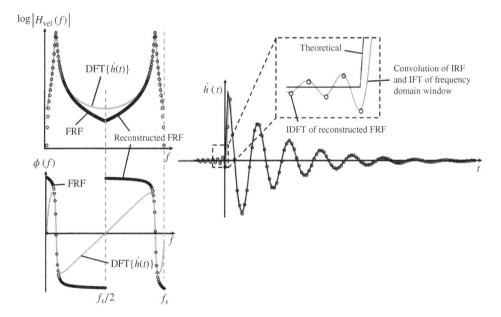

Figure 4.10 Illustration of the effects of windowing data in the frequency domain on the velocity IRF of an SDOF system, calculated using the IDFT.

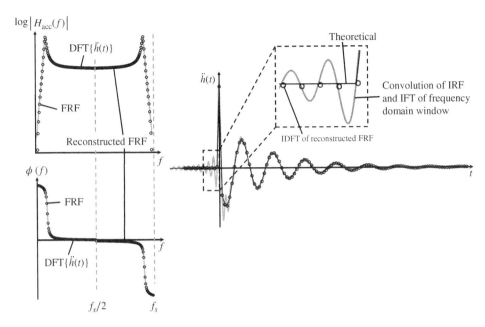

Figure 4.11 Illustration of the effects of windowing data in the frequency domain on the acceleration IRF of an SDOF system, calculated using the IDFT.

frequency content than the receptance and mobility (it is constant at high frequency); thus, the windowing effect in the frequency domain is much greater. The acausal effects are clear in Figure 4.11, especially in the inset. However, it is also clear that the acausal effects are largely hidden in the IRF calculated using the IDFT, because the samples occur approximately when the ripples pass through zero as there is a shift a quarter of a cycle (or 90°) between acceleration and velocity.

MATLAB Example 4.6

In this example, the effect of windowing the FRF on the IRF of an SDOF system, calculated using the IDFT, is illustrated.

```
clear all

%% Parameters                              % see MATLAB example 3.1
m = 1;                        % [kg]
k = 10000;                    % [N/m]
z = 0.1; c = 2*z*sqrt(m*k);   % [Ns/m]
wn=sqrt(k/m); wd=sqrt(1-z^2)*wn;   % [rad/s]

%% Time and frequency parameters           % see MATLAB example 3.1
fs=200; T=1; dt=1/fs; t=0:dt:T;    % [s]
a=0.05; dtt=a*dt; tt=0:dtt:T;      % [s]
ts=-T/2:dtt:T/2;                   % [s]
df=1/T; f=0:df:fs;                 % [Hz]

%% Theoretical IRFs
hdt=1/(m*wd)*exp(-z*wn*t).*sin(wd*t);  % [m/Ns]      % displacement IRF
hv1=1/(m*wd)*exp(-z*wn*t);
```

(Continued)

MATLAB Example 4.6   (Continued)

```
hv2=(wd*cos(wd*t)-z*wn*sin(wd*t));
hvt=hv1.*hv2;                        % [m/Ns^2]    % velocity IRF
ha1=1/(m*wd)*exp(-z*wn*t);
ha2=-(wd^2*sin(wd*t)+z*wn*wd*cos(wd*t));
ha3=-z*wn*(wd*cos(wd*t)-z*wn*sin(wd*t));
hat=ha1.*(ha2+ha3);
hat(1)=1/(dt*m)+hat(1);              % [m/Ns^3]    % acceleration IRF

%% Calculation of DFTs
Hd=dt*fft(hdt);                                    % receptance
Hv=dt*fft(hvt);                                    % mobility
Ha=dt*fft(hat);                                    % accelerance

%% Theoretical FRFs
fr=0:df:fs/2;                        % [Hz]        % frequency vector
w=2*pi*fr;                           % [rad/s]
HD=1./(k-w.^2*m+j*w*c);              % [m/N]       % receptance
HV=j*w.*HD;                          % [m/Ns]      % velocity
HA=j*w.*HV;                          % [m/Ns^2]    % accelerance

%% Form double-sided spectra
Hda=[HD fliplr(conj(HD))];HDd=Hda(1:length(t));    % receptance
Hva=[HV fliplr(conj(HV))];HVd=Hva(1:length(t));    % velocity
Haa=[HA fliplr(conj(HA))];HAd=Haa(1:length(t));    % accelerance

%% inverse Fourier transforms
hdi=fs*ifft(HDd);                                  % displacement IRF
hvi=fs*ifft(HVd);                                  % velocity IRF
hai=fs*ifft(HAd);                                  % acceleration IRF

%% calc, of convolved sinc function with IRF
W=fs*sinc(fs*ts);
hd=1/(m*wd)*exp(-z*wn*tt).*sin(wd*tt);             % displacement IRF
hdc=conv(hd,W)*dtt;hdcc=hdc(1:length(tt));         % convolved IRF with sinc func.
hv1=1/(m*wd)*exp(-z*wn*tt);
hv2=(wd*cos(wd*tt)-z*wn*sin(wd*tt));
hv=hv1.*hv2;                                       % velocity
hvc=conv(hv,W)*dtt;hvcc=hvc(1:length(tt));         % convolved IRF with sinc func.
ha1=1/(m*wd)*exp(-z*wn*tt);
ha2=-(wd^2*sin(wd*tt) + z*wn*wd*cos(wd*tt));
ha3=-z*wn*(wd*cos(wd*tt) - z*wn*sin(wd*tt));
ha=ha1.*(ha2+ha3); ha(1)=1/(dtt*m)+ha(1);          % acceleration
hac=conv(ha,W)*dtt;hacc=hac(1:length(tt));         % convolved IRF with sinc func.

%% shifting the IRFs to see the acausality
hdcirc=circshift(hdt',100);
hdicirc=circshift(hdi',100);
hdcccirc=circshift(hdcc',(length(tt)+1)/2+100/a);
hvcirc=circshift(hvt',100);
hvicirc=circshift(hvi',100);
hvcccirc=circshift(hvcc',(length(tt)+1)/2+100/a);
hacirc=circshift(hat',100);
haicirc=circshift(hai',100);
hacccirc=circshift(hacc',(length(tt)+1)/2+100/a);

%% plot the results
figure(1)                                          % receptance
plot(f,20*log10(abs(HDd)),'ok'),hold on            % modulus
```

(Continued)

MATLAB Example 4.6   (Continued)

```
plot(f,20*log10(abs(Hd)),'k')
axis([0,fs,-120,-60])
xlabel('frequency (Hz)');
ylabel('|receptance| (dB ref 1 m/N)');
figure (2)                                  % phase
plot(f,180/pi*(angle(HDd)),'ok'),hold on
plot(f,180/pi*(angle(Hd)),'k')
axis([0.1,fs,-200,200])
xlabel('frequency (Hz)');
ylabel('phase (degrees)');
figure (3)                                  % displacement IRF
plot(tt,hdcccirc),hold on
plot(t,hdicirc,'ok'),hold on
plot(t,hdcirc,'k')
axis([0.4,1,-0.01,0.01])
xlabel('time(s)');
ylabel('displacement IRF (m/Ns)');

figure (4)                                  % mobility
plot(f,20*log10(abs(HVd)),'ok'),hold on     % modulus
plot(f,20*log10(abs(Hv)),'k')
axis([0,fs,-70,-20])
xlabel('frequency (Hz)');
ylabel('|mobility| (dB ref 1 m/Ns)');
figure (5)                                  % phase
plot(f,180/pi*(angle(HVd)),'ok'),hold on
plot(f,180/pi*(angle(Hv)),'k','Linewidth',2)
axis([0,fs,-90,90])
xlabel('frequency (Hz)');
ylabel('phase (degrees)');
figure (6)                                  % velocity IRF
plot(tt,hvcccirc),hold on
plot(t,hvicirc,'ok'),hold on
plot(t,hvcirc,'k')
axis([0.4,1,-1,1.2])
xlabel('time(s)');
ylabel('velocity IRF (m/Ns^2)');

figure (7)                                  % accelerance
plot(f,20*log10(abs(HAd)),'ok'),hold on     % modulus
plot(f,20*log10(abs(Ha)),'k')
axis([0,fs,-40,20])
xlabel('frequency (Hz)');
ylabel('|accelerance| (dB ref 1 m/Ns^2)');
figure (8)                                  % phase
plot(f,180/pi*(angle(HAd)),'ok'),hold on
plot(f,180/pi*unwrap(angle(Ha)),'k')
axis([0,fs,-200,200])
xlabel('frequency (Hz)');
ylabel('phase (degrees)');
figure (9)                                  % acceleration IRF
plot(tt,hacccirc),hold on
plot(t,haicirc,'ok'),hold on
plot(t,hacirc,'k')
axis([0.4,1,-150,200])
xlabel('time(s)');
ylabel('acceleration IRF (m/Ns^3)');
```

(Continued)

MATLAB Example 4.6   (Continued)

Results

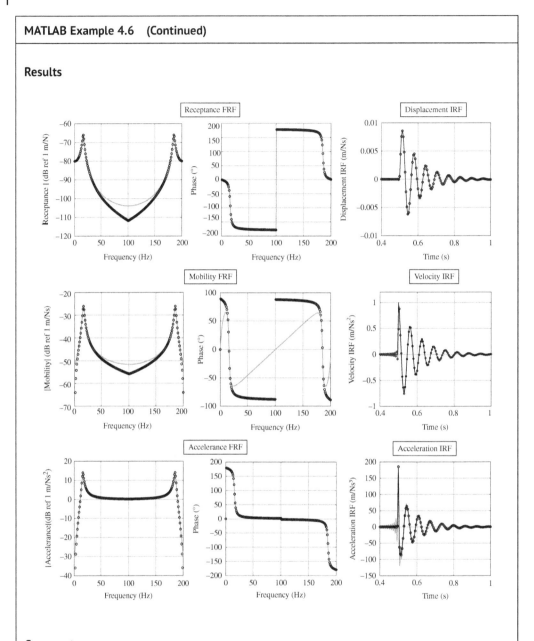

Comments

1. The figures are for the same parameters used in Figures 4.9, 4.10, and 4.11.
2. An exercise for the reader is to explore what happens when the sampling frequency is changed.

4.5   Summary

The process of sampling a signal, either in the time or frequency domain, has the effect of distorting the transformed signal (in either the frequency or the time domain) using the DFT or the IDFT. This has been explored in detail for the displacement, velocity, and acceleration impulse response functions (IRFs) for an SDOF vibrating system. In transforming a time domain signal to the frequency domain, the main effects are to add artefacts to the frequency domain representation signal. This cannot be avoided in numerical simulations of vibrating systems, because continuous time models have an infinite bandwidth, and hence have an infinite number of frequency components. When a signal is sampled, a finite sampling frequency must be used and hence there are always aliased components. There are also additional components if the time signal is not zero at the beginning and the end of the time window in which the signal is sampled, which always occurs for velocity and acceleration IRFs. The sum of these two effects at zero frequency and half the sampling frequency for the receptance, mobility, and accelerance FRFs is tabulated in Table 4.1 to show how these effects can be minimised by careful choice of the sampling frequency and consideration of the system parameters.

The FRF can be further distorted if the 'tail' of the IRF is not captured correctly by having a time window of insufficient length. This means that the calculated FRF is then effectively the actual FRF convolved with the spectrum of the time window. This has been illustrated for a rectangular time window, which has a spectrum described by a sinc function.

When transforming an FRF to an IRF using the IDFT, data truncation always occurs because of a finite sampling frequency. This effectively means that the actual FRF is multiplied by a rectangular window in the frequency domain. As mentioned above when a rectangular window is transformed from one domain to another, the result is a sinc function. Therefore, the transformed IRF is effectively the actual FRF convolved with a sinc function, which results in an additional acausal component.

Table 4.1   Effects on the FRFs of an SDOF system using the IDFT of their respective IRFs.

Frequency	Signal type		Receptance	Mobility	Accelerance
$f = 0$	Continuous	Modulus	$\dfrac{1}{k}$	0	0
		Phase	$0°$	$90°$	$180°$
	Sampled	Modulus	$\dfrac{1}{k} - \dfrac{1}{12f_s^2 m}$	$\dfrac{1}{2f_s m}$	$\left\| \dfrac{\pi}{m}\dfrac{f_n}{f_s}\left(\dfrac{\pi}{3}\dfrac{f_n}{f_s} - 2\zeta\right)\right\|$
		Phase	$0°$	$0°$	$0°$ for $\zeta \le \dfrac{\pi}{6}\dfrac{f_n}{f_s}$, $-180°$ for $\zeta > \dfrac{\pi}{6}\dfrac{f_n}{f_s}$
$f = \dfrac{f_s}{2}$	Continuous	Modulus	$\dfrac{1}{(\pi f_s)^2 m}$	$\dfrac{1}{\pi f_s m}$	$\dfrac{1}{m}$
		Phase	$-180°$	$-90°$	$0°$
	Sampled	Modulus	$\dfrac{1}{4f_s^2 m}$	$\dfrac{1}{2f_s m}$	$\dfrac{1}{m}$
		Phase	$-180°$	$0°$	$0°$

References

Abramowitz, M. and Stegun, I.A. (2014). *Handbook of Mathematical Functions with Formulas, Graphs, and Mathematical Tables.* Dover Publications.

Iwanaga, M.K., Brennan, M.J., Tang, B., et al. (2021). Some features of the acceleration impulse response function. *Meccanica*, 56, 169–177. https://doi.org/10.1007/s11012-020-01265-4.

Shin, K. and Hammond, J.K. (2008). *Fundamentals of Signal Processing for Sound and Vibration Engineers.* Wiley.

5

Vibration Excitation

5.1 Introduction

As mentioned in Chapter 1, most vibration testing in the laboratory is carried out using electrodynamic shakers or instrumented impact hammers. Other means of excitation can of course be used, such as hydraulic shakers, piezoelectric actuators, pyrotechnic excitation, and acoustic excitation (Ewins, 2000), but the details of these are not covered here. The type of device used for vibration excitation influences the time history of the force applied to the structure. This chapter describes two common ways of exciting a structure, namely an electrodynamic shaker and an instrumented hammer. In particular, the signals representing typical forces generated by these devices are examined, including their characteristics in the time and frequency domains. The aim of the chapter is not to give detailed knowledge on how to use specific vibration excitation devices, as this can be found in specialised texts on vibration testing, for example McConnell and Varoto (2008) and Waters (2013). Rather, it is to introduce some specific signals used in vibration testing, which are then used in subsequent chapters in virtual vibration experiments.

5.2 Vibration Excitation Devices

5.2.1 Electrodynamic Shaker

An example of a structure being excited by an electrodynamic shaker is shown in Figure 5.1. The shaker is supplied with an oscillating current $i_s(t)$, causing oscillatory motion of the moving part of the shaker, which consists of the coil and the armature, as shown in the figure. The shaker is connected to the structure through a thin rod, called a stinger, to ensure a point connection which minimises moment excitation of the structure. Careful design of the stinger is required to minimise its effect on vibration measurements, details of which are given in Waters (2013) and the references therein. The force applied is measured using a force gauge placed between the stinger and the structure. The output from the sensor and associated conditioning amplifier is a voltage which is proportional to the force applied $f_e(t)$. Some features of excitation using electrodynamic shakers are as follows:

- They allow a controlled force to be applied to a structure both in terms of frequency content and force amplitude.
- The force generated by a shaker is proportional to the oscillating current supplied to the shaker. The current is supplied by a power amplifier, which is fed by a signal generated by a computer or

Virtual Experiments in Mechanical Vibrations: Structural Dynamics and Signal Processing,
First Edition. Michael J. Brennan and Bin Tang.
© 2023 John Wiley & Sons Ltd. Published 2023 by John Wiley & Sons Ltd.
Companion website: www.wiley.com/go/brennan/virtualexperimentsinmechanicalvibrations

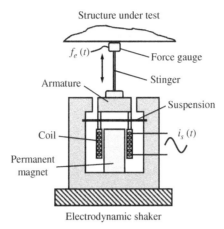

Figure 5.1 Vibration excitation of a structure using an electrodynamic shaker.

a signal generator. Note that some amplifiers can be operated in either voltage or current mode, in which the output is either a constant voltage or a constant current that is independent of the dynamics of the structure under test. If a voltage amplifier is used, the shaker movement generates a back electromotive force which causes additional damping to be added to the structure under test. If a constant current amplifier is used then the force applied to the structure is much less affected by the dynamics of the vibrating structure, and no additional damping is added. More information on the types of amplifiers and their influence on vibration test results can be found in McConnell and Varoto (2008) and Waters (2013).

- The shaker is a mass-spring-damper system, in which the stiffness is due to the suspension and the mass is due to the moving mass, which comprises the armature, the coil, the stinger, and part of the force gauge. This can have some effect on the vibration of the structure, and there may be issues of shaker–structure interaction.
- The frequency range, in which a constant force can be applied to the structure, is limited. At low frequencies this limitation is related to the maximum stroke of the shaker. In the mid-frequency range, the force is governed by the product of the magnetic field strength, the coil length, and the current supplied. At high frequencies the force is limited by an internal resonance of the shaker, and hence smaller shakers generally have a higher frequency capability.

5.2.2 Instrumented Impact Hammer

Figure 5.2 shows a structure being excited by an instrumented impact hammer. In this type of vibration test, the force gauge is not attached to the structure, but it is an integral part of the impact hammer. The output from the sensor and associated conditioning amplifier is a voltage which is proportional to the force applied to the structure $f_e(t)$. A hammer is normally supplied with a variety of tips, each of which has a different stiffness that helps to control the frequency content of the force generated. To illustrate the type of force applied to the structure by the impact hammer, the situation shown in Figure 5.3 is considered. An elastic body of mass m, which represents the mass of an impact hammer, impacts a rigid structure with a velocity $\dot{x}_0$. There is a local strain because of the contact stiffness k (which in practice is a combination of the tip and local stiffness of the structure). Whilst the body is in contact with the structure it can be modelled simply by combination of m and k, as shown in the top right part of Figure 5.3. The system has a natural frequency of $\omega_n = \sqrt{k/m}$

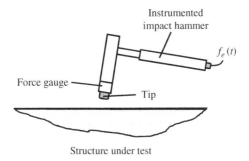

Figure 5.2 Vibration excitation of a structure using an instrumented impact hammer.

and hence a natural period of $T_n = 2\pi\sqrt{m/k}$. The time that the body is in contact is half of this period. Thus, the contact time is $T_c = \pi\sqrt{m/k}$. The force applied to the structure (assuming no damping) is given by $f_e(t) = kx$ where $x = (\dot{x}_0/\omega_n)\sin\omega_n t$. The peak force is thus given by $f_{peak} = \dot{x}_0\sqrt{km}$. The idealised force time history for a force generated by an instrumented impact hammer is illustrated in the lower left part of Figure 5.3. It can be seen that the duration of the impact is a function of the mass of the hammer and the stiffness of the tip, and the maximum force is also a function of these parameters, as well as the velocity of the impact. The frequency content of the force can be determined by calculating the $\mathcal{F}\{f_e(t)\}$ to give

$$F(j\omega) = T_c\dot{x}_0\sqrt{km}\frac{2\pi\cos(\omega T_c/2)}{\pi^2 - (\omega T_c)^2}e^{-j\omega T_c/2}. \tag{5.1}$$

The modulus of $F(j\omega)$ is shown in the lower right part of Figure 5.3. It can be seen that the frequency content of the applied force is limited by the contact time – a shorter contact time results

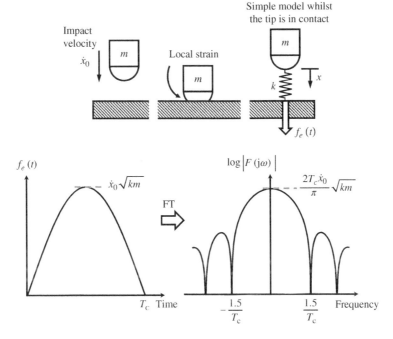

Figure 5.3 Simple model of a force applied to a structure by an instrumented impact hammer.

in a higher frequency content. This simple analysis shows how the stiffness of the hammer tip and the mass of the hammer influence the frequency content of the force applied with an instrumented impact hammer. Although this type of device is very convenient for measuring FRFs, especially if many FRFs have to be measured on a structure, it does require some skill to use correctly, and it is difficult to control the level of force excitation.

5.3 Vibration Excitation Signals

Some typical vibration signals used for vibration testing are shown in Figure 5.4. It can be seen that when a shaker is used, several types of signals are available for a vibration test, whereas when an instrumented impact hammer is used then the type of signal is limited to a half-sine impulse (in the ideal case). Although there are some other types of signals used for FRF estimation (Brandt, 2011; Waters, 2013), only the signals pictured in Figure 5.4 are considered in this book as they are in common use, and an understanding of these signals will enable the reader to follow more specialised texts on practical vibration testing. The time and frequency domain characteristics of the signals are described, as is the relationship between them in the two domains.

A fundamental relationship between any signal in the time domain and its transformation to the frequency domain is the equivalence of energy or power contained in the signal in both domains.

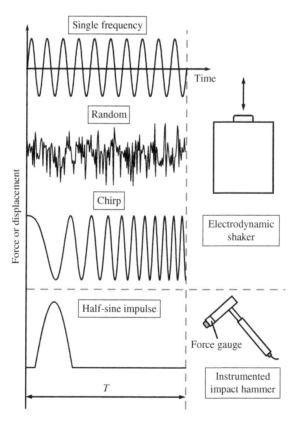

Figure 5.4 Signals supplied to an electrodynamic shaker or generated by an instrumented impact hammer. (The signals are denoted by force or displacement for convenience).

This relationship is given by Parseval's theorem[1] (Shin and Hammond, 2008). The choice of using power or energy depends on whether a signal is considered to act over all time, i.e. is persistent, or whether it is a transient so that it acts over a finite time period. In Figure 5.4, the single frequency and random excitation signals are assumed to be persistent, and the chirp signal and the half-sine impulse are transient signals. If a signal is persistent, the total energy in the signal increases with increasing time, so characterising this type of signal in terms of energy is not helpful as it is unbounded. It is better to calculate its power, which is defined by

$$\text{signal power} = \frac{1}{T}\int_{-T/2}^{T/2} x_T^2(t)\mathrm{d}t, \tag{5.2}$$

where $x_T(t)$ is the signal $x(t)$ in the time period $-T/2 \leq t \leq T/2$. This is illustrated in Figure 5.5, which shows an arbitrary stationary displacement signal that acts for an infinite period of time. Note that the term stationary refers to the statistical properties of the signal, i.e. its mean and variance. These should be constant for any period over which they are calculated for the signal to be considered stationary. It should also be noted that the term signal power does not necessarily refer to power in the physical sense, but it is defined according to Eq. (5.2), and is the *mean square value* of the signal in the time domain. If the 'windowed' part of the time series $x_T(t)$ is transformed to the frequency domain, Parseval's theorem determines that

$$\frac{1}{T}\int_{-T/2}^{T/2} x_T^2(t)\mathrm{d}t = \int_{-\infty}^{\infty}\frac{1}{T}|X_T(f)|^2\mathrm{d}f, \tag{5.3}$$

where $X_T(f)$ is the Fourier transform of $x_T(t)$. On the right-hand side of Eq. (5.3) the term $\frac{1}{T}|X_T(f)|^2 = \frac{1}{T}X_T(f)X_T^*(f)$ is the *power spectral density* (PSD) of the signal (Note that PSD is normally defined for stochastic processes, where an averaging procedure is required. When no averaging as carried out then it is often called the raw, or sample PSD. This is further discussed in Section 5.3.2). Note also that PSD is a real quantity. If the measurement is displacement with units of metres, then as discussed in Chapter 3, the unit of $X_T(f)$ is m/Hz, so the unit of PSD, in this case, is m²/Hz. Thus, the power in the signal, in the frequency domain, is the integral of the PSD over all negative and positive frequencies, i.e. the area under the PSD plot.

If a signal is transient, then the total energy in the signal can be captured within a single window. Provided that the signal is zero at each end of the window, then if the window size is increased, the signal energy does not increase. Note, however, that the signal power would decrease if the time period is increased because of the term $1/T$. As an example, consider the half-sine impulse signal shown in Figure 5.4. Provided that the pulse is captured by the window, the energy does not change

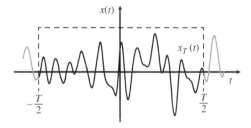

Figure 5.5 A windowed section of a signal used for analysis.

---

1 The theorem is named after Marc-Antoine Parseval (1755–1836). Parseval stated the theorem, but did not prove it, claiming it to be self-evident.

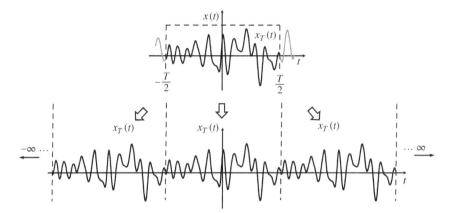

Figure 5.6 Illustration of the assumption of periodicity in the time domain that occurs with Fourier transformation to the frequency domain.

if the window length increases, but the signal power would decrease. The definition of signal energy for a transient signal is given by

$$\text{signal energy} = \int_{-T/2}^{T/2} x_T^2(t)\mathrm{d}t, \tag{5.4}$$

where the time duration T is of arbitrary length provided that the transient is not truncated. In this case Parseval's theorem means that

$$\int_{-T/2}^{T/2} x_T^2(t)\mathrm{d}t = \int_{-\infty}^{\infty} |X_T(f)|^2 \mathrm{d}f, \tag{5.5}$$

where the term $|X_T(f)|^2$ on the right-hand side of Eq. (5.5) is called the Energy Spectral Density (ESD), which for a displacement signal has units of $(\text{m/Hz})^2$, or $\text{m}^2\text{s/Hz}$ as is more commonly found in the literature. Note that, similar to the definition of signal power, signal energy does not necessarily refer to energy in the physical sense, but it is defined according to Eq. (5.4). When units of $\text{m}^2\text{s/Hz}$ are used, however, it is clear how the ESD relates to the PSD, which has units of m^2/Hz, for a displacement signal.

The application of Parseval's theorem is explored for the vibration signals shown in Figure 5.4, in which the differences when dealing with sampled rather than continuous data are also discussed. Note that when processing the data using the Fourier transform it is assumed that the truncated time series $x_T(t)$ is repeated infinitely for both negative and positive time as shown in Figure 5.6.

5.3.1 Excitation at a Single Frequency

Perhaps the simplest way to excite a structure using a shaker is to use a harmonic signal, which is described by a sine or a cosine function. This is frequently carried out in a laboratory setting to check the test set up and to determine whether its behaviour is linear or nonlinear. A linear system will only respond at the excitation frequency, and this is easy to check by using single-frequency excitation and an oscilloscope, whereas a nonlinear system may respond at multiple frequencies, which manifests itself as a distorted sine wave. Furthermore, there is an established test procedure to measure an FRF, frequency by frequency, using a method called the *stepped-sine* approach.

Because a sine or cosine wave is a periodic function, it can be completely described by a single period. However, analysis is often carried out over several periods. Such a signal is shown at the top of Figure 5.7. Provided that the time over which such a signal is analysed, is equal to an integer

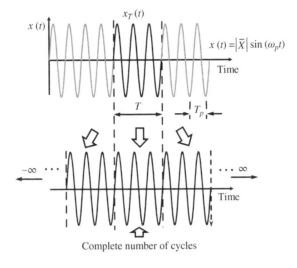

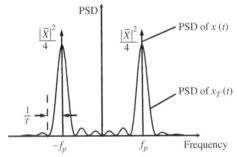

Figure 5.7 Power spectral density of a truncated sine wave with a window of time duration T that captures a complete number of cycles.

multiple of cycles, and the signal is given by, for example, $x(t) = |\overline{X}| \sin(\omega_p t)$, in which $|\overline{X}|$ is the amplitude, the mean square value of a signal is given by

$$x_{\text{mean}}^2 = \frac{1}{T} \int_{-T/2}^{T/2} x^2(t) \mathrm{d}t = \frac{1}{T} \int_{-T/2}^{T/2} (|\overline{X}| \sin(\omega_p t))^2 \mathrm{d}t = \frac{|\overline{X}|^2}{2}. \tag{5.6}$$

The root mean square (rms) value of the signal is determined by simply taking the square root of Eq. (5.6) to give $x_{\text{rms}} = \sqrt{x_{\text{mean}}^2} = |\overline{X}|/\sqrt{2}$. To calculate the PSD, the FT is required, and this results in some complications due to the finite duration of the window. First consider the idealised situation when the window tends to infinity, then the PSD is given by $\frac{1}{T}|X_T(f)|^2 = \frac{1}{T}X_T(f)X_T^*(f)$, where $T \to \infty$ and

$$X(f) = \int_{-\infty}^{\infty} x(t) e^{-\mathrm{j}2\pi f t} \mathrm{d}t. \tag{5.7}$$

Substituting for $x(t) = |\overline{X}| \sin(\omega_p t)$, and noting that $\sin(\omega_p t) = \left(e^{\mathrm{j}2\pi f_p t} - e^{-\mathrm{j}2\pi f_p t}\right)/\mathrm{j}2$, Eq. (5.7) evaluates to

$$X(f) = \frac{|\overline{X}|}{\mathrm{j}2}[\delta(f - f_p) - \delta(f + f_p)], \quad \text{for } -\infty < f < \infty. \tag{5.8}$$

Thus, the PSD $= [\delta(f - f_p) + \delta(f + f_p)]|\overline{X}|^2/4$, which is plotted in the bottom part of Figure 5.7 as arrows at $f = \pm f_p$. The energy, given by $|\overline{X}|^2/4$ at frequencies $\pm f_p$, sums to give the mean square value in Eq. (5.6).

Now, consider the case where the FT is carried out on the windowed section of data $x_T(t)$, shown in the top part of Figure 5.7, but now T is finite rather than infinite as discussed previously. Note that one effect of windowing is that the effective time series is then a repetition of the sections of data in the window as shown in the central part of Figure 5.7. If the window encompasses a complete number of cycles of $x(t)$, then the effective time series is the same as the actual time series. However, this is not the case if an incomplete number of cycles are captured by the window – this is discussed later. The finite length window also has the effect of smearing the energy in the frequency domain, so that it is not concentrated at the excitation frequency. This windowing effect was discussed in detail in Chapter 4. In the time domain, the data to be transformed to the frequency domain are the product of the window and the signal. In the frequency domain, this becomes convolution of the window and the signal. If a rectangular time domain window is used as shown in Figure 5.7, this transforms to a sinc function in the frequency domain. Thus, the FT of $x_T(t) = |\overline{X}| \sin(\omega_p t)$ becomes

$$X_T(f) = \frac{|\overline{X}|}{j2}[\delta(f - f_p) - \delta(f + f_p)] * T\frac{\sin(\pi f T)}{\pi f T}e^{-j\pi f T}, \quad \text{for} \quad -\infty < f < \infty, \tag{5.9a}$$

where '*' denotes the convolution operation. Equation (5.9a) evaluates to

$$X_T(f) = \frac{|\overline{X}|T}{j2}\left(\frac{\sin(\pi(f - f_p)T)}{\pi(f - f_p)T}e^{-j\pi(f - f_p)T} - \frac{\sin(\pi(f + f_p)T)}{\pi(f + f_p)T}e^{-j\pi(f + f_p)T}\right), \quad \text{for} \quad -\infty < f < \infty. \tag{5.9b}$$

The PSD is given by $\frac{1}{T}|X_T(f)|^2$, so that

$$\mathrm{PSD}(x_T(t)) = \frac{|\overline{X}|^2 T}{4}\left|\frac{\sin(\pi(f - f_p)T)}{\pi(f - f_p)T} - \frac{\sin(\pi(f + f_p)T)}{\pi(f + f_p)T}e^{-j2\pi f_p T}\right|^2, \quad \text{for} \quad -\infty < f < \infty. \tag{5.10}$$

This is plotted at the bottom part of Figure 5.7. It is clear that the effect of windowing the data in the time domain, as shown in the top part of Figure 5.7, is to cause a distribution of energy from the excitation frequency to other frequencies, and this is described by the combination of sinc functions. The term 'leakage' is the term used to describe this effect. Leakage can be reduced by using a longer duration time window, which can be seen in the lower part of Figure 5.7, as the width of the main lobe is $2/T$. Note that the area under the PSD plot is equal to the mean square value of the time history, and this is independent of the duration of the time window. If either the frequency or the time window is large enough, the PSD is dominated by the main lobes in Figure 5.7, centred around the excitation frequency at $\pm f_p$. The peaks in the PSD at $\pm f_p$ are then given approximately by $|\overline{X}|^2 T/4$. It can thus be seen, that as the time window is increased, the amplitude of the PSD at the excitation frequency increases proportionately, and because of Parseval's theorem, the signal power becomes more concentrated around the excitation frequency.

Now, consider the situation where the time window does not capture an integer number of cycles, which is shown in the top part of Figure 5.8. It is clear that the assumed signal is no longer an actual sine wave in this case, as shown in the central part of Figure 5.8, as there are discontinuities at each end of the time window. The mean square value of the assumed signal is no longer necessarily equal to $|\overline{X}|^2/2$. There is increased leakage of the signal to other frequencies, including a DC component. This can be seen in the lower part of Figure 5.8. However, Parseval's theorem still holds, with the mean square value of the time domain signal being equal to the area under the PSD plot. Thus, because there is increased leakage to frequencies other than the excitation frequency, the amplitude of the PSD at $\pm f_p$ is reduced.

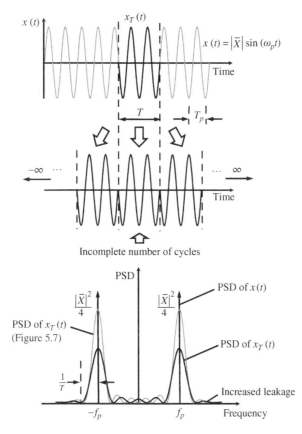

Figure 5.8 Power spectral density of a truncated sine wave with a window of time duration T that captures an incomplete number of cycles.

If sampled data are considered, there are further considerations. For the excitation frequency to coincide with a sample in the frequency domain, the excitation frequency needs to be an integer multiple of the frequency resolution, i.e. $f_p/\Delta f$ must be an integer. This can be written in alternative ways by noting that $\Delta f = 1/(N\Delta t) = f_s/(N)$. An example to illustrate the effects of sampling is given in Figure 5.9. In the top left part of the figure, three cycles of a sine wave with a frequency of f_p are sampled with a sampling frequency of f_s Hz, such that there are N samples. The PSD is calculated both analytically and numerically using the FT to give PSD $= |X_T(f)|^2/T_1$ and is plotted in the lower left part of Figure 5.9 for the frequency range $0-f_s/2$. Note that the area under the PSD in this frequency range is 50% of the mean square value of the signal, which is $|\overline{X}|^2/2$ in this case. As the excitation frequency f_p is an integer multiple of Δf then there is a sample at f_p, which can be clearly seen in the figure. Moreover, as the other samples occur at multiples of $\Delta f = 1/(T)$ in this case, all the other samples are zero. The PSD at f_p is $|\overline{X}|^2 T/4$. In the example shown in the upper right part of Figure 5.9, two and a half cycles of the same sine wave as that in the upper left part of the figure are captured using the same sampling frequency, so that there are P samples. The PSD is again calculated both analytically and numerically to give PSD $= |X_T(f)|^2/T_2$ and is plotted in the lower right part of Figure 5.9 for the frequency range $0-f_s/2$. In this case, it is clear that because the excitation frequency f_p is not an integer multiple of $\Delta f = 1/(P\Delta t)$ there is no sample at f_p. The samples in the PSD calculated using the DFT still coincide with the analytical curve as can be seen in the lower right part of Figure 5.9, but they have a much smaller level than the peak of

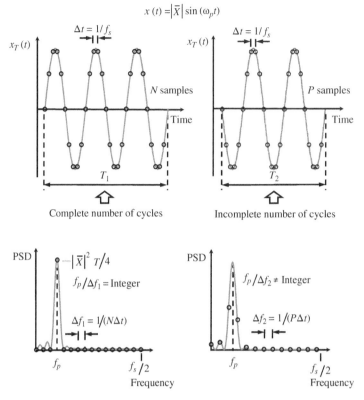

Figure 5.9 Power spectral densities of sampled sine waves (from 0 to $f_s/2$).

the PSD of the continuous time signal. Furthermore, it can be seen that the other samples are no longer zero, which means that there is significant leakage to frequencies other than the excitation frequency (including a DC component).

MATLAB Example 5.1

In this example, the phenomenon of leakage is explored, and Parseval's theorem is applied to a sine wave.

```
clear all

%% Time and frequency parameters
fn=2; fs=40;                    % [Hz]           % excitation and sampling frequencies
Nc=2.5; A=1;                    % [ ],[m]        % no. of cycles; amplitude of signal
T=Nc/fn; dt=1/fs; t=0:dt:T-1/fs; % [s]           % time vector
N=length(t);                                     % number of points
TT=N*dt;                        % [s]

%% Sine wave
x = A*sin(2*pi*fn*t);           % [m]            % displacement sine wave

%% Calculation of PSD
xdft=fft(x)*dt;                                  % DFT of sine wave
psd=abs(xdft).^2/TT;                             % PSD of sine wave
df=1/TT;f=0:df:(N-1)*df;                         % freq. resolution and freq. vector
```

(Continued)

MATLAB Example 5.1   (Continued)

```
%% Pareseval's theorem
MSV=sum((x).^2)/TT*dt                      % mean square value
PSD=2*trapz(psd(1:round(N/2+1)))*df        % area under the PSD plot

%% Plot the results
figure (1)                                 % time domain
plot(t,x,'k','linewidth',2)
hold on
plot(t,x,'ok','linewidth',2,'MarkerSize',10)
xlabel('time(s)');
ylabel('displacement (m)');
grid;axis square

figure (2)                                 % frequency domain
plot(f,psd,'k','linewidth',2)
hold on
plot(f,psd,'ok','linewidth',2,'MarkerSize',10)
xlim([0 fs])
xlabel('frequency (Hz)');
ylabel('PSD (m^2/Hz)');
grid;axis square
```

Results

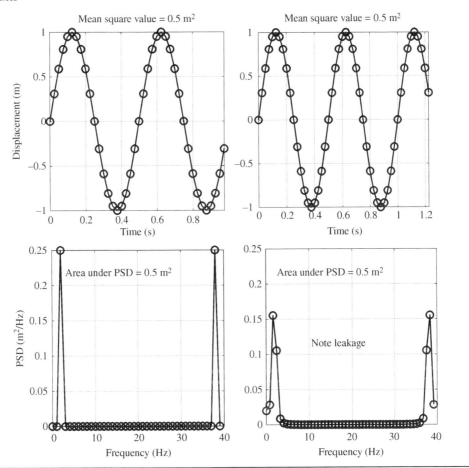

MATLAB Example 5.1   (Continued)

Comments

1. An exercise for the reader is to explore what happens when:
   (a) the sampling frequency is changed,
   (b) the time period over which the analysis is conducted is changed,
   (c) the frequency of excitation is changed.

5.3.2 Excitation Using a Random Signal

A common way to excite a structure during a vibration test using a shaker is to use a random, or pseudo random signal. This type of signal is easily generated using a computer, and the signals are relatively easy to process. As mentioned previously, this type of signal is assumed to last for all time, so signal power rather than signal energy is of interest. In an experiment, the bandwidth of such a signal is generally limited to the frequency range of interest using a low-pass or band-pass filter. This can also be done in simulations using sampled data, but if a filter is not applied, then the maximum frequency content of the random signal is limited to half of the sampling frequency. A typical random signal, which has a time duration of T_c seconds, is shown in Figure 5.10. To calculate the PSD of this signal, it is split into segments of length T and the raw PSD of each segment is calculated. Note that in the case considered, the segments are non-overlapping. The raw PSDs are then averaged as indicated in Figure 5.10 to give an estimate of the true PSD of the signal. This method is called segment averaging or Welch's method (Welch, 1967). The data are sampled (which is not explicitly shown in Figure 5.10), and there are N points in each segment, such that there are N data points in the estimated PSD, and if there are P segments there are NP points in the complete time history. As in the previous case discussed, the sampling frequency is f_s, the time resolution is $\Delta t = 1/f_s$, and the frequency resolution is $\Delta f = 1/(N\Delta t)$. The actual PSD is given by Shin and Hammond (2008)

$$S_{xx}(f) = \lim_{T \to \infty} \frac{E\left[X_p^*(f)X_p(f)\right]}{T}, \tag{5.11}$$

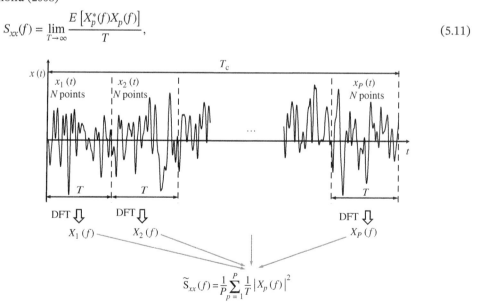

Figure 5.10 A typical random signal split into time segments for averaging purposes.

where $X_p(f)$ is the Fourier transform of the time series in the p-th segment. As T is finite, only an estimate of the PSD is possible. It is denoted as $\tilde{S}_{xx}(f)$ and is given by

$$\tilde{S}_{xx}(f) = \frac{1}{P}\sum_{p=1}^{P}\frac{1}{T}|X_p(f)|^2. \tag{5.12}$$

This is only an estimate of the actual PSD $S_{xx}(f)$, so there is an error at each frequency, which can be reduced by the averaging process. Because $\tilde{S}_{xx}(f)$ is a random variable, the error is quantified by the standard deviation $\sigma(\tilde{S}_{xx}(f))$ of the estimate. The relationship between the error and the actual PSD for a time series consisting of P uncorrelated sections of data is given by Shin and Hammond (2008)

$$\frac{\sigma(\tilde{S}_{xx}(f))}{S_{xx}(f)} \approx \frac{1}{\sqrt{P}}. \tag{5.13}$$

Equation (5.13) is plotted in dB in Figure 5.11. It can be seen that it is worthwhile to apply the averaging procedure, but in practice there is a limit to the improvement in the estimate that can be achieved. For example, if only a raw estimate is calculated, i.e. $T = T_c$, then $P = 1$, and the standard deviation is the same as the estimate! The estimate is improved by 3 dB for each doubling of the number of averages. For example, if there are 16 averages there is an improvement of 12 dB, but this only improves by a further 3 dB if there are an additional 16 averages. This behaviour follows the 'law of diminishing returns', such that beyond an acceptable level of accuracy, a very large number of additional averages are needed to obtain a relatively small improvement in the estimate.

Previously in this book, a rectangular window has been used to extract data from a time history before transformation into the frequency domain. As shown in Section 5.3.1, there can be abrupt discontinuities in the data at the edge of the windowed data using this type of window, which results in leakage in the frequency domain. To overcome this problem, many types of windows have been proposed, each of which has different characteristics (Harris, 1978; Shin and Hammond, 2008). In this book, only one of these windows is considered, as this is used most often in vibration testing when using random signals. This is the *Hanning* window, which has a smooth characteristic, reducing to zero at each end of the window. It is given by

$$\begin{aligned} w_{\text{Hanning}}(t) &= \cos^2\left(\frac{\pi t}{T}\right) \quad \text{for} -\frac{T}{2} \le t \le \frac{T}{2} \\ &= 0 \qquad\qquad \text{otherwise} \end{aligned} \tag{5.14a}$$

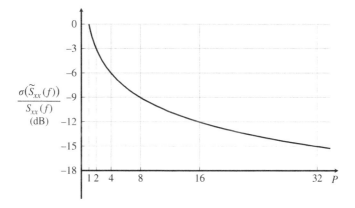

Figure 5.11 Effect of averaging on the variance of the estimate of the PSD.

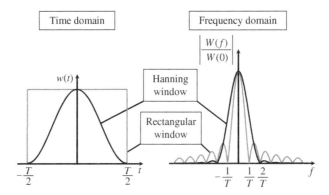

Figure 5.12   Rectangular and Hanning windows.

This can be contrasted with the rectangular window, which is given by

$$w_{\text{rectangular}}(t) = 1 \quad \text{for} -\frac{T}{2} \le t \le \frac{T}{2}.$$
$$= 0 \quad \text{otherwise}$$

(5.14b)

These windows are plotted in the left part of Figure 5.12, where the difference between them is obvious. A correction factor is needed for the Hanning window as there is a loss of 'energy' compared to the rectangular window. This is given by

$$\text{scaling factor} = \sqrt{\frac{\displaystyle\int_{-T/2}^{T/2} w_{\text{rectangular}}^2(t)}{\displaystyle\int_{-T/2}^{T/2} w_{\text{Hanning}}^2(t)}} = \sqrt{\frac{8}{3}}.$$

(5.15)

The Fourier transforms for the two windows are given by[2]

$$W_{\text{Hanning}}(f) = \frac{T}{2} \frac{\sin(\pi f T)}{\pi f T [1 - (fT)^2]} = \frac{T}{2[1 - (fT)^2]} \text{sinc}(fT)$$

(5.16a)

and

$$W_{\text{rectangular}}(f) = T \frac{\sin(\pi f T)}{\pi f T} = T\text{sinc}(fT).$$

(5.16b)

To illustrate the different spectral shapes of the windows, the normalised amplitude for the FT of each window is plotted in the right part of Figure 5.12. Note that the actual amplitude for the rectangular window at $f = 0$ is twice that of the Hanning window. It can be seen that the main lobe is broader for the Hanning window compared to the rectangular window, but the amplitudes of the side lobes are much smaller. The asymptotic roll-off for these lobes is 6 dB/octave for the rectangular window and 18 dB/octave for the Hanning window (Shin and Hammond, 2008). Hence, if the Hanning window is used, the 'energy' is much more confined to frequencies close to the actual frequency, compared to when a rectangular window is used.

An illustration of the use of the Hanning window applied to a random signal is shown in the top part of Figure 5.13. The smooth transition between adjacent windows is apparent. It is also clear that the penalty paid for the smooth transition is the loss of some data. To capture the data, another set of windows is applied as shown in the lower part of Figure 5.13. The amount of overlap between

---

2 $\text{sinc}(fT) = \sin(\pi f T)/(\pi f T)$, which is the normalised sinc function.

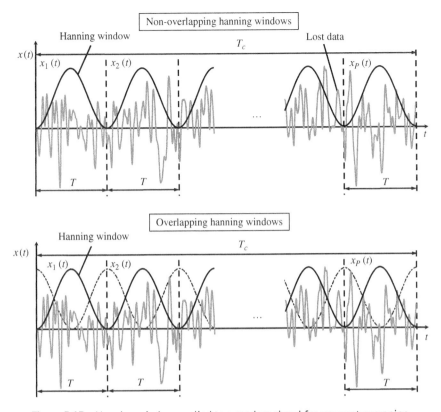

Figure 5.13 Hanning window applied to a random signal for segment averaging.

the windows can be chosen by the analyst, but in the case shown in Figure 5.13, the overlap is set at 50% of the window length (which is commonly used in practice). Of course, this means that some of the data appears in more than one window, so there is degree of correlation between each overlapping window, which means that Eq. (5.13) does not strictly apply. Further details on this can be found in Brandt (2011).

A random time signal and its PSD estimate calculated using segment averaging with a Hanning window and 50% overlap are shown in Figure 5.14. The PSD differs from that shown for the sine wave in two respects. The first is that it is shown on a logarithmic scale (which is the usual case for a PSD), and the second is that it is a single-sided PSD estimate $\tilde{G}_{xx}(f)$, which is given by

$$\begin{aligned} \tilde{G}_{xx}(f) &= 2\tilde{S}_{xx}(f) & f > 0 \\ &= \tilde{S}_{xx}(f) & f = 0. \\ &= 0 & f < 0 \end{aligned} \tag{5.17}$$

Thus, the area under the single-sided PSD plot up to half the sampling frequency is approximately equal to the mean square value of the time domain signal, $(x_{\mathrm{rms}})^2$. In the example shown in Figure 5.14, the area under the PSD is approximately a rectangle, so the average value of the PSD is constant with frequency and is given approximately by $(x_{\mathrm{rms}})^2/(f_s/2)$. The effects of averaging can also be seen in the lower part Figure 5.14. The PSD estimate calculated with 4 averages is much less smooth than the PSD estimate calculated with 32 averages. Also note that for a fixed data length of T_c seconds (see Figure 5.10 or 5.13), then as the number of averages increases, the

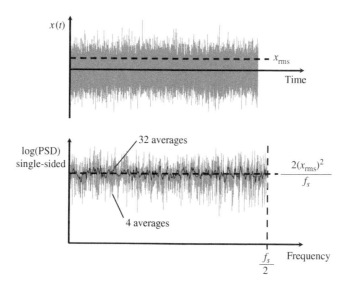

Figure 5.14 Time history of a random signal and its single-sided PSD estimate.

random error reduces but the frequency resolution (related to the bias error) worsens (Shin and Hammond, 2008).

MATLAB Example 5.2

In this example, the spectral characteristics of a random signal are investigated, as is the trade-off between frequency resolution and random error, and Parseval's theorem is verified for a random signal.

```
clear all

%% Time and frequency parameters
fs=200;                          % [Hz]        % sampling frequency
T=5; dt=1/fs; t=0:dt:T;          % [s]         % signal length; time res. time vector
N=length(t); A=2;                             % no. of points; amplitude of signal

%% Random signal
x=A*randn(length(t),1);          % [m]         % random signal
x=x-mean(x);                                  % set mean of signal to zero

%% Calculation of PSD
Na = 4;                                       % number of averages = 4
nfft=round(N/Na);                             % number of points in the DFT
noverlap=round(nfft/2);                       % number of points in the overlap
psd=pwelch(x,hann(nfft),noverlap,nfft,fs);    % calculation of single-sided PSD
df=1/(nfft*dt);                               % frequency resolution
f=0:df:fs/2;                                  % frequency vector

%% Calculation of PSD
Na = 32;                                      % number of averages = 32
nfft=round(N/Na);
noverlap=round(nfft/2);
psd2=pwelch(x,hann(nfft),noverlap,nfft,fs);
df2=1/(nfft*dt);
f2=0:df2:fs/2;
```

(Continued)

MATLAB Example 5.2 (Continued)

```
%% Parseval's theorem
MSV=sum((x).^2)/T*dt                    % mean square value

STD=sqrt(MSV);                          % standard deviation
PSD=trapz(psd2)*df2                     % area under the PSD plot
amp=PSD/(fs/2)                          % average value of PSD

%% Plot the results
figure(1)                               % time domain
t1=0.001:0.1:T;
plot(t,x,'linewidth',2,'Color',[.6 .6 .6])
hold on
plot(t1,STD*t1./t1,'--k','linewidth',3)
xlabel('time(s)'); ylabel('displacement (m)');
grid, axis square

figure(2)                               % frequency domain plot
fa=0:600;
plot(f,10*log10(psd),'Color',[.6 .6 .6])
hold on
plot(f,10*log10(psd),'o','Color',[.6 .6 .6])
hold on
plot(f2,10*log10(psd2),'k','linewidth',2)
hold on
plot(f2,10*log10(psd2),'sk','linewidth',2)
plot(fa,10*log10(fa./fa*amp),'--k')
xlabel('frequency (Hz)');
ylabel('PSD (m^2/Hz)');
grid; axis square
```

Results

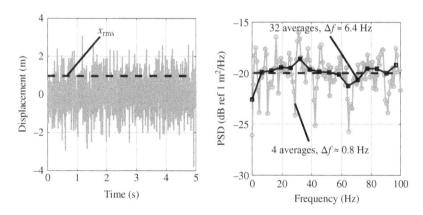

Comments

1. An exercise for the reader is to explore what happens when:
   (a) the sampling frequency is changed,
   (b) the number of averages is changed,
   (c) the amount of window overlap is changed,
   (d) a rectangular window is used.
2. An exercise for the reader is to investigate the difference between a single-sided PSD and a two-sided PSD.

5.3.3 Excitation Using a Chirp or Swept Sine

A chirp signal is described by a sine function where the instantaneous frequency changes with time. There are two ways in which a chirp signal can be used to excite a structure. One is to increase the frequency of excitation very slowly such that the duration of the chirp signal is much greater than the fundamental natural period of the structure. In this case the response of the structure is quasi steady-state at any instant during the sweep. The other way is to use a fast sweep (White and Pinnington, 1982), where the duration of the chirp signal is much less than the fundamental natural period of the structure. In this case, the effect of the excitation is very similar to an impact being applied to the structure. As mentioned previously, a single chirp signal is a transient signal, so signal energy rather than power is of interest. A chirp signal, in which the frequency of excitation increases linearly with time, is given by

$$x(t) = |\overline{X}| \sin(\phi(t)), \tag{5.18}$$

where $\phi(t)$ is the instantaneous phase of the signal, given by

$$\phi(t) = \frac{(\omega_2 - \omega_1)}{2T} t^2 + \omega_1 t, \tag{5.19}$$

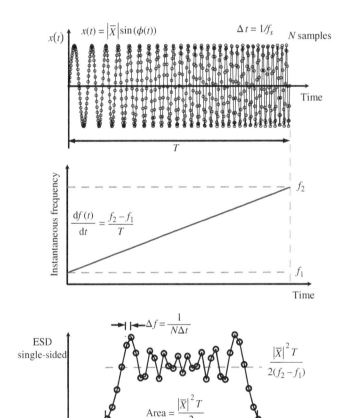

Figure 5.15 A linear chirp and its single-sided ESD estimate.

in which $\omega_1 = 2\pi f_1$ and $\omega_2 = 2\pi f_2$ are the starting and finishing frequencies of the sweep, respectively. Note that the instantaneous frequency is given by

$$\omega(t) = \frac{\mathrm{d}\phi(t)}{\mathrm{d}t} = \frac{(\omega_2 - \omega_1)}{T}t + \omega_1, \tag{5.20}$$

and the rate of change of frequency is given by $\frac{\mathrm{d}\omega(t)}{\mathrm{d}t} = (\omega_2 - \omega_1)/T$.

A typical linear chirp signal described by Eq. (5.18) is shown in the top part of Figure 5.15. The frequency increases from f_1 to f_2 over a period of T seconds. Also shown are samples of the continuous signal sampled at a frequency of f_s samples per second, such that the time resolution is $\Delta t = 1/f_s$ seconds. The excitation frequency is thus increased at a rate of $(f_2 - f_1)/T$ Hz/s, which is shown in the central part of Figure 5.15. The single-sided ESD of the chirp signal is calculated from $|X(f)|^2$ in a similar way to that for the single-sided PSD given in Eq. (5.17), and is plotted in the lower part of Figure 5.15. The frequency resolution is given by $1/(N\Delta t)$, which is approximately given by $1/T$. It can be seen that the energy of the chirp signal is broadly contained within the frequencies f_1 and f_2. An approximate value of the average amplitude of the single-sided ESD within this frequency range can be determined by noting that the area under the two-sided ESD is equal to the mean square value of the signal multiplied by T, i.e. $|\overline{X}|^2 T/2$ (if the chirp contains an integer number of cycles). As this area is equal to the product of $f_2 - f_1$ and the amplitude of the single-sided ESD, within this frequency range the approximate average amplitude of the single-sided ESD within this frequency range is given by $|\overline{X}|^2 T/[2(f_2 - f_1)]$. This is shown in the lower part of Figure 5.15.

MATLAB Example 5.3

In this example, the spectral characteristics of a linear chirp signal are investigated and Parseval's theorem is verified for a chirp signal.

```
clear all

%% Time and frequency parameters
fs=40;                        % [Hz]          % sampling frequency
T = 8;dt=1/fs;t=0:dt:T;       % [s]           % chirp length; time res. time vector
N=length(t);A=2;                              % no. of points; amplitude of chirp
f1=1; f2=5;                   % [Hz]          % upper and lower frequencies
TT=N*dt;                      % [s]

%% Chirp signal
a=2*pi*(f2-f1)/(2*T);b=2*pi*f1;               % coefficients
x=A*sin(a*t.^2+b*t);          % [m]           % chirp signal

%% Calculation of ESD
X=fft(x)*dt;                  % [m/Hz]        % DFT of chirp signal
esd = 2*abs(X).^2;            % [m²s/Hz]      % ESD of chirp
df=1/TT;f=0:df:fs/2;          % [Hz]          % freq. resolution; freq. vector
esds=2*abs(X(1:round(N/2))).^2;               % ESD of chirp (single-sided)
esds(1)=abs(X(1)).^2;

%% Parseval's theorem
e=trapz((x).^2)*dt            % [m²s]         % energy in the time domain
E=2*trapz(esd(1:round(N/2+1)))*df  % [m²s]    % energy in the frequency domain
amp=A^2*T/2/(f2-f1)           % [m²s/Hz]      % average amplitude of ESD

%% Plot the results
figure (1)                                    % time domain
plot(t,x,'k','linewidth',2)
hold on
```

(Continued)

MATLAB Example 5.3   (Continued)

```
plot(t,x,'ok','linewidth',2,'MarkerSize',10)
xlabel('time(s)');
ylabel('displacement (m)');
grid;

figure (2)                              % frequency domain
plot(f,esds,'k','linewidth',2)
hold on
plot(f,esds,'ok','linewidth',2,'MarkerSize',10)
hold on
plot(f,amp*f./f,'--k','linewidth',2)
axis([0,6,0,inf]);grid
xlabel('frequency (Hz)');
ylabel('ESD (m^2s/Hz)');
```

Results

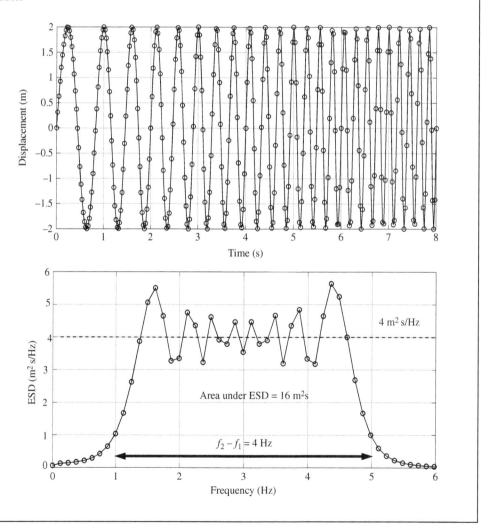

MATLAB Example 5.3   (Continued)

Comments

1. An exercise for the reader is to explore what happens when:
   (a) the sampling frequency is changed,
   (b) the time duration of the chirp signal is changed,
   (c) the frequency range of excitation is changed.

5.3.4   Excitation Using a Half-Sine Pulse

The final vibration excitation signal discussed in this chapter is the signal generated by an impact hammer, which is idealised as a half-sine impulse as discussed in Section 5.2.2. This type of signal is shown in Figures 5.3 and 5.4. It is a transient signal, so signal energy rather than signal power is considered. In Figure 5.4, it can be seen that following the half-sine impulse, there is a significant time period during which the signal is zero. This is needed, because in a vibration test the force and the response signals should have the same time duration for subsequent frequency domain analysis. After the half-sine pulse force has been applied to a structure using an impact hammer, the resulting free vibration of the structure takes some time to decay away (this time is, of course, dependent upon the amount of damping in the structure), which needs to be accounted for when setting the overall duration of the force signal. The time history of a half-sine pulse force of duration T_1 and its sampled form is shown in the top left part of Figure 5.16, and the time history for the half-sine pulse followed by an extended time period when the signal is zero such that the total time duration is T_2, is shown in the upper right part of Figure 5.16. The modulus of the FT of the half sine pulse is given by Eq. (5.1), which is plotted for positive frequencies only, as the continuous line in the lower part of Figure 5.16. Also shown in this plot are the moduli of the DFTs for the two sampled time histories shown in the upper part of Figure 5.16. It is clear that the duration of the signal used

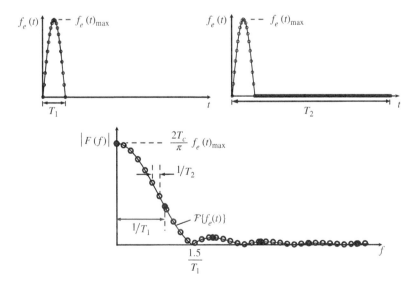

Figure 5.16   Time history of a half-sine pulse, and its FT and DFT (from 0 to $f_s/2$).

to calculate the DFT plays an important role in terms of the number of frequency points in the spectrum. The effect of adding the extra signal (of zero values) to the half-sine pulse does not add additional information, but it does add more frequency points. It effectively interpolates the data between the frequency points, which occur at frequencies of integer multiples of $1/T_1$. The effect is the same as *zero padding* a signal which is described in Shin and Hammond (2008).

The ESD of the half-sine pulse can be found in the same way as for the chirp signal described in Section 5.3.3. This is carried out in MATLAB Example 5.4.

MATLAB Example 5.4

In this example, the ESD of a half-sine pulse is explored and Parseval's theorem is verified for a half-sine pulse. Also, the effect of extending the duration of the signal by adding zeros is investigated.

```
clear all

%% Time and frequency parameters
fs=2000;                               % [Hz]      % sampling frequency
T=0.01;                                % [s]       % contact time
dt=1/fs;t=0:dt:T-dt;                   % [s]       % time resolution; time vector
N=length(t);                                       % number of points
TT=N*dt;

%% Half-sine pulse
x = sin(pi*t/T);                       % [m]       % displacement half-sine pulse

%% Extended half-sine pulse signal
Nz=5*N;                                            % number of zeros to add
xe = [x zeros(Nz,1)'];                 % [m]       % extended signal
te = 0:dt:(length(xe)-1)*dt;           % [s]       % time vector for extended signal
Ne=length(te);                                     % number of points in extended signal
TTe=Ne*dt;

%% Calculation of ESDs
X=fft(x)*dt;                           % [m/Hz]    % DFT of half-sine pulse
esd=conj(X).*X;                        % [m²s/Hz]  % ESD of half-sine pulse
df=1/TT;fx=0:df:fs/2;                  % [Hz]      % freq. resolution; freq. vector
esds=2*abs(X(1:round(N/2)+1)).^2;                  % single sided ESD
esds(1)=abs(X(1)).^2;

Xe=fft(xe)*dt;                         % [m/Hz]    % DFT of extended signal
esde=conj(Xe).*Xe;                     % [m²s/Hz]  % ESD of extended signal
dfe=1/TTe;fxe=0:dfe:fs/2   ;           % [Hz]      % freq. resolution; freq. vector
esdes=2*abs(Xe(1:round(Ne/2)+1)).^2;               % single sided ESD
esdes(1)=abs(Xe(1)).^2;

%% Theory
ft=0:1:250;w=2*pi*ft;                              % frequency vector
Xt=T*2*pi*cos(w*T/2)./(pi^2-(w*T).^2);             % spectrum
esdts=2*abs(Xt).^2;                                % ESD of half-sine pulse (single-sided)
esdts(1)=abs(Xt(1)).^2;

%% Parseval's theorem
e=trapz(x.^2)*dt                       % [m²s]     % energy in the time domain
E1=2*trapz(esd(1:round(N/2+1)))*df     % [m²s]     % energy in the frequency domain
E2=2*trapz(esde(1:round(Ne/2+1)))*dfe  % [m²s/Hz]

%% Plot the results
figure (1)                                         % time domain
plot(te,xe,'k')
hold on
```

(Continued)

MATLAB Example 5.4   (Continued)

```
plot(te,xe,'ok')
hold on
plot(t,x,'ok','MarkerFaceColor',[.1 .1 .1])
grid;xlim([0 0.06])
xlabel('time(s)');
ylabel('displacement (m)');

figure (2)                              % frequency domain
plot(ft,10*log10(abs(esdts)),'k')
hold on
plot(fx,10*log10(abs(esds)),'ok',...
'MarkerFaceColor',[.1 .1 .1])
hold on
plot(fxe,10*log10(abs(esdes)),'ok')
grid;axis([0, 250, -90, -40])
xlabel('frequency (Hz)');
ylabel('ESD (m^2s/Hz)');
```

Results

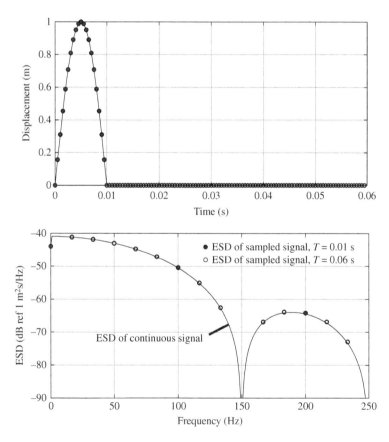

Comments

1. An exercise for the reader is to explore what happens when:
   (a) the sampling frequency is changed,
   (b) the contact time is changed,
   (c) the time period over which the analysis is conducted is changed.

Table 5.1 Some characteristics of excitation signals.

Persistent signals		
x(t) (m)	**PSD (m²/Hz)**	**Parseval's theorem**

Single frequency

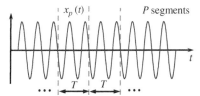

$$\tilde{S}_{xx}(\omega) = \frac{1}{P}\sum_{p=1}^{P}\frac{1}{T}|X_p(f)|^2$$

where
$X_p(f) = \mathcal{F}\{x_p(t)\}$
assumption of stationarity

Random excitation

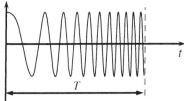

$$\frac{1}{T}\int_{-T/2}^{T/2} x_p^2(t)\mathrm{d}t = \int_{-\infty}^{\infty}\tilde{S}_{xx}(\omega)\mathrm{d}f$$

Transient signals		
x(t) (m)	**ESD (m²s/Hz)**	**Parseval's theorem**

Chirp or swept sine

Half-sine pulse

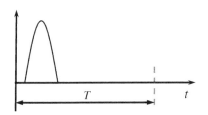

$$E_{xx}(\omega) = |X(f)|^2$$
where
$X(f) = \mathcal{F}\{x(t)\}$

$$\int_{-T/2}^{T/2} x^2(t)\mathrm{d}t = \int_{-\infty}^{\infty} E_{xx}(\omega)\mathrm{d}f$$

5.4   Summary

Four signals commonly used in vibration testing shown in Figure 5.4 have been discussed in this chapter. Three of these are used to drive an electrodynamic shaker, and one is the force generated by an instrumented impact hammer. For convenience, the characteristics of the signals have been considered using displacement signals to illustrate the units of the signals in both the time and frequency domains. The way in which the signals are processed depends upon the type of signal, which are classified as either persistent, where they are assumed to act over infinite time, or transient, where they act for a limited time. Of course, no signal acts for an infinite time, but the persistent signals last for longer than the time window in which a segment of the data is transformed to the frequency domain. A fundamental assumption is that the signal has the same mean and variance in each segment – that the signal is stationary in the statistical sense. For persistent signals the energy in the signal increases as the signal length increases, but the power is constant, so the frequency domain quantity of power spectral density (PSD) is calculated. For a transient signal, the signal energy is independent of the window length provided that the signal is completely captured by the window. Thus, in this case, the frequency domain quantity of energy spectral density (ESD) is calculated. Table 5.1 summarises the way in which the four signals are processed.

References

Brandt, A. (2011). *Noise and Vibration Analysis: Signal Analysis and Experimental Procedures*. Wiley.

Ewins, D.J. (2000). *Modal Testing: Theory, Practice and Application*, 2nd Edition. Research Studies Press.

Harris, F.J. (1978). On the use of windows for harmonic analysis with the discrete Fourier transform. *Proceedings of the IEEE*, 66(1), 51-83. https://doi.org/10.1109/PROC.1978.10837.

McConnell, K.G. and Varoto, P.S. (2008). *Vibration Testing: Theory and Practice*, 2nd Edition. Wiley.

Shin, K. and Hammond, J.K. (2008). *Fundamentals of Signal Processing for Sound and Vibration Engineers*. Wiley.

Waters, T.P. (2013). *Vibration Testing, Chapter 9 in Fundamentals of Sound and Vibration,* (eds F.J. Fahy and D.J. Thompson). Taylor & Francis Group.

Welch, P.D. (1967). The use of fast Fourier transform for the estimation of power spectra: A method based on time averaging over short, modified periodograms. *IEEE Transactions on Audio and Electroacoustics*, 15(2), 70–73. https://doi.org/10.1109/TAU.1967.1161901.

White, R.G. and Pinnington, R.J. (1982). Practical application of the rapid frequency sweep technique for structural frequency response measurement. *Aeronautical Journal*, 86(855), 179–199. https://doi.org/10.1017/S0001924000018741.

6

Determination of the Vibration Response of a System

6.1 Introduction

In Chapter 5 some common types of signals used for vibration testing were discussed. When carrying out a real experiment, these signals would be used to excite the structure under test and the response (displacement, velocity, or acceleration) would be measured. The force input and the vibration response signals could then be used to determine the dynamic behaviour of the system, and important physical parameters that influence this behaviour could be identified. In a virtual experiment, such as that discussed in this book, the vibration response of the system has to be determined numerically. Three ways of doing this are discussed in this chapter. They are convolution, calculation of the response via transformation to the frequency domain, and numerical integration of the equation of motion. The focus of the study is the input–output relationship for the simple vibrating system illustrated in Figure 6.1.

6.2 Determination of the Vibration Response

6.2.1 Convolution in the Time Domain

The relationship between the displacement response of a vibrating system and the force input (which is zero for $t < 0$), such as that shown in Figure 6.1, is given by Shin and Hammond (2008)

$$x(t) = \int_0^\infty h(t - \tau) f_e(\tau) \mathrm{d}\tau. \tag{6.1a}$$

This is called the convolution integral and is derived in Appendix G, along with an historical discussion on the operation. An alternative way of writing Eq. (6.1a) is by

$$x(t) = f_e(t) * h(t), \tag{6.1b}$$

where $*$ denotes convolution. Note that $f_e(t)$ denotes force and has units of N, $x(t)$ is the displacement response and has units of m, and $h(t)$ is the displacement impulse response function (IRF) and has units of m/Ns. This is derived in Chapter 2 for an SDOF system and is given by

$$h(t) = \frac{1}{m\omega_d} e^{-\zeta\omega_n t} \sin(\omega_d t), \tag{6.2}$$

where $\omega_n = \sqrt{k/m}$ is the undamped natural frequency, $\zeta = c/(2m\omega_n)$ is the viscous damping ratio, and $\omega_d = \omega_n \sqrt{1 - \zeta^2}$ is the damped natural frequency. In practice, the output is calculated using

Virtual Experiments in Mechanical Vibrations: Structural Dynamics and Signal Processing,
First Edition. Michael J. Brennan and Bin Tang.

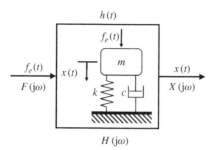

Figure 6.1 Block diagram of a simple vibrating system.

force input data that are sampled at a frequency of $f_s = 1/\Delta t$, so that $t = n\Delta t$, where Δt is the time resolution of the sampled signals. Accordingly, the IRF is also sampled, and the convolution operation given in Eq. (6.1a) becomes

$$x(n\Delta t) = \sum_{m=0}^{n} h(n\Delta t - m\Delta t)f_e(m\Delta t),$$ (6.3a)

or

$$x(n\Delta t) = f_e(n\Delta t) * h(n\Delta t).$$ (6.3b)

Examples of convolution are given in Appendix G, together with an animation to illustrate the process for each time step Δt. Determination of the vibration response of the system in Figure 6.1 is illustrated in Section 6.3 for different types of excitation force.

6.2.2 Calculation of the Response via the Frequency Domain

It is possible to determine the response of a vibrating system by transforming the force input to the frequency domain using the Fourier transform (FT), multiplying by the frequency response function (FRF) to give the displacement response in the frequency domain before transforming back to the time domain using the inverse Fourier transform (IFT). This is illustrated by

$$\begin{array}{cc} f_e(t) & x(t) \\ \text{FT} & \uparrow \\ \downarrow & \text{IFT} \\ F(\text{j}\omega) \times H(\text{j}\omega) & = X(\text{j}\omega). \end{array}$$ (6.4)

The vibration response is often determined in this way because it is faster than using convolution in the time domain. As with convolution, calculation of the vibration response via the frequency domain is usually carried out using sampled data. Thus, the process of transforming to the frequency domain and back to the time domain is carried out using the DFT and the IDFT, respectively. This brings an additional problem in which the sampled response $x(n\Delta t)$, calculated in this way, is periodic, as discussed in Chapter 3. When the response is calculated via the frequency domain using sampled data it is called *circular convolution*. This is illustrated in Figure 6.2 and is discussed in Appendix G. To avoid the issue of wrap around with circular convolution, which is discussed in Appendix G, the vibration response should be zero at the end of the time period over which the analysis is conducted. This ensures that there is an accurate representation of $x(n\Delta t)$, comparable with that calculated by convolution in the time domain.

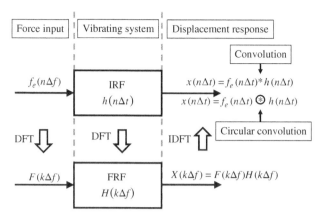

Figure 6.2 Determination of the response of a vibrating system by convolution in the time domain, and by circular convolution via the frequency domain.

To achieve this, the duration of the force time history is doubled, by adding zeros to the original sampled time history. This is illustrated in MATLAB Example G3.

6.2.3 Numerical Integration of the Equation of Motion

If the equation of motion for the vibrating system is known, as it is for the SDOF system considered in this chapter, the response can be determined by numerical integration of the equation of motion for the system. This is easily carried out in MATLAB using one of the functions designed for this purpose, for example `ode45`. This approach is illustrated in Appendix D using the fourth-order Runge–Kutta method. For the SDOF system shown in Figure 6.1, the equation of motion is given by

$$m\ddot{x} + c\dot{x} + kx = f_e. \tag{6.5}$$

This is written in terms of two first-order equations as $\dot{x} = y$ and $\dot{y} = \frac{1}{m}(f_e - c\dot{x} - kx)$, which can be combined and written in vector-matrix (state-space) form as

$$\dot{\mathbf{x}} = \mathbf{A}\mathbf{x} + \mathbf{b}, \tag{6.6}$$

where $\mathbf{A} = \begin{bmatrix} 0 & 1 \\ -\frac{k}{m} & -\frac{c}{m} \end{bmatrix}$, $\mathbf{b} = \begin{Bmatrix} 0 \\ \frac{f_e}{m} \end{Bmatrix}$, $\mathbf{x} = \begin{Bmatrix} x \\ y \end{Bmatrix}$. This is now in a form which can be solved numerically for each time step Δt as described in Appendix D. Note that the vibration response calculated using numerical integration is an approximation. The accuracy can be improved by reducing the time step, but there may be some issues in achieving accurate results using some numerical equation solvers. This is beyond the scope of this book, but the interested reader can find further information in Lambert (1991) and Hairer et al. (1993).

6.3 Calculation of the Vibration Response of an SDOF System

In Chapter 5 four signals commonly used for vibration testing were considered. They are single-frequency excitation, a chirp, random excitation, and a half-sine impulse. The response due to a single frequency is trivial to calculate as the response is simply the amplitude of the force input multiplied by the FRF evaluated at the excitation frequency, to give the amplitude and relative phase of the displacement response at that frequency. The remaining cases are considered in this section.

6.3.1 Impulsive Force

Before discussing the remaining cases, an interesting benchmark case to consider is an impulsive force, which is represented as a force vector in which all the elements are zero except one, which has a finite value. In this case the response is simply the IRF of the system, delayed with respect to the time that the force is applied, multiplied by the magnitude of the force impulse, i.e.

$$
\begin{aligned}
x(t) &= \frac{\hat{f}_e}{m\omega_d} e^{-\zeta\omega_n(t-t_0)} \sin(\omega_d(t-t_0)) \quad \text{for } t \geq t_0 \\
&= 0 \qquad\qquad\qquad\qquad\qquad\qquad\quad t < t_0
\end{aligned}
\tag{6.7}
$$

where the force impulse is applied at $t = t_0$, and the time resolution is Δt so that $t = n\Delta t$. If the force has a value of $f_{e(max)}$, then the magnitude of the force is simply $\hat{f}_e = f_{e(max)}\Delta t$. The calculation of the displacement response in the time domain using Eq. (6.7) and the response calculated via the frequency domain are illustrated in Figure 6.3. Note that although the plots appear to be continuous lines, they are, in fact, sampled data. This means that the DFT and the IDFT are used to transform data between the time and frequency domains, which results in double-sided spectra. At each frequency the data is a complex number, and are plotted in terms of modulus and phase, for ease of interpretation. The displacement response calculated by convolution and by the frequency domain approach give identical results. The response calculated by numerical integration of the equation of motion is not shown. This is calculated and compared with the other approaches in MATLAB Example 6.1a.

6.3.2 Half-sine Force Impulse

As discussed in Chapter 5, when an instrumented hammer is used to excite a structure the force time history is approximately a half-sine pulse. It is described by

$$
\begin{aligned}
f_e(t) &= f_{e(max)} \sin(\pi t/T_c) \quad \text{for } 0 \leq t < T_c \\
&= 0 \qquad\qquad\qquad\quad \text{otherwise}
\end{aligned}
\tag{6.8}
$$

where T_c is the time period during which the impact hammer is in contact with the structure under test, and $f_{e(max)}$ is the maximum force applied. If $T_c \ll T_n$, where $T_n = 2\pi\sqrt{m/k}$ is the natural period of the system (which is the period corresponding to the natural frequency) then the approximate displacement response can be calculated in the same way as for the pure impulse, i.e. $x(t) \approx \hat{f}_e \times h(t)$, where $\hat{f}_e = 2f_{e(max)}T_c/\pi$, which is the area under the half-sine force impulse. If T_c is not small compared with T_n, the simple approach to determine the displacement response is not applicable and the output needs to be calculated numerically using one of the three methods described in Section 6.2. An example of the process to calculate the displacement response of an SDOF system to a half-sine force impulse using convolution and via the frequency domain is given in Figure 6.4. The comments concerning sampled data made for Figure 6.3 are also applicable to Figure 6.4. Note that zeros have been added to the force time history to ensure that it has the same length as the IRF. Although this is not necessary to determine the displacement response using convolution, it is necessary to calculate the displacement response via the frequency domain and to compare the two approaches. Examining the displacement response in Figure 6.4, it can be seen that the

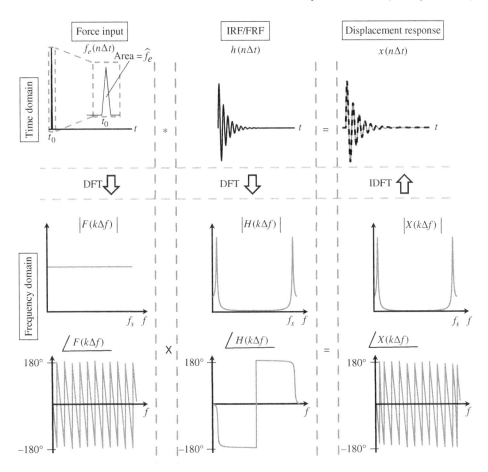

Figure 6.3 Calculation of the displacement response of an SDOF system excited by an impulsive force using convolution and via the frequency domain (circular convolution).

time duration of the impulse is much longer than the natural period of the system. The response follows the pattern of the impulse force with some additional dynamics until the end of the force impulse, and then the mass undergoes damped free vibration. In MATLAB Example 6.1b the effect on the displacement response of changing the system properties and the time duration of the sine impulse is investigated. The response is also calculated by numerical integration of the equation of motion.

6.3.3 Chirp (Swept Sine) Force Input

A signal often used to excite a structure using a shaker is a chirp. The characteristics of this signal were discussed in Chapter 5, the force time history of which is given by

$$f_e(t) = |\overline{F}| \sin \left(\frac{(\omega_2 - \omega_1)}{2T} t^2 + \omega_1 t \right), \tag{6.9}$$

where $\omega_1 = 2\pi f_1$ and $\omega_2 = 2\pi f_2$ are starting and finishing frequencies of the chirp, respectively, and T is the duration of the chirp. If the frequency range is wide enough and the chirp duration

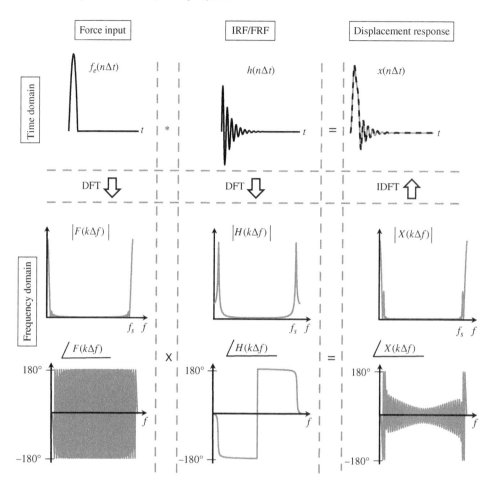

Figure 6.4 Calculation of the displacement response of an SDOF system excited by a half-sine impulse force using convolution, and via the frequency domain (circular convolution).

is small compared to natural period of the system, the effect of using a chirp excitation is similar to that for impact excitation, discussed in the previous sub-sections. If the duration of the chirp is very long compared to the natural period, the dynamic response is quite different. In this case the effect of a slow sweep excitation is similar to that for stepped sine excitation in which the system is excited frequency by frequency, and the system is allowed to reach steady-state between the frequency changes.

To ensure that the response is zero at the end of the period of excitation, so that the method of calculating the response by convolution and by the frequency domain method can be compared, the duration of the IRF is doubled by adding zeros. The chirp signal can then be any length up to the duration of the original IRF, and then zero-padded so that it has the same time duration as the modified IRF. This ensures that there is no wraparound which distorts the beginning of the displacement response when calculated using the frequency domain method. Examples of how to calculate the displacement response of an SDOF system to a slow and a fast chirp using the three methods discussed in Section 6.2 are given in MATLAB Example 6.1c.

6.3.4 Random Force Input

A common signal used to excite a shaker in vibration testing is a random signal with a Gaussian distribution as discussed in Chapter 5. This is easily generated in MATLAB using the `randn` function. To ensure that there is no wraparound in the displacement response calculated using the frequency domain method, both the force time history and the IRF are padded with zeros, to double their original time duration, in the same way as described for the chirp force input. An example of how to calculate the displacement response of an SDOF system to random excitation using the three methods discussed in Section 6.2 is given in MATLAB Example 6.1d.

MATLAB Example 6.1

In this example, the displacement response of an SDOF system is calculated for several types of input forces using convolution in the time domain, via the frequency domain, and by numerical integration of the equation of motion.

```
clear all

%% Parameters
m=1;                               % [kg]        % mass
k=1000;                            % [N/m]       % stiffness
z=0.1;c=2*z*sqrt(m*k);             % [Ns/m]      % damping
wn=sqrt(k/m);wd=sqrt(1-z^2)*wn;    % [rad/s]     % (un)damped natural frequency
fn=wn/(2*pi);                      % [Hz]        % natural frequency
Tn=1/fn;                           % [s]         % natural period

%% Example 6.1a Impulsive force
TT=2;                              % [s]         % time duration of force
fs=100;                            % [Hz]        % sampling frequency
dt=1/fs; t=0:dt:TT-dt;             % [s]         % time vector
f=zeros(1,length(t));f(12)=1;      % [N]         % force vector
[xc,xf,xn]=calculate(f,m,k,c,wd,wn,z,t,fs);      % function to calculate response
plots(t,f,TT,xc,xf,xn)                           % plots

%% Example 6.1b. Half-sine pulse
Tc=Tn*0.5;                         % [s]         % time duration of half-sine pulse
fs=100;                            % [Hz]        % sampling frequency
dt=1/fs; t=0:dt:Tc;                % [s]         % time vector
f = sin(pi*t/Tc);                  % [N]         % half-sine pulse
Nz=20*length(t);                                 % number of zeros to add
f = [f zeros(Nz,1)'];              % [N]         % zero-padded force signal
t = 0:dt:(length(f)-1)*dt;         % [s]         % time vector for extended signal
N=length(t);TT=N*dt;
[xc,xf,xn]=calculate(f,m,k,c,wd,wn,z,t,fs);      % function to calculate response
plots(t,f,TT,xc,xf,xn)                           % plots

%% Example 6.1c. Chirp
fs=1000;                           % [Hz]        % sampling frequency

% slow chirp
T=60;dt=1/fs;t=0:dt:T;             % [s]         % chirp duration; time vector
f1=1;f2=10;                        % [Hz]        % upper and lower frequencies
[f,t,TT]=chrp(f1,f2,t,T,dt);                     % function to calculate chirp force
[xc,xf,xn]=calculate(f,m,k,c,wd,wn,z,t,fs);      % function to calculate response
plots(t,f,TT,xc,xf,xn)                           % plots
```

(Continued)

MATLAB Example 6.1   (Continued)

```
% fast chirp
T=1.25;t=0:dt:T;                              % [s]      % chirp duration; time vector
f1=1;f2=100;                                  % [Hz]     % upper and lower frequencies
[f,t,TT]=chrp(f1,f2,t,T,dt);                            % function to calculate chirp force
[xc,xf,xn]=calculate(f,m,k,c,wd,wn,z,t,fs);             % function to calculate response
plots(t,f,TT,xc,xf,xn)                                  % plots

%% Example 6.1d. Random excitation
fs=100;                                                 % sampling frequency
T=10;dt=1/fs;t=0:dt:T;                                  % signal duration; time vector
fc = randn(1,length(t));                                % random signal
fc=fc-mean(fc);                                         % set the mean to zero
f=[fc zeros(1,length(fc))];                             % zero padded force signal
t=0:dt:2*T+dt;                                          % time vector for extended signal
N=length(t);TT=N*dt;
[xc,xf,xn]=calculate(f,m,k,c,wd,wn,z,t,fs);             % function to calculate response
plots(t,f,TT,xc,xf,xn)                                  % plots

%% Functions
function[xc,xf,xn]=calculate(f,m,k,c,wd,wn,z,t,fs)      % function to calculate the response
%% Impulse response
h=1/(m*wd)*exp(-z*wn*t).*sin(wd*t); % [m/Ns]            % IRF
%% Convolution
xc = conv(h,f)/fs;                                      % displacement response
xc = xc(1:length(f));                                   % displacement response
%% Frequency domain method
H = fft(h)/fs;                                          % FRF
F = fft(f)/fs;                                          % DFT of force
X = H.*F;                                               % DFT of displacement response
xf = ifft(X)*fs;                                        % displacement response
%% Numerical integration
A=[0 1;-k/m -c/m];                                      % system matrix
B=[0; 1/m];                                             % system matrix
n=t;                                                    % dummy variable
[t,xn]=ode45(@(t,x) imp(A,B,x,f,t,n),t,[0 0]);          % numerical integration using ode45
xn=xn(:,1);                                             % displacement response
end

function [f,t,TT] = chrp(f1,f2,t,T,dt);                 % function for chirp force
a=2*pi*(f2-f1)/(2*T); b=2*pi*f1;                        % coefficients
fc=sin(a*t.^2+b*t);                                     % chirp signal
f=[fc zeros(1,length(fc))];                             % zero padded force signal
t=0:dt:2*T+dt;                                          % time vector for extended signal
N=length(t);TT=N*dt;
end

function dxdt=imp(A,B,x,f,t,n)                          % used in numerical solution
f = interp1(n,f,t);
dxdt=A*x+B*f;
end

function plots(t,f,TT,xc,xf,xn)                         % function for plots
figure
subplot(2,1,1)                                          % force input
plot(t,f,'-k','linewidth',3),grid
axis([0,TT,1.1*min(f),1.1*max(f)])
xlabel('time (s)');
ylabel('force (N)');
```

MATLAB Example 6.1   (Continued)

```
subplot(2,1,2)                                      % displacement response
plot(t,xc,'linewidth',3,'Color',[.7 .7 .7]),grid
hold on
plot(t,xf,'-k','linewidth',3)
hold on
plot(t,xn,'k','linewidth',1)
axis([0,TT,1.1*min(xc),1.1*max(xc)])
xlabel('time (s)');
ylabel('displacement (m)');
end
```

Results

Example 6.1a

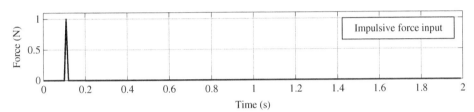

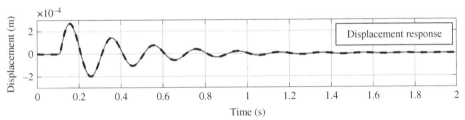

Example 6.1b

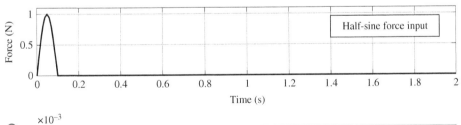

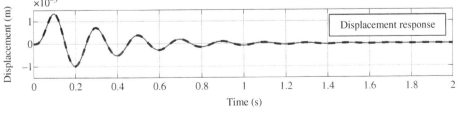

(Continued)

MATLAB Example 6.1   (Continued)

Example 6.1c

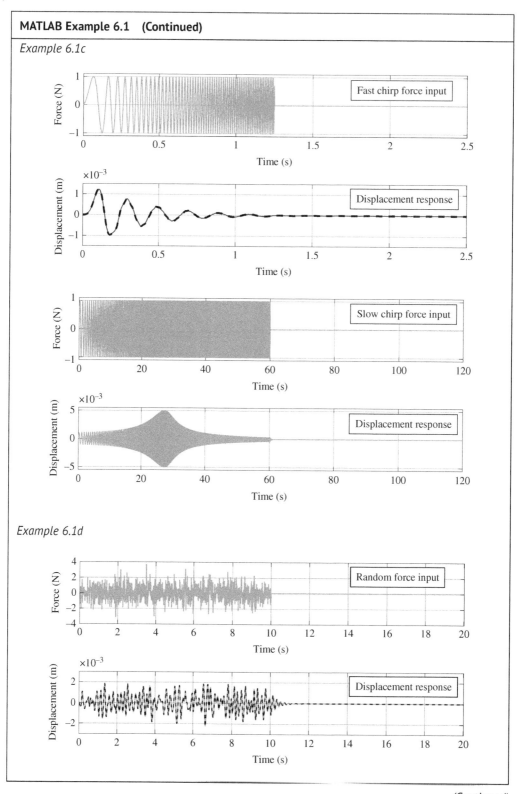

Example 6.1d

(Continued)

MATLAB Example 6.1   (Continued)

Comments:

1. Exercises for the reader are to:
    (a) compare the results in Figure 6.1a,b with a scaled IRF,
    (b) change the duration of the half-sine force input,
    (c) change the duration of the chirp force input,
    (d) reduce the number of zeros added to the chirp and the random signal and observe the effects at the beginning of the response time histories when calculated via the frequency domain.

6.4 Summary

This chapter has shown how to calculate of the displacement response of a vibrating system due to a force input. Three methods have been described:

(a) Convolution of the force time history and the impulse response function (IRF) of the system.
(b) Transformation of the force time history to the frequency domain and then multiplying by the frequency response function (FRF) before transforming back to the time domain. If this is done using sampled data, it is equivalent to circular convolution.
(c) Numerical integration of the equation of motion.

These operations are summarised for an SDOF system in Table 6.1.

Table 6.1   Some ways to calculate the output of a linear SDOF force-excited vibrating system ($t = n\Delta t, f = k\Delta f$).

Input	System description	Operation	Output
$f_e(t)$	Impulse response function (IRF) $$h(t) = \frac{1}{m\omega_d}e^{-\zeta\omega_n t}\sin(\omega_d t)$$	Convolution $$x(t) = f_e(t) * h(t)$$	$x(t)$
	Frequency response function $$H(\mathrm{j}\omega) = \frac{1}{k - \omega^2 m + \mathrm{j}\omega c}$$	Circular convolution $f_e(t) \qquad x(t)$ $\mathrm{DFT}\Downarrow \qquad \mathrm{IDFT}\Uparrow$ $F(k\Delta f) \times H(k\Delta f) = X(k\Delta f)$	
	Equation of motion $$m\ddot{x}(t) + c\dot{x}(t) + kx(t) = f_e(t)$$	Numerical integration Runge–Kutta method (Appendix D)	

References

Hairer, E., Nørsett, S.P., and Wanner, G (1993). *Solving Ordinary Differential Equations I: Nonstiff Problems*, 2nd Edition. Springer-Verlag.

Lambert, J.D. (1991). *Numerical Methods for Ordinary Differential Systems: The Initial Value Problem*. Wiley.

Shin, K. and Hammond, J.K. (2008). *Fundamentals of Signal Processing for Sound and Vibration Engineers*. Wiley.

7

Frequency Response Function (FRF) Estimation

7.1 Introduction

In Chapter 5 the signals used to excite a structure under test were discussed in terms of their temporal and spectral characteristics. The concept of spectral density was introduced to describe the spectral characteristics of a signal, and the way in which spectra are calculated from time domain data was presented for both transient and persistent signals. In Chapter 6, the process of calculating the resulting vibration from a model structure was described. Once both input and output signals have been recorded, they can be used to determine some characteristics of the structure under test. As mentioned in Chapter 1, this is often done by estimating the frequency response function (FRF), and the way this is accomplished is the subject of this chapter. The procedure is described, as are some practical issues, in particular the way in which contaminating noise either in the measured input or output signals is dealt with. Finally, a virtual experiment concerning vibration isolation is described, in which the stiffness and damping of the isolator are estimated from measured data. As in previous chapters, the discussion is centred around the SDOF system shown in Figure 7.1.

7.2 Transient Excitation

Consider a structure which is excited by a transient force such as a half-sine pulse delivered by an instrumented impact hammer, or a fast chirp delivered by an electrodynamic shaker. Provided that the force input and the response time histories are each captured entirely within single windows, the receptance FRF is given by

$$H(j\omega) = \frac{X(j\omega)}{F(j\omega)} = |H(j\omega)|e^{j\phi(j\omega)}, \tag{7.1}$$

where $F(j\omega)$ and $X(j\omega)$ are the Fourier transforms of the force and the displacement time histories, respectively, and $|H(j\omega)|$ and $\phi(j\omega)$ are the modulus and phase of the FRF, respectively. To minimise windowing effects, the force and the displacement response should be zero at the beginning and at the end of their respective time histories. Rectangular windows should then be used. Although Eq. (7.1) describes the FRF of the system, it is rarely calculated this way in practice. The reason for this is that all measurements contain some noise, and there are better ways to calculate the FRF, such that some of the noise is removed or at least attenuated. The general situation, where noise is added to both the force and the displacement response signals, is depicted in Figure 7.2. In this chapter, it is assumed that the noise $n_f(t)$ in the measured force signal is uncorrelated with the noise $n_x(t)$ in the measured displacement signal, and that both noise signals are random and

Virtual Experiments in Mechanical Vibrations: Structural Dynamics and Signal Processing,
First Edition. Michael J. Brennan and Bin Tang.
© 2023 John Wiley & Sons Ltd. Published 2023 by John Wiley & Sons Ltd.
Companion website: www.wiley.com/go/brennan/virtualexperimentsinmechanicalvibrations

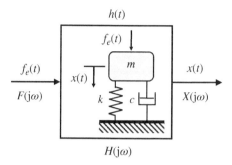

Figure 7.1 Block diagram of a simple vibrating system.

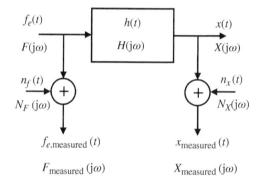

Figure 7.2 Block diagram of a force-excited system with added noise in the measured force and displacement data.

are uncorrelated with the signals $f_e(t)$ and $x(t)$. To suppress this noise in the FRF estimate, an averaging procedure is carried out. To achieve this, several measurements must be made, which for an SDOF system excited by an impact hammer, typically result in the signals shown in Figure 7.3, which depicts the situation where P measurements are made. The force time histories for each measurement are shown in the top row, and the corresponding displacement responses are shown in the second row. The bottom two rows of Figure 7.3 depict the FRFs in terms of the magnitude and phase for each measurement, calculated using Eq. (7.1).

To ensure that the force signal is relatively flat over the frequency range of interest, the contact time of the impact hammer is set to be much smaller than the natural period of the structure. The spectrum due to a half-sine pulse is discussed in detail in Chapter 5. As the added noise is random, it is attenuated by the averaging procedure, resulting in a 'smoother' FRF estimate. The FRF of a single p-th measurement, including the measurement noise, is given by

$$H_{\text{measured},p} = \left(\frac{X_{\text{measured}}}{F_{\text{measured}}} \right)_p = \left(\frac{X + N_X}{F + N_F} \right)_p, \tag{7.2}$$

where N_F and N_X are the Fourier transforms of the noise in the measured force and displacement signals, respectively, as shown in Figure 7.2. Note that the frequency dependence of these quantities (together with others) is assumed, but omitted for clarity. One way of averaging the results could be to calculate the FRF from a single record of the measured data as given in Eq. (7.2) and then average the FRFs for P measurements, so that

$$\tilde{H} = \frac{1}{P} \sum_{p=1}^{P} \left(\frac{X_p + N_{Xp}}{F_p + N_{Fp}} \right). \tag{7.3}$$

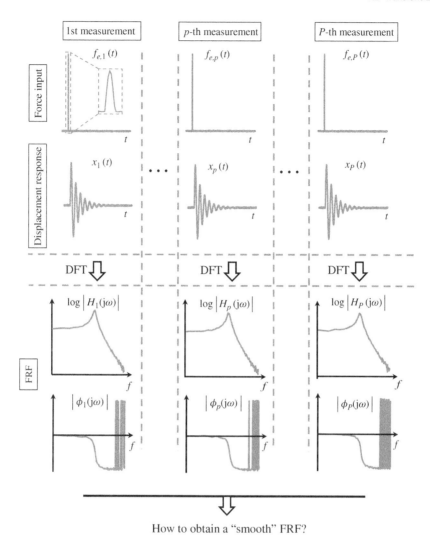

Figure 7.3 Individual measurements to determine the FRF of an SDOF system from transient force excitation.

To investigate the effects of noise on the FRF estimate, it is assumed that $P \to \infty$, so $\lim_{P \to \infty} \frac{1}{P} \sum_{p=1}^{P}(\bullet) = \mathrm{E}[\bullet]$, in which E is the expectation operator. Thus, the averaged quantities are written as expectations for simplicity. Therefore, $\tilde{H} = \mathrm{E}\left[\frac{X + N_X}{F + N_F}\right]$, where N_F and N_X are random variables, and X and F can be considered to be constants (at each frequency), which leads to

$$\tilde{H} = H \times \mathrm{E}\left[\frac{1 + N_X/X}{1 + N_F/F}\right], \tag{7.4}$$

where $H = X/F$. Letting $N = \mathrm{E}[(1 + N_X/X)/(1 + N_F/F)]$ and noting that $H = |H|e^{\mathrm{j}\phi}$ and $N = |N|e^{\mathrm{j}\phi_N}$, Eq. (7.4) can be written as $\tilde{H} = |H|e^{\mathrm{j}\phi}|N|e^{\mathrm{j}\phi_N}$, or

$$\tilde{H} = |NH|e^{\mathrm{j}(\phi + \phi_N)}, \tag{7.5}$$

which shows that the noise, which although is diminished by the averaging process, contaminates both the modulus and the phase of the FRF estimate.

7.2.1 H_1 and H_2 Estimators

There are two principal alternative ways to calculate the FRF. In the first, for each measurement the ratio of the Fourier transform of the displacement to the Fourier transform of the force is formed, and the numerator and the denominator are multiplied by the complex conjugate of the Fourier transform of the force. The numerator and denominator are then averaged separately to give an estimate of the FRF. This is generally called the H_1 estimate[1] (Bendat and Piersol, 1980, 2000), and if measurement noise is included, is given by

$$H_1 = \frac{\mathrm{E}\left[(F + N_F)^*(X + N_X)\right]}{\mathrm{E}\left[(F + N_F)^*(F + N_F)\right]}, \tag{7.6a}$$

where * denotes the complex conjugate. Equation (7.6a) expands to

$$H_1 = \frac{\mathrm{E}\left[F^*X + N_F^*X + F^*N_X + N_F^*N_X\right]}{\mathrm{E}\left[F^*F + N_F^*F + F^*N_F + N_F^*N_F\right]}. \tag{7.6b}$$

Now, all the products of uncorrelated terms tend to zero as the number of averages increases, and $\mathrm{E}[F^*X]/\mathrm{E}[F^*F] = X/F = H$. As mentioned above, it is assumed that the noise in the input and output measured signals is uncorrelated, and this noise is also uncorrelated with the force and displacement response signals. Noting that $H = |H|e^{\mathrm{j}\phi}$, and applying the assumptions, Eq. (7.6b) reduces to

$$H_1 = \left|\frac{H}{N_1}\right| e^{\mathrm{j}\phi}, \tag{7.6c}$$

where $N_1 = 1 + \tilde{S}_{nfnf}/S_{\overline{ff}}$, in which $\tilde{S}_{nfnf} = \mathrm{E}\left[N_F^*N_F\right]/T$ is an estimate of the power spectral density (PSD) of the noise in the force signal, and $S_{\overline{ff}} = F^*F/T$ is the PSD of the force signal without added noise; T is the time duration of a single measurement. It can be seen from Eq (7.6c) that in the limit of an infinite number of averages, the H_1 estimate is not affected (biased) by output noise. Furthermore, only the modulus is affected by noise in the measured force signal, with the phase being an unbiased estimate. As $N_1 \geq 1$, and because both $\tilde{S}_{nfnf}$ and $S_{\overline{ff}}$ are positive real quantities, the estimate of the modulus is an underestimate. In the signal processing literature this is called a bias error. Note that the numerator in Eq. (7.6a) divided by T is generally written as $\tilde{S}_{fx}$, which is an estimate of the cross power spectral density (CPSD) between the measured force and displacement signals, and is analogous to the PSD, but is complex, i.e. it has both modulus and phase. In fact, the phase of the CPSD is the same as the phase of the FRF. The process of calculating the CPSD effectively acts as a filter, removing uncorrelated contaminating noise from both the input and the output. The denominator $\tilde{S}_{ff}$ is the PSD of the measured force signal, discussed in Chapter 5, so that

$$H_1(\mathrm{j}\omega) = \frac{\tilde{S}_{fx}(\mathrm{j}\omega)}{\tilde{S}_{ff}(\omega)}. \tag{7.6d}$$

The second method of calculating the FRF is similar to the first, but both the numerator and denominator of Eq. (7.2) are now multiplied by the complex conjugate of the Fourier transform of

---

1 FRF estimators are generally defined in terms of power spectral densities (PSDs), which are simply scaled versions of the energy spectral densities (ESDs) for transient signals. The scaling factor is $1/T$, which is the time duration of a single transient measurement. Note that because both numerator and denominator of an FRF estimator are scaled by $1/T$, it does not matter whether PSDs or ESDs are used. For simplicity of notation, and to provide consistency with FRF estimation using random data, which is discussed later in this chapter, PSDs rather than ESDs are used.

the displacement, to give

$$H_2 = \frac{E\left[(X + N_X)^*(X + N_X)\right]}{E\left[(X + N_X)^*(F + N_F)\right]}.$$ (7.7a)

Note that the numerator and denominator are again averaged separately to give an FRF estimate, which is generally called the H_2 estimate (Bendat and Piersol, 1980, 2000). Equation (7.7a) expands to

$$H_2 = \frac{E\left[X^*X + N_X^*X + X^*N_X + N_X^*N_X\right]}{E\left[X^*F + N_X^*F + X^*N_F + N_X^*N_F\right]}.$$ (7.7b)

As with the H_1 estimate, all products of uncorrelated terms tend to zero as the number of averages increases, so that Eq. (7.7b) reduces to

$$H_2 = |N_2 H|e^{i\phi},$$ (7.7c)

where $N_2 = 1 + \tilde{S}_{nxnx}/S_{\overline{xx}}$, in which $\tilde{S}_{nxnx} = E\left[N_X^*N_X\right]/T$ is an estimate of the PSD of the noise in the displacement signal, and $S_{\overline{xx}} = X^*X/T$ is the PSD of the displacement signal without added noise. It can be seen from Eq. (7.7c), that in the limit of an infinite number of averages, the H_2 estimate is unaffected by noise in the force signal. Furthermore, the phase is not affected by noise in the measured displacement signal, but the modulus is affected by noise in the measured displacement signal. The estimate of the modulus is an overestimate because $N_2 \geq 1$. Note that the denominator (divided by T) in Eq. (7.7a) is generally written as $\tilde{S}_{xf}$, which is an estimate of the CPSD between the measured displacement and force signals. It is the complex conjugate of $\tilde{S}_{fx}$, i.e. $\tilde{S}_{xf} = \tilde{S}_{fx}^*$, so it has the same modulus but the phase is of opposite sign. The denominator $\tilde{S}_{xx}$ is the PSD of the measured displacement signal, so that

$$H_2(j\omega) = \frac{\tilde{S}_{xx}(\omega)}{\tilde{S}_{xf}(j\omega)}.$$ (7.7d)

The process of estimating an FRF using impact hammer excitation and averaging over a finite number of measurements is illustrated in Figure 7.4. The PSDs for the measured force and displacement signals are shown in the top two rows, and the corresponding modulus and phase of the CPSDs between these signals are shown for each measurement in the bottom two rows. Also shown at the bottom of the figure is the averaging procedure used to calculate the H_1 and H_2 estimators. At the lower right part of Figure 7.4 is the coherence function. This is an extremely important function, when estimating FRFs, and is used to estimate the quality of a measurement. It is discussed in the following subsection.

If there is noise in both the force and displacement measured data, there are other FRF estimators that could be used (Shin and Hammond, 2008). However, they are not as simple as the H_1 and H_2 estimators and require some knowledge of the noise characteristics, for example the ratio of the noise in the force and displacement spectra. As this book is an introductory text, these are not covered here.

7.2.2 Coherence Function

The coherence function between the measured force and the measured displacement signals is equal to the ratio of the H_1 and H_2 estimators (Brandt, 2011). It is thus a function of frequency and is given by

$$\gamma_{fx}^2(\omega) = \frac{H_1(j\omega)}{H_2(j\omega)},$$ (7.8a)

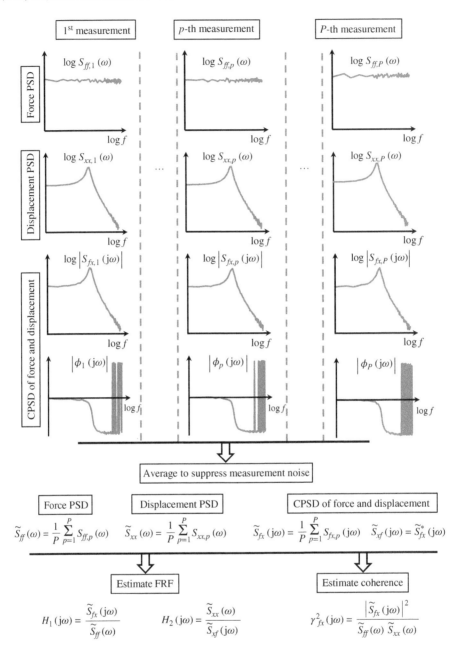

Figure 7.4 Process of FRF estimation of an SDOF system due to transient force excitation in the presence of measurement noise.

where H_1 is given in Eq. (7.6d) and H_2 is given in Eq. (7.7d). Note that because $\tilde{S}_{xf} = \tilde{S}_{fx}^*$ the coherence function is real-valued at each frequency. Moreover, it was shown above that H_1 underestimates the modulus of the actual FRF and is insensitive to noise in the measured displacement response, and H_2 overestimates the modulus of the actual FRF and is insensitive to noise in the measured input force. Thus, $0 \leq \gamma_{fx}^2(\omega) \leq 1$. Substituting for H_1 and H_2 from Eqs. (7.6d) and (7.7d) into Eq. (7.8a) results in the more usual expression for the coherence function given by

$$\gamma_{fx}^2(\omega) = \frac{|\tilde{S}_{fx}(j\omega)|^2}{\tilde{S}_{ff}(\omega)\tilde{S}_{xx}(\omega)}. \tag{7.8b}$$

The coherence function is a measure of the degree of linear relationship between the two measured signals at each frequency. If it is less than unity it could be due to one or more of the following reasons:

1. Noise may be present in the force and/or the displacement signals.
2. The system being measured may have some nonlinearity. This could be the structure under test and/or the measurement system.
3. The displacement measurement is not only due to the force input. There may be other inputs, which manifest themselves as noise in the output signal.

It is interesting to examine the coherence function in the special cases when noise is only in the force signal or only in the displacement response signal. In the first case, when the noise is only in the measured force signal, from Eq. (7.6c), $H_1 = H/(1 + \tilde{S}_{n_f n_f}/S_{ff})$ and $H_2 = H$. Substituting for H_1 and H_2 in Eq. (7.8a) results in

$$\gamma_{fx}^2(\omega) = \frac{1}{1 + \tilde{S}_{n_f n_f}(\omega)/S_{\overline{ff}}(\omega)}. \tag{7.9}$$

Note that $S_{\overline{ff}}(\omega)/\tilde{S}_{n_f n_f}(\omega)$ is the signal-to-noise ratio (SNR) of the force signal *at each frequency*. As the SNR tends to infinity, $\tilde{S}_{n_f n_f}(\omega)/S_{\overline{ff}}(\omega) \to 0$, and $\gamma_{fx}^2(\omega) \to 1$. When the noise is only in the measured displacement response $H_1 = H$ and $H_2 = H(1 + \tilde{S}_{n_x n_x}/S_{\overline{xx}})$, so that

$$\gamma_{fx}^2(\omega) = \frac{1}{1 + \tilde{S}_{n_x n_x}(\omega)/S_{\overline{xx}}(\omega)}. \tag{7.10}$$

Note that in this case $S_{\overline{xx}}(\omega)/\tilde{S}_{n_x n_x}(\omega)$ is the SNR of the displacement signal *at each frequency*, and as the SNR tends to infinity, the coherence tends to unity. Equations (7.9) and (7.10) have the same form and are plotted in Figure 7.5 to show the relationship between the SNR at each frequency and the coherence function. The SNR is plotted in dB (calculated as $10\log_{10}(\text{SNR}(\omega))$) as this is the more usual way of quoting an SNR, and the coherence has a value between 0 and 1, as mentioned above. It can be seen that the coherence increases rapidly as the SNR increases and has a value greater than 0.9 when the SNR is greater than approximately 9 dB.

7.2.3 Examples

To illustrate the effect of averaging the measured data to give a smoother FRF, the PSDs and CPSDs shown in Figure 7.4 are considered, but with noise added. A set of P measurements is made of the input force and the resulting displacement response of an SDOF system. The FRF is calculated for two cases (i) when random noise is added to the force signal and (ii) when random noise is added to the displacement response signal. The amount of random noise added to the time domain signal (either force or displacement) is determined by specifying an overall SNR for the signal, which is defined by

$$\text{SNR (dB)} = 10\log_{10}\left(\frac{\text{signal energy}}{\text{noise energy}}\right). \tag{7.11}$$

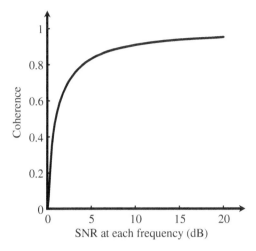

Figure 7.5 Relationship between the coherence and SNR at each frequency for a system with measurement noise in *either* the force or displacement data.

Recall from Chapter 5 that in the time domain the signal energy is calculated by squaring the signal and integrating the result. In the frequency domain, this is equivalent to the product of the PSD integrated over frequency, and the time duration for a measurement sample. The time domain data for a single measurement with the addition of noise in the force measurement are shown in the top part of Figure 7.6. The averaged PSDs and CPSDs from a set of P measurements as shown in Figure 7.4, and are used to determine the H_1 and H_2 estimators. The modulus and phase of the FRFs calculated in this way are shown in the lower part of Figure 7.6, together with the FRFs calculated for each of the individual measurements (all samples), and the theoretical FRF. It is clear that the averaging process significantly improves the FRF estimates, reducing the effect of the measurement noise in both the modulus and phase. It is also clear that, for the simulation carried out, the H_2 estimator gives a better estimate of the modulus of the FRF compared to the H_1 estimator. The bias error due to the measurement noise in the modulus of the H_1 estimator is evident. There is no bias error in the phase estimate, as discussed above. The coherence is also shown in Figure 7.6. It is about 0.9 over the frequency range shown, which means that the force SNR is about 9 dB at each frequency. Note that this SNR is not particularly high, and that generally it would be higher in a real experiment. A small SNR was used in the simulation to illustrate the clear differences between the H_1 and H_2 estimators.

The second example concerns the SDOF system considered previously, but now with measurement noise contaminating the displacement response signal instead of the measured force signal. The results are shown in Figure 7.7, in which the random noise contaminating the displacement signal is evident in the time domain signal. The H_1 and H_2 estimators are calculated in the same way as in the previous example. Again, it is clear that the averaging process significantly improves the FRF estimates, but it does not remove the noise completely. The added noise has a different effect than when it is added to the force signal. This is because the displacement response is significantly attenuated at high frequencies, above the resonance frequency of the system, and the noise then dominates the response, because the system displacement reduces with the square of frequency, whereas the noise has a flat spectrum. This effect is particularly evident in the coherence

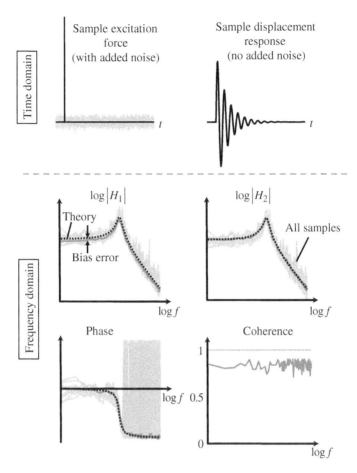

Figure 7.6 Estimates of the FRF of an SDOF system excited with a half-sine impulse which has added noise in the measured force signal.

plot, also shown in Figure 7.7. The bias error in the H_2 estimate manifests itself when the SNR is poor, as can be seen in the figure.

An alternative way to tackling the measurement noise problem, which is sometimes used in practice, is to apply a window to the half-sine force impulse alone, setting the remainder of the force time history to zero, and applying an exponential window to the response. The exponential window effectively adds damping to the system response, which can be compensated for in post processing of the data if modal models are used as discussed in Chapter 8. These approaches are not considered in this book, but the interested reader can consult (Brandt, 2011; Avitabile, 2017), for further information.

Although the simulations presented so far in this chapter have involved a half-sine pulse excitation, similar to that delivered by an impact hammer, the approach is applicable to any type of transient excitation. In MATLAB Example 7.1, the FRF of an SDOF system is estimated using a fast chirp signal as the force signal, and this is compared with the estimated FRF when a half-sine pulse force signal is used to excite the system.

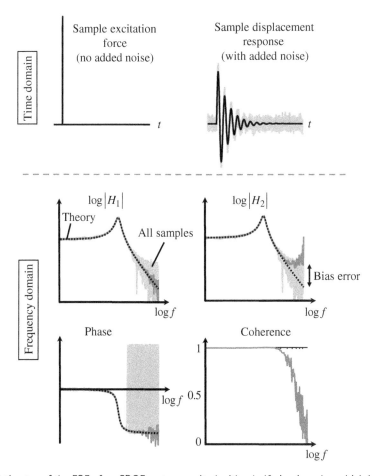

Figure 7.7 Estimates of the FRF of an SDOF system excited with a half-sine impulse which has added noise in the measured displacement signal. The frequency axes are logarithmic as are the moduli axes.

MATLAB Example 7.1

In this example, the FRF of an SDOF system is estimated using the H_1 and H_2 estimators for half-sine pulse and chirp excitation signals with added random noise.

```
clear all

%% Parameters
m=1;k=10000;                    % [kg,N/m]      % mass and stiffness
z=0.1;c=2*z*sqrt(m*k);          % [Ns/m]        % damping
wn=sqrt(k/m);wd=sqrt(1-z^2)*wn; % [rad/s]       % natural frequency
fn=wn/(2*pi);Tn=1/fn;           % [Hz,s]        % natural frequency/period
SNRf=20;SNRx=100;                              % SNRs for added random noise

%% Half-sine pulse
Tc=Tn*0.02;fs=5000;             % [s,Hz]        % contact time/sampling frequency
dt=1/fs; t1=0:dt:Tc;            % [s]           % time vector
fp=sin(pi*t1/Tc);               % [N]           % half sine pulse
Nz=600*length(t1);                             % Number of zeros to add
```

(Continued)

MATLAB Example 7.1   (Continued)

```
f=[zeros(Nz/10,1)',fp zeros(Nz,1)'];  % [N]    % zero-padded force signal
t = 0:dt:(length(f)-1)*dt;                      % time vector for extended signal
N=length(t);TT=N*dt;Tm=max(t);
[xc]=calc(f,m,k,c,wd,wn,z,t,fs);       % [m]    % function to calculate displ.
[fwn,xwn,Sff,Sxx,H1,H1a,H2,H2a,coh]=...         % function to calculate frequency
FRF(f,xc,dt,Tm,SNRf,SNRx);                      %  domain quantities
Sffs=Sff;Sxxs=Sxx;H1s=H1;H1as=H1a;              % frequency domain quantities
H2s=H2;H2as=H2a;cohs=coh;


plots(t,fwn,TT,xwn)                             % function for time domain plots

%% Chirp
T=TT/2;tt=0:dt:T;                      % [s]    % time vector
f1=1;f2=200;                           % [Hz]   % upper and lower frequencies
a=2*pi*(f2-f1)/(2*T); b=2*pi*f1;                % coefficients
fc=sin(a*tt.^2+b*tt);                  % [N]    % chirp signal
f=[fc zeros(1,length(f)-length(fc))]; % [N]     % zero padded force signal
[xc]=calc(f,m,k,c,wd,wn,z,t,fs);       % [m]    % function to calculate displ.
[fwn,xwn,Sff,Sxx,H1,H1a,H2,H2a,coh]=...         % function to calculate frequency
FRF(f,xc,dt,Tm,SNRf,SNRx);                      %  domain quantities
Sffc=Sff;Sxxc=Sxx;H1c=H1;H1ac=H1a;              % frequency domain quantities
H2c=H2;H2ac=H2a;cohc=coh;


plots(t,fwn,TT,xwn)                             % function for time domain plots

%% Theoretical FRF
df=1/(N*dt);ff=0:df:fs-df;             % [Hz]   % frequency vector
w=2*pi*ff;
Ht=1./(k-w.^2*m+j*w*c);                % [m/N]  % theoretical FRF

%% Plots of frequency domain quantities
figure                                          % plot of force PSD
semilogx(ff,10*log10(mean(Sffc)))
hold on
semilogx(ff,10*log10(mean(Sffs)))
axis square; grid; axis([1,1000,-80,-30])
xlabel('frequency (Hz)');
ylabel('force PSD (dB ref 1 N^2/Hz)');


figure                                          % plot of displ. PSD
semilogx(ff,10*log10(mean(Sxxc))); hold on
semilogx(ff,10*log10(mean(Sxxs)))
axis square; grid; axis([1,1000,-200,-80])
xlabel('frequency (Hz)');
ylabel('displ, PSD (dB ref 1 m^2/Hz)');


figure                                          % plot of H1 estimator
semilogx(ff,20*log10(abs(H1a))); hold on
semilogx(ff,20*log10(abs(H1c))); hold on
semilogx(ff,20*log10(abs(H1s))); hold on
semilogx(ff,20*log10(abs(Ht)))
axis square; grid; axis([1,1000,-120,-60])
xlabel('frequency (Hz)');
ylabel('displ./force (dB ref 1 m/N)');


figure                                          % plot of H2 estimator
semilogx(ff,20*log10(abs(H1a))); hold on
semilogx(ff,20*log10(abs(H2c))); hold on
semilogx(ff,20*log10(abs(H2s))); hold on
```

(Continued)

MATLAB Example 7.1   (Continued)

```
semilogx(ff,20*log10(abs(Ht)))
axis square; grid; axis([1,1000,-120,-60])
xlabel('frequency (Hz)');
ylabel('displacement/force (dB ref 1 m/N)');

figure                                    % plot of phase
semilogx(ff,180/pi*angle(H1a)); hold on
semilogx(ff,180/pi*unwrap(angle(H1c))); hold on
semilogx(ff,180/pi*unwrap(angle(H1s))); hold on
semilogx(ff,180/pi*unwrap(angle(Ht)))
axis square; grid; axis([1,1000,-200,200])
xlabel('frequency (Hz)');
ylabel('phase (degrees)');

figure                                    % plot of coherence
semilogx(ff,cohc,ff,cohs)
axis square; grid; axis([1,1000,0,1])
xlabel('frequency (Hz)'); ylabel('coherence');

%% Functions                              % calculate displacement
function [xc]=calc(f,m,k,c,wd,wn,z,t,fs)
 % Impulse response
 h=1/(m*wd)*exp(-z*wn*t).*sin(wd*t);      % IRF
 % Convolution
 xc=conv(h,f)/fs;                         % displacement response
 xc=xc(1:length(f));
end

function[fwn,xwn,Sff,Sxx,H1,H1a,H2,H2a,coh]=...  % calculate frequency domain
FRF(f,xc,dt,Tm,SNRf,SNRx)                         properties
 for n=1:16
   fwn=awgn(f,SNRf,'measured','dB');      % add random noise
   xwn=awgn(xc,SNRx,'measured','dB');     % add random noise
   F=fft(fwn)*dt;         % [N/Hz]        % fft of force
   X=fft(xwn)*dt;         % [m/Hz]        % fft of displacement
   Sff(n,:)=F.*conj(F)/Tm;   % [N²/Hz]    % force PSD
   Sxx(n,:)=X.*conj(X)/Tm;   % [m²/Hz]    % displacement PSD
   Sfx(n,:)=X.*conj(F)/Tm;   % [Nm/Hz]    % CPSD fx
   Sxf(n,:)=F.*conj(X)/Tm;   % [Nm/Hz]    % CPSD xf
 end

 coh=abs(mean(Sxf)).^2./(mean(Sxx).*mean(Sff));  % coherence
 H1=mean(Sfx)./mean(Sff);H1a=Sfx./Sff;  % [m/N]  % H1 estimator/all samples
 H2=mean(Sxx)./mean(Sxf);H2a=Sxx./Sxf;  % [m/N]  % H2 estimator/all samples
end

function plots(t,fwn,TT,xwn)                     % plots time domain quantities
 figure
 subplot(2,1,1)
 plot(t,fwn); grid
 axis([0,TT,1.1*min(fwn),1.1*max(fwn)])
 xlabel('time (s)');ylabel('force (N)');
 subplot(2,1,2)
 plot(t,xwn); grid
 axis([0,TT,1.1*min(xwn),1.1*max(xwn)])
 xlabel('time (s)');ylabel('displ. (m)');
end
```

(Continued)

MATLAB Example 7.1   (Continued)

Results

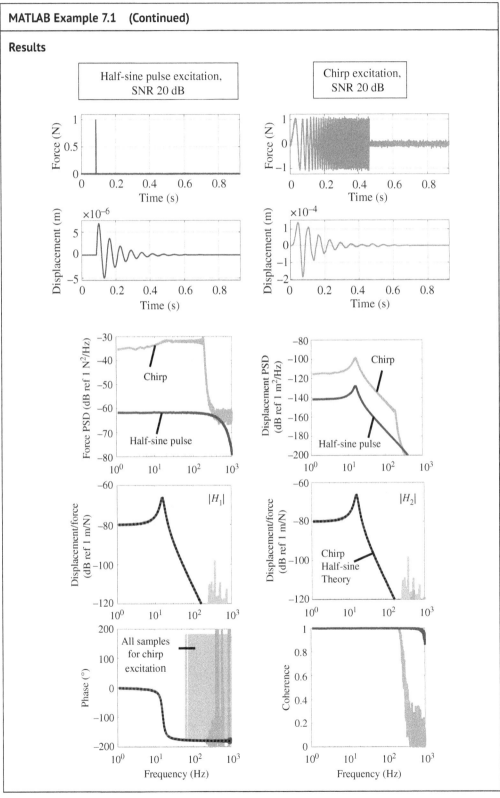

MATLAB Example 7.1   (Continued)

Comments:

1. The figures show the situation when the SNR for the measured force and displacement signals are 20 and 100 dB, respectively. An exercise for the reader is to explore what happens when the SNRs are changed for both input and output. Also explore the effects of increasing the number of averages.
2. From the results above, it is clear that the spectra of the force excitation signals are very different in terms of both level and frequency range, depending on the type of excitation. An exercise for the reader is to explore the effects of changing the excitation parameters, such as
   (a) the duration of the half-sine pulse,
   (b) the lower and upper frequencies for the chirp excitation, and the rate at which the frequency is increased.

7.3   Random Excitation

As mentioned previously, a common way to excite a structure to measure an FRF is to use a shaker supplied with random noise passed through a power amplifier. The way in which random signals are treated in terms of estimating their spectra is given in Chapter 5, and the same approach is followed in this section. A typical measured force input to an SDOF system, without measurement noise, is shown in Figure 7.8, as is the measured displacement response. Note the difference in the force and displacement signals. The force signal contains a wide-range of frequencies – it has a flat spectrum up to half the sampling frequency, and the displacement response has dominant low-frequency content. Close examination of the displacement response shows that there is a periodic component, which corresponds to the natural frequency of the system. Also shown in Figure 7.8 are Hanning windows of length T, which are used in the transformation of the data to the frequency domain. Note, that in practice, overlapping windows are generally used, but these are not shown in the figure to avoid clutter. The effect of using Hanning windows with 50% overlap is discussed later. In Figure 7.8, the total time over which the data are collected is T_c, and there are P averages. The process of estimating the FRF is similar to that discussed for transient excitation, and is shown at the bottom of Figure 7.8, but because the signals are random and persistent, there are leakage issues (which are minimised by using Hanning windows as discussed in Chapter 5), bias errors, and random errors. The frequency domain quantities calculated from the time domain data are shown in Figure 7.9. They are the spectral densities of the force and the displacement, the magnitude and phase of the FRF and the coherence. Note that because there is no measurement noise in this simulation, then $H = H_1 = H_2$, and the FRFs are calculated by using either the PSD of the force input or the displacement response, and the CPSD, as with transient excitation. The theoretical FRF is also shown in Figure 7.9 as a dashed line.

A summary of the random errors associated with each quantity is given in Table 7.1 (Shin and Hammond, 2008). It can be seen that only the number of averages govern the error for the PSDs. However, the coherence between the input and output is also important for the FRF (modulus and phase) and the coherence estimators. The smaller the coherence, the larger the random error. Note that in the derivation of the expressions for the errors in Table 7.1, it is assumed that the data segments are mutually uncorrelated. If overlapping windows are used, then more averages

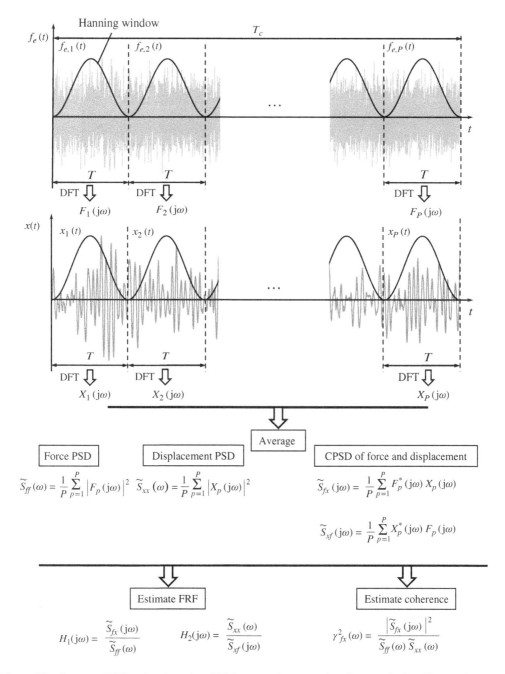

Figure 7.8 Process of FRF estimation of an SDOF system due to random force excitation. The overlapped Hanning windows are not shown for clarity.

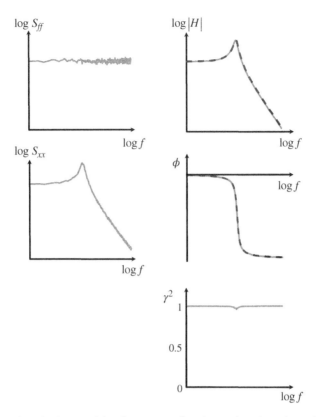

Figure 7.9 Frequency domain data resulting from processing the random data shown in Figure 7.8. The frequency axes are logarithmic as are the axes for the PSDs and the modulus.

Table 7.1 Random errors for the various estimators.

Estimator	Random error
$\tilde{S}_{ff}(\omega)$ or $\tilde{S}_{xx}(\omega)$	$\dfrac{\sigma(\tilde{S}_{ff}(\omega))}{S_{ff}(\omega)} = \dfrac{\sigma(\tilde{S}_{xx}(\omega))}{S_{xx}(\omega)} = \dfrac{1}{\sqrt{P}}$
$|\tilde{H}_1(j\omega)|$ or $|\tilde{H}_2(j\omega)|$	$\dfrac{\sigma(|\tilde{H}_1(j\omega)|)}{|H_1(j\omega)|} = \dfrac{\sigma(|\tilde{H}_2(j\omega)|)}{|H_2(j\omega)|} = \sqrt{\dfrac{1-\gamma^2(\omega)}{2\gamma^2(\omega)P}}$
$\tilde{\phi}(j\omega)$	$\sigma(\tilde{\phi}(j\omega)) = \sqrt{\dfrac{1-\gamma^2(\omega)}{2\gamma^2(\omega)P}}$
$\tilde{\gamma}^2(\omega)$	$\dfrac{\sigma(\tilde{\gamma}^2(\omega))}{\gamma^2(\omega)} = (1-\gamma^2(\omega))\sqrt{\dfrac{2}{\gamma^2(\omega)P}}$

are needed to maintain the error at the same value. For example, if a Hanning window with 50% overlap is used, the number of averages has to be increased by a factor of 1.89 (Brandt, 2011).

Returning to the coherence function, shown in Figure 7.9, it can be seen that there is a very slight reduction in the value from unity at the resonance frequency. This feature is commonly seen in the coherence function when random excitation is used. The cause of this phenomenon is

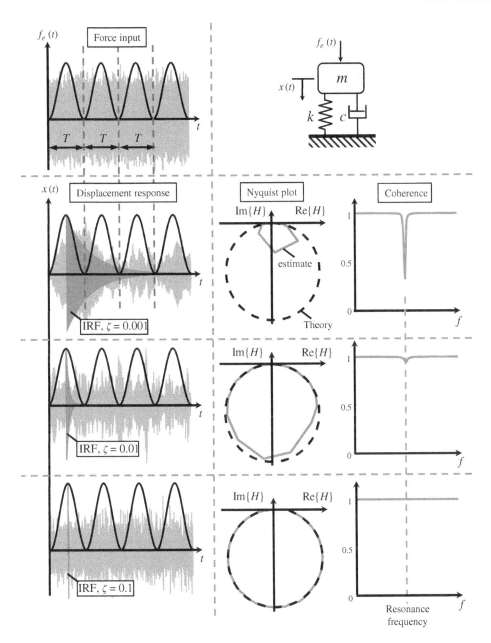

Figure 7.10 Plots showing the effect of the window size compared to on the length of the IRF, on the FRF, and the coherence.

illustrated in Figure 7.10. Three simulations are shown in this figure for an SDOF with three values of damping ratio, $\zeta = 0.001, 0.01$, and 0.1. This means that the IRFs for each system have different time durations. The IRFs are plotted together with the time histories of the corresponding displacement responses to the random force input so that the time duration of the IRFs can be clearly seen. What is important is the time duration of the IRF compared to the time length of the segment T. Also plotted in Figure 7.10 in the top left part of the figure is the force time history for each system. Now, the coherence is essentially a measure of how much of the displacement response in each time segment is linearly related to the force input in that segment. For most frequencies this is fine, but close to the resonance frequency the response is dominated by damping, and if the damping is light, it can be seen that the IRF takes more than one segment to decay to a very small level. In this case, the displacement response close to the resonance frequency is due to the force input in preceding segments as well as the current segment, and hence the coherence function is less than unity, even though there is no extraneous measurement noise. It is clear from Figure 7.10 that if the length of the segment T is much greater (smaller) than the duration of the IRF, then the effect on the coherence at the resonance frequency is small (large). This bias error also manifests itself in the FRF at frequencies close to the resonance frequency. The effect is to underestimate the amplitude of the FRF close to the resonance frequency. This is most clearly seen in the Nyquist plot, which emphasises the region around the resonance frequency compared to other frequencies. The Nyquist plots for all three damping values are shown in Figure 7.10. The correlation between the error in the estimate of the FRF compared to the actual FRF, and the duration of the IRF compared to the segment length as well as the drop in the coherence can be clearly seen.

To reduce the bias error for a lightly damped system, it is evident that the window length needs to be increased. However, for a given length of data this means that there would be fewer averages, and hence the random error would increase in the FRF estimate, as can be seen in Table 7.1. Thus, there is a trade-off between bias error and random error. It is clear from the above discussion that for a lightly damped system, bias error can be a significant problem, and to maintain acceptable levels of bias and random errors in the estimated FRF for such as system, large data segments and hence long duration time histories of force and response data are necessary. In some situations, this can possibly result in another problem. For example, if the temperature changes, this could mean that the dynamic properties of the system under test may vary, especially if it contains visco-elastic materials, which would violate the assumption that the system has time-invariant properties.

MATLAB Example 7.2

In this example, the FRF of an SDOF system excited by random force excitation is estimated.

```
clear all

%% Parameters
m=1;k=10000;                      % [kg,N/m]     % mass and stiffness
z=0.01;c=2*z*sqrt(m*k);           % [Ns/m]       % damping
wn=sqrt(k/m);wd=sqrt(1-z^2)*wn;   % [rad/s]      % natural frequency
fn=wn/(2*pi);Tn=1/fn;             % [Hz,s]       % natural frequency/period
SNRf=1000;SNRx=1000;                            % SNRs for added random noise

%% Random force signal
fs=2000;                          % [Hz]         % sampling frequency
dt=1/fs; T=120; t=0:dt:T;         % [s]          % time vector
f=randn(length(t),1);             % [N]          % random force signal
N=length(f);

%% Calculation of displacement response
h=1/(m*wd)*exp(-z*wn*t).*sin(wd*t); % [m/Ns]     % IRF
xc=conv(h,f)/fs;                  % [m]          % convolution
xc=xc(1:length(f));                             % displacement response

%% Theoretical FRF
df=0.001; ft=0:df:fs-df;          % [Hz]         % frequency vector
w=2*pi*ft;                        % [rad/s]
Ht=1./(k-w.^2*m+j*w*c);           % [m/N]        % theoretical FRF
```

(Continued)

MATLAB Example 7.2   (Continued)

```
%% Frequency domain calculations
fwn=awgn(f,SNRf,'measured','dB');                          % add random noise to force signal
xwn=awgn(xc,SNRx,'measured','dB');                         % add random noise to displ. signal

Na = 32;                                                   % number of averages
nfft=round(N/Na);                                          % number of points in the DFT
noverlap=round(nfft/2);                                    % number of points in the overlap
Sff=cpsd(fwn,fwn,hann(nfft),noverlap,nfft,fs);            % force PSD
Sxx=cpsd(xwn,xwn,hann(nfft),noverlap,nfft,fs);            % displacement PSD
Sfx=cpsd(xwn,fwn,hann(nfft),noverlap,nfft,fs);            % CPSD between force and displ.
Tfx=tfestimate(fwn,xwn,hann(nfft),noverlap,nfft,fs);      % FRF between force and displ.
Coh=mscohere(fwn,xwn,hann(nfft),noverlap,nfft,fs);        % coherence
df=1/(nfft*dt);                                            % frequency resolution
ff = 0:df:fs/2;                                            % frequency vector

%% Time domain plots                                       % time domain plots
figure
subplot(2,1,1)
plot(t,fwn); grid
axis([0,T,1.1*min(fwn),1.1*max(fwn)])
xlabel('time (s)');ylabel('force (N)');
subplot(2,1,2)
plot(t,xwn); grid
set(gca,'fontsize',16)
axis([0,T,1.1*min(xwn),1.1*max(xwn)])
xlabel('time (s)'); ylabel('displacement (m)');

%% Frequency domain plots                                  % frequency domain plots
figure
semilogx(ff,10*log10(Sff)),grid                           % plot of force PSD
axis square; axis([1,1000,-60,-20])
xlabel('frequency (Hz)');
ylabel('force PSD (dB ref 1 N^2/Hz)');

figure
semilogx(ff,10*log10(Sxx)),grid                           % plot of displ. PSD
axis square; axis([1,1000,-180,-60])
xlabel('frequency (Hz)');
ylabel('displacement PSD (dB ref 1 m^2/Hz)');

figure
semilogx(ff,20*log10(abs(Tfx)))                           % plot of FRF
hold on
semilogx(ft,20*log10(abs(Ht))),grid
axis square; axis([1,1000,-140,-40])
xlabel('frequency (Hz)');
ylabel('displacement/force (dB ref 1 m/N)');

figure
semilogx(ff,180/pi*angle(Tfx))                            % plot of phase
hold on
semilogx(ft,180/pi*unwrap(angle(Ht))),grid
axis square,axis([1,1000,-200,0])
xlabel('frequency (Hz)');ylabel('phase (degrees)');

figure
semilogx(ff,Coh); grid                                     % plot of coherence
axis square, axis([1,1000,0,1])
xlabel('frequency (Hz)');ylabel('coherence');
```

(Continued)

MATLAB Example 7.2   (Continued)

Results

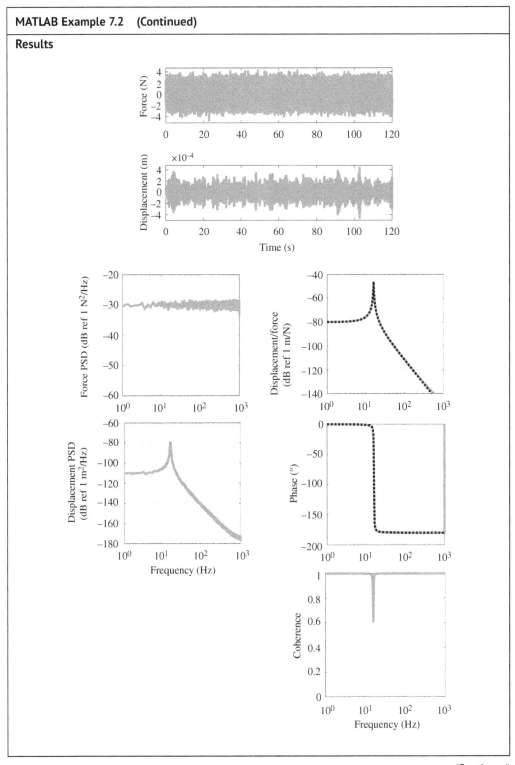

(Continued)

MATLAB Example 7.2 (Continued)

Comments:

1. The figures show the situation when the SNR for the measured force and displacement signals are very large such that the measurement noise is negligibly small. An exercise for the reader is to explore what happens when the SNRs are changed for the input and output. Try using the H_2 estimator.
2. Other exercises for the reader are to:
   (a) explore the effects on the FRF and coherence of increasing (reducing) the number of averages by reducing (increasing) the segment size,
   (b) change the damping in the system and investigate the effects on the FRF and the coherence for different segment sizes (number of points in the DFT),
   (c) increase/decrease the overall time in which the system is excited, and the force and response data are collected, and investigate the effects of changing the number of averages on the quality of the FRF.

7.4 Comparison of Excitation Methods and Effects of Shaker–Structure Interaction

As discussed in Chapter 5, a shaker that is driven by a chirp signal or random noise can be used to excite a structure, and an impact hamper can be used to deliver a force similar to a half-sine impulse. It was shown previously in this chapter that there are differences in processing data from tests using transient excitation, such as a chirp or an impulse, compared to persistent excitation from random noise. Thus, it is reasonable to ask what is the best method of excitation? As with many simple questions, there is not necessarily a simple answer, but some features of each approach can be articulated. Before doing this, however, it is of interest to investigate the effects that a shaker may have on test data due to dynamic interaction between the shaker and the structure under test.

Consider the test set-up shown in the left part of Figure 7.11, and a simple model for an SDOF test structure shown in the right part of Figure 7.11. A shaker, which is connected to a structure through a stinger and a force gauge, is driven by an oscillating current $i_s(t)$. The force gauge measures the force applied to the structure $f_e(t)$, which is not equal to the force $f_s(t)$ that is proportional to the current supplied to the shaker. This is because part of the force generated is needed to drive the shaker, which is given by $f_{shaker}(t)$, with the remainder being delivered to the structure. This force relationship can be written as

$$f_s(t) = f_e(t) + f_{shaker}(t). \tag{7.12}$$

Now it is assumed that the force gauge and the stinger are very stiff, so that displacements of the structure and the shaker are the same. If the excitation is harmonic at angular frequency ω then the forces have the form $f_e(t) = \overline{F}e^{j\omega t}$ and the displacement is given by $x(t) = \overline{X}e^{j\omega t}$, where $\overline{F}$ and $\overline{X}$ are complex functions of frequency. As the shaker and the structure are modelled as parallel combinations of lumped parameter elements, they can be described in a simple manner by their dynamic stiffness as described in Chapter 2. Now,

$$F_{shaker}(j\omega) = K_{shaker}(j\omega)X(j\omega) \tag{7.13a}$$

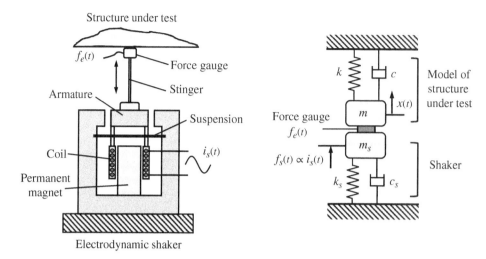

Figure 7.11 Structure excited by an electrodynamic shaker and a simple model of the test.

and

$$F(j\omega) = K(j\omega)X(j\omega), \tag{7.13b}$$

where $K_{\text{shaker}}(j\omega) = k_s - \omega^2 m_s + j\omega c_s$ and $K(j\omega) = k - \omega^2 m + j\omega c$ are the dynamic stiffnesses of the shaker and structure under test, respectively. Eqs. (7.12), (7.13a), and (7.13b) can be combined to give

$$\frac{F(j\omega)}{F_s(j\omega)} = \frac{K(j\omega)}{K(j\omega) + K_{\text{shaker}}(j\omega)}. \tag{7.14a}$$

Substituting for the dynamic stiffness of the structure and the shaker results in

$$\frac{F(j\omega)}{F_s(j\omega)} = \frac{k - \omega^2 m + j\omega c}{(k + k_s) - \omega^2(m + m_s) + j\omega(c + c_s)}. \tag{7.14b}$$

It can be seen that if F_s has a constant value independent of frequency, then the force applied to the structure is not constant. Recall that, in the simulations hitherto in this chapter, the excitation force due to a random signal had a flat spectrum, which was also broadly the case for chirp excitation between the lower and upper cut-off frequencies. However, as shown in Eq. (7.14a,b), if the amplitude of the current supplied to the shaker is kept constant, which is the case if the associated power amplifier is used in current mode, the amplitude of the force applied to the structure changes with frequency. In particular, at the resonance frequency of the structure, when $k \approx \omega^2 m$, the force applied to the structure becomes very small. This is known as '*force drop out*'. The effect is illustrated in Figure 7.12. Also shown in Figure 7.12 is the force that drives the shaker, which can be determined by simply subtracting the force applied to the structure from the generated force to give

$$\frac{F_{\text{shaker}}(j\omega)}{F_s(j\omega)} = \frac{K_{\text{shaker}}(j\omega)}{K(j\omega) + K_{\text{shaker}}(j\omega)} \tag{7.15a}$$

or

$$\frac{F_{\text{shaker}}(j\omega)}{F_s(j\omega)} = \frac{k_s - \omega^2 m_s + j\omega c_s}{(k + k_s) - \omega^2(m + m_s) + j\omega(c + c_s)}. \tag{7.15b}$$

In Figure 7.12, it can be seen that provided that the shaker stiffness is much less than the stiffness of the structure, then at low frequencies the force that drives the shaker is small compared

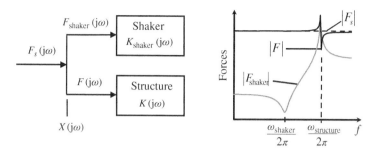

Figure 7.12 Frequency domain representation of the division of the generated force between the shaker and the structure. The force and frequency axes are logarithmic.

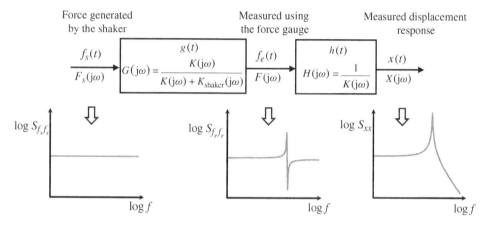

Figure 7.13 Block diagram showing the relationship between the generated force, the force applied to the structure, and its response.

to the force applied to the structure. It reduces at the natural frequency of the shaker, then it increases whilst the shaker operates in its mass-controlled regime until a peak occurs at the natural frequency of the combination of the shaker and the structure, after which it decreases. This is because above this frequency, the dynamic stiffness of the structure is greater than that of the shaker.

A block diagram of an SDOF system excited by a shaker is shown in Figure 7.13. Also shown in this figure are the PSD of the force generated by the shaker, the PSD of the force applied to the structure, and the PSD of the mass displacement of the SDOF structure. It is clear that the force applied to the structure does not have a flat spectrum. The force peaks at the resonance frequency of the combined shaker–structure system, and 'drops out' at the resonance frequency of the structure. The PSD of the displacement response has a peak, but this peak does not occur at the resonance frequency of the structure. It peaks at the resonance frequency of the combined shaker–structure system. This means that care should be taken when determining the natural frequency of a structure when excited by a shaker. This should be done by examining the peak of the modulus of the FRF rather than the PSD of the response.

A comparison between two types of excitation – random noise and a chirp – is illustrated in MATLAB Example 7.3.

MATLAB Example 7.3

In this example, the FRF of a shaker-excited SDOF system is investigated for random and chirp excitation.

```
clear all

%% Parameters
% structure
m=1;k=10000;                        % [kg,N/m]      % mass and stiffness
z=0.01;c=2*z*sqrt(m*k);             % [Ns/m]        % damping
wn=sqrt(k/m);wd=sqrt(1-z^2)*wn;     % [rad/s]       % natural frequency
SNRf=1000;SNRx=1000;                               % SNRs for random excitation
SNRf=20;SNRx=100;                                  % SNRs for chirp excitation

% shaker
ms=0.1;ws=2*pi*10; ks=ws^2*ms;      % [kg,rad/s,N/m]% mass, nat. freq. and stiffness
zs=0.1;cs=2*zs*sqrt(ms*ks);         % [Ns/m]        % damping

%% Random force signal
fs=2000;                            % [Hz]          % sampling frequency
dt=1/fs;T=120;t1=0:dt:T;            % [s]           % time vector
f=randn(length(t1),1);              % [N]           % random force signal
N=length(f);TT=N*dt;
t=0:dt:(length(f)-1)*dt;            % [s]           % time vector

%% Calculation of frequency domain quantities
Na=32;nfft=round(N/Na);                            % number of averages, points in DFT
noverlap=round(nfft/2);                            % number of points in the overlap
df=1/(nfft*dt);                     % [Hz]          % frequency resolution
ff=0:df:fs/2;w=2*pi*ff;ft=ff;       % [Hz]          % frequency vector


[fe,xc,Ht]=calc(f,m,k,c,ms,ks,cs,w,wd,wn,z,t,fs);  % calculate force and displacement

Sfsfs=cpsd(f,f,hann(nfft),noverlap,nfft,fs);       % PSD of force generated by shaker
Sff=cpsd(fe,fe,hann(nfft),noverlap,nfft,fs);       % PSD of force applied to structure
Sxx=cpsd(xc,xc,hann(nfft),noverlap,nfft,fs);       % PSD of displacement response
H=tfestimate(fe,xc,hann(nfft),noverlap,nfft,fs);   % FRF
coh=mscohere(fe,xc,hann(nfft),noverlap,nfft,fs);   % coherence

plots(ft,ff,Sfsfs,Sff,Sxx,H,Ht,coh)               % function to plot the results
clear Sfsfs; clear Sff; clear Sxx; clear Sfx

%% chirp force signal
T=5;t1=0:dt:T;                      % [s]           % time vector
f1=1;f2=300;                        % [Hz]          % upper and lower frequencies
a=2*pi*(f2-f1)/(2*T); b=2*pi*f1;                    % coefficients
fc=sin(a*t1.^2+b*t1);               % [N]           % chirp force signal
f=[fc zeros(1,length(fc))];         % [N]           % zero padded force signal
N=length(f);
t=0:dt:(N-1)*dt; Tm=max(t);         % [s]           % time vector
df=1/(N*dt);ft=0:df:fs/2;w=2*pi*ft;                 % frequency vector

[fe,xc,Ht]=calc(f,m,k,c,ms,ks,cs,w,wd,wn,z,t,fs);  % calculate force and displacement

%% Calculation of frequency domain quantities
for n=1:16
  fwn=awgn(fe,SNRf,'measured','dB');               % add random noise
  xwn=awgn(xc,SNRx,'measured','dB');               % add random noise
  Fs=fft(f)*dt;                     % [N/Hz]        % DFT of generated force
```

(Continued)

MATLAB Example 7.3   (Continued)

```
F=fft(fwn)*dt;               % [N/Hz]      % DFT of force applied to structure
X=fft(xwn)*dt;               % [m/Hz]      % DFT of displacement
Sfsfs(n,:)=Fs.*conj(Fs)/Tm;  % [N^2/Hz]    % PSD of force generated by shaker
Sff(n,:)=F.*conj(F)/Tm;      % [N^2/Hz]    % PSD of force applied to structure
Sxx(n,:)=X.*conj(X)/Tm;      % [m^2/Hz]    % PSD of displacement response
Sfx(n,:)=X.*conj(F)/Tm;      % [Nm/Hz]     % CPSD of force and displacement
end


Sfsfs=mean(Sfsfs);Sff=mean(Sff);Sxx=mean(Sxx);   % averaging the results
Sfx=mean(Sfx);


H=Sfx./Sff;                  % [m/N]       % FRF
coh=abs(Sfx).^2./(Sxx.*Sff);              % coherence
ff=0:df:fs-df;               % [Hz]        % frequency vector
plots(ft,ff,Sfsfs,Sff,Sxx,H,Ht,coh)       % function to plot the results


%% Functions
%% Calculation of the force and displacement
function [fe,xc,Ht]=...                   % function to calc. force and displ.
calc(f,m,k,c,ms,ks,cs,w,wd,wn,z,t,fs)
  K=k-w.^2*m+j*w*c;          % [N/m]       % dynamic stiffness of structure
  Ks=ks-w.^2*ms+j*w*cs;      % [N/m]       % dynamic stiffness of shaker
  Fe=K./(K+Ks);              % [N]         % force applied to structure
  G=[Fe,fliplr(conj(Fe(1:length(Fe)-1)))]; % force FRF
  g=fs*ifft(G);              % [1/s]       % force IRF
  fe = conv(real(g),f)/fs;   % [N]         % applied force
  fe = fe(1:length(f));      % [N]         % applied force
  Ht=1./K;                   % [m/N]       % theoretical FRF


  % impulse response
  h=1/(m*wd)*exp(-z*wn*t).*sin(wd*t); % [m/Ns]  % IRF
  xc=conv(h,fe)/fs;          % [m]         % Convolution
  xc=xc(1:length(fe));       % [m]         % displacement response
end


%% Plot the results
function plots(ft,ff,Sfsfs,Sff,Sxx,H,Ht,coh)   % Frequency domain plots
  figure
  semilogx(ff,10*log10(Sfsfs));           % plot of force PSD
  hold on
  semilogx(ff,10*log10(Sff)); axis square; grid
  xlabel('frequency (Hz)');
  ylabel('force PSD (dB ref 1 N^2/Hz)');


  figure
  semilogx(ff,10*log10(Sxx)); axis square; grid   % plot of displ. PSD
  xlabel('frequency (Hz)');
  ylabel('displacement PSD (dB ref 1 m^2/Hz)');


  figure
  semilogx(ff,20*log10(abs(H)));          % plot of FRF
  hold on
  semilogx(ft,20*log10(abs(Ht))); axis square; grid
  xlabel('frequency (Hz)');
  ylabel('displacement/force (dB ref 1 m/N)');


  figure
  semilogx(ff,180/pi*unwrap(angle(H)));   % plot of phase
  hold on
  semilogx(ft,180/pi*unwrap(angle(Ht)));
  axis square; grid
  xlabel('frequency (Hz)');ylabel('phase...
 (degrees)');
```

(Continued)

MATLAB Example 7.3   (Continued)

```
figure                                    % plot of coherence
semilogx(ff,coh); axis square; grid;
xlabel('frequency (Hz)');ylabel('coherence');
end
```

Results

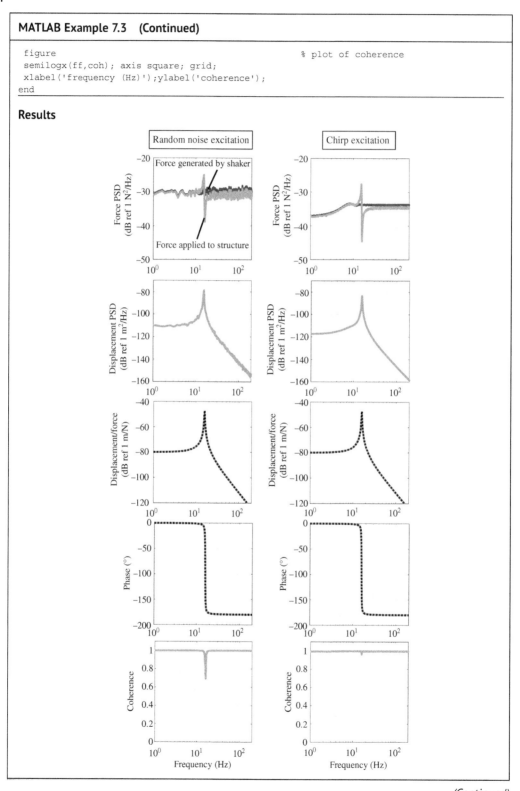

MATLAB Example 7.3 (Continued)

Comments:

1. The figures show the frequency domain plots when the SNRs for the measured force and displacement signals are very large. An exercise for the reader is to explore what happens when the SNRs are changed for the input and output. Try using the H_2 estimator.
2. Although the PSD of the force applied to the structure exhibits a peak and a trough, compared to the generated force, there is no trough in the PSD of the displacement response. The trough coincides with the resonance frequency of the structure, and the peak coincides with the resonance frequency of the combined structure and shaker system.
3. Even though the force input and displacement response are different for random noise and chirp excitation, the estimated modulus and the phase for both cases are broadly similar. Note that a Hanning window is used for random excitation, and a rectangular window is used for chirp excitation. There is a small amount of bias error at frequencies close to the resonance frequency for the FRF estimated using random excitation. An exercise for the reader is to adjust the segment window size to reduce this error.
4. An exercise for the reader is to adjust the rate of change of frequency (chirp rate) and see if this makes a difference to the estimate of the FRF.
5. Other exercises for the reader are to:
   (a) change the damping in the system and investigate the effects on the force drop out, the FRF estimates and the coherence for different segment sizes (number of points in the DFT), different chirp rate, etc.
   (b) by changing the SNRs and other processing parameters, investigate the accuracy of the FRF estimates, and speculate on which is the better way to excite the system.

7.5 Virtual Experiment – Vibration Isolation

A ubiquitous problem in vibration engineering is the dynamic decoupling of a vibrating system from its surroundings. This is called vibration isolation. It is often achieved by using a resilient element that is modelled as a parallel combination of a stiffness and a viscous damper. Vibration isolation occurs above the resonance frequency when the stiffness of the isolator interacts with the mass of the vibration source. To calculate the frequency at which the resonance occurs and the subsequent level of vibration isolation, the stiffness and the damping of the isolator need to be known, and in some cases, these are estimated from experimental data. The way in which this can be achieved is described in this chapter, but before this, the principal mechanisms of vibration are discussed in Section 7.5.1.

7.5.1 The Physics of Vibration Isolation

To illustrate the fundamental mechanisms of vibration isolation, consider the two systems shown in Figure 7.14. The system on the left shows a force-excited system, which could represent, for example, the isolation of a vibrating machine from its host structure, and the system on the right shows a base-excited system, which could represent the isolation of an item of equipment from a vibrating host structure. These two systems involve idealised models in which many assumptions

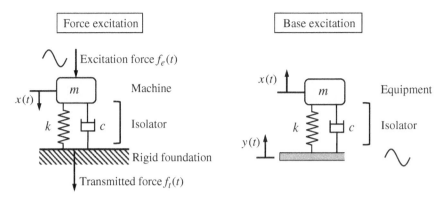

Figure 7.14 Two basic vibration isolation situations.

are made such as the host structure for the force-excited system is rigid, and there is only an SDOF system in each case. Nonetheless, the systems illustrate the basic physical principles involved in vibration isolation. The efficacy of a vibration isolator is assessed at each frequency. Thus, harmonic excitation is assumed, so that $f_e(t) = \overline{F}e^{j\omega t}$, $f_t(t) = \overline{F}_te^{j\omega t}$, $y(t) = \overline{Y}e^{j\omega t}$, and $x(t) = \overline{X}e^{j\omega t}$. In the force-excited system the ratio of the amplitude of the transmitted force to the amplitude of the excitation force is used to quantify vibration isolation. This is called the force transmissibility and is given by $|T_{\text{force}}| = |\overline{F}_t/\overline{F}|$. Likewise, in the base-excited case, vibration isolation is quantified by the displacement transmissibility given by $|T_{\text{displ.}}| = |\overline{X}/\overline{Y}|$. If the isolator comprises linear stiffness and damping, the force transmissibility is equal to the displacement transmissibility, so the term transmissibility is simply used and is represented by $|T|$. The equation of motion for the force-excited system is given by

$$m\ddot{x}(t) + c\dot{x}(t) + kx(t) = f_e(t), \tag{7.16}$$

and the force transmitted to the rigid foundation is given by

$$f_t(t) = c\dot{x}(t) + kx(t). \tag{7.17}$$

Assuming harmonic excitation, as discussed above, results in

$$k\overline{X} - \omega^2 m\overline{X} + j\omega c\overline{X} = \overline{F} \tag{7.18}$$

and

$$\overline{F}_t = k\overline{X} + j\omega c\overline{X}. \tag{7.19}$$

Note that in Eq. (7.19) the term $k\overline{X}$ is the force that is transmitted to the foundation through the spring and the term $j\omega c\overline{X}$ is the force transmitted to the foundation through the damper. It can be seen that relative to the force transmitted through the spring, the force transmitted through the damper increases with frequency, and therefore can be dominant at high frequencies. Combining Eqs. (7.18) and (7.19) results in

$$|T| = \left|\frac{\overline{F}_t}{\overline{F}}\right| = \left|\frac{k + j\omega c}{k - \omega^2 m + j\omega c}\right| = \left|\left(\frac{1}{k - \omega^2 m + j\omega c}\right) \times (k + j\omega c)\right|. \tag{7.20}$$

The first term in brackets in the right-hand side of Eq. (7.20) is the receptance of the SDOF system and the second term in brackets corresponds to the transmitted force per unit displacement amplitude of the mass. The damping term in these terms has different physical effects. In the first term the damping has a beneficial effect in that it reduces the displacement of the mass, and hence the transmitted force, at the resonance frequency. However, the damping in the second

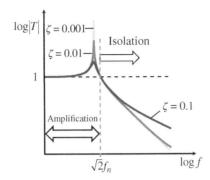

Figure 7.15 Force or displacement transmissibility of an SDOF system.

term is undesirable, as an increase in damping results in a higher force being transmitted at high frequencies.

The modulus of the transmissibility (for both force and displacement) is plotted in Figure 7.15, for three values of damping ratio, $\zeta = 0.001, 0.01$, and 0.1. Note that both axes have logarithmic scales. It can be seen that there is a frequency region below the isolation frequency, given by $f = \sqrt{2}f_n$, in which f_n is the resonance frequency, that is independent of the level of damping, where there is force or displacement amplification rather than attenuation. In the frequency range above the isolation frequency, i.e. where $f > \sqrt{2}f_n$, the transmitted force (displacement) is less than the excitation force (displacement), so vibration isolation occurs. Note that an increase in damping results in a reduction in the response at the resonance frequency, but an increase in the transmitted force (displacement) at high frequencies. Thus, there is a trade-off between reducing the transmitted force (displacement) at the resonance frequency and improving vibration isolation at high frequencies. The phase of the transmissibility is not shown as it is of little interest in vibration isolation unless there are multiple sources. If the machine to be isolated operates at constant speed, such that it generates a force at a frequency corresponding to the operating speed of the machine (and possibly harmonics of this frequency), the isolation system should be designed so that the resonance frequency of the system is much less than the operating frequency of the machine. Provided that the machine can be run-up to speed so that it transverses the resonance frequency rapidly, then it is better to have an isolation system with light damping. This is investigated in MATLAB Example 7.5.

7.5.2 Experimental Determination of the Stiffness and Damping of a Vibration Isolator

Sometimes it is necessary to determine the stiffness and damping properties of an isolator by experiment. This can be carried out using specialised test equipment in which the force and relative displacement across the isolator can be measured as it is cycled at very low frequency. The dynamic stiffness of the isolator may also be measured (ISO 10846, 2008). In many cases, however, the test equipment to carry out such measurements is not available, so an improvised experiment needs to be conducted. An example of such an experiment is shown in Figure 7.16. The isolator and suspended mass are placed on a shaker. The mass should be compact and representative of the amount of mass that the isolator will support in practice, so that the static equilibrium position of the isolator is correct. Either a chirp or a random noise signal can be used to drive the shaker via a power amplifier. Using two accelerometers, one positioned on the shaker and one positioned on the mass, the displacement transmissibility can be measured (note that lasers can also be used to measure velocity or displacement instead of acceleration). The acceleration signals are passed

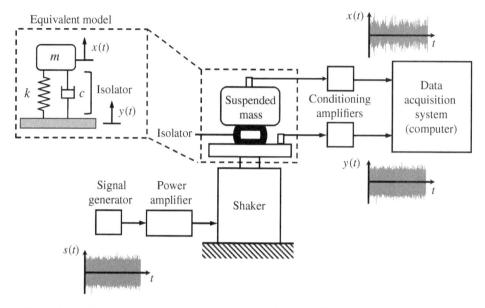

Figure 7.16 Schematic diagram of an experiment to determine the stiffness and damping characteristics of a vibration isolator, and its simplified equivalent model.

through conditioning amplifiers and captured by a data acquisition system, which converts the signals from analogue to digital format.

In a real experiment the transmissibility can be calculated by using the Fourier transforms of the acceleration signals, which is also the case in the virtual experiment described in MATLAB Example 7.4. For simplicity, here it is assumed that time history of the shaker acceleration – the base of the isolator – is simply a scaled version of the current supplied to the shaker. This is considered to be a reasonable assumption, if the shaker is large enough such that the dynamics of the isolator and suspended mass do not affect the vibration of the shaker, or a control system is used to maintain a prescribed level of shaker vibration. Once the transmissibility has been estimated with sufficient frequency resolution to capture the peak at the resonance frequency, the stiffness and damping can be estimated. If the mass is measured before it is attached to the isolator, the stiffness can be determined by estimating the natural frequency from the transmissibility plot, and noting that

$$k = (2\pi f_n)^2 m, \tag{7.21}$$

and the damping can be determined by noting that, for light damping, at the resonance frequency, Eq. (7.20) becomes

$$|T|_{max} \approx \frac{k}{2\pi f_n c} = \frac{1}{2\zeta}. \tag{7.22}$$

Note that it is extremely important to check that the frequency resolution is adequate, so that damping value is estimated accurately. The choice of frequency resolution can be guided by noting that the damping ratio is related to the half-power point frequencies f_1 and f_2, and the resonance frequency by $\zeta = (f_2 - f_1)/(2f_n)$. To ensure that there are at least three frequency points between the half-power point frequencies the frequency resolution should satisfy the inequality $\Delta f \leq \zeta f_n$. This means that the segment time duration should satisfy the inequality $T \geq 1/(\zeta f_n)$. It can thus be seen that the choice of signal processing parameters is dependent upon the properties of the system. If the natural frequency is low and the damping in the isolator is thought to be small, then a large time segment is required. Further, if random noise is used to excite the system, then long duration time histories should be captured to ensure that enough averages so that there is a small random error.

MATLAB Example 7.4

In this example, an experiment to estimate the stiffness and damping properties of a vibration isolator is simulated.

```
clear all

% Suspended mass
m=1;                                % [kg]              % suspended mass
% Isolator
k=10000;                            % [N/m]             % stiffness
z=0.01;c=2*z*sqrt(m*k);             % [Ns/m]            % damping

%% Random acceleration signal
fs=2000;                            % [Hz]             % sampling frequency
dt=1/fs;T=240;t1=0:dt:T;            % [s]              % time vector
y = randn(length(t1),1);            % [m/s^2]          % random acceleration signal
N=length(y);TT=N*dt;
t=0:dt:(length(y)-1)*dt;            % [s]              % time vector

%% Parameters for processing the data
Na = 16;nfft=round(N/Na);                              % number of averages, points in DFT
noverlap=round(nfft/2);                                % number of points in the overlap
df=1/(nfft*dt);                                        % frequency resolution
ff=0:df:fs/2;w=2*pi*ff;ft=ff;                          % frequency vector

%% Calculation of the acceleration of the mass
Tf=(k+j*w*c)./(k-w.^2*m+j*w*c);                        % acceleration transmissibility
TTf=[Tf,fliplr(conj(Tf(1:length(Tf)-1)))];
ttf=fs*ifft(TTf);                                      % IDFT of the transmissibility
x=conv(real(ttf),y)/fs;                                % acceleration response
x=x(1:length(y));                                      % acceleration response

%% Calculation of the frequency domain quantities
Syy=cpsd(y,y,hann(nfft),noverlap,nfft,fs);%[(m^2/s^2)/Hz] % PSD of shaker acceleration
Sxx=cpsd(x,x,hann(nfft),noverlap,nfft,fs);%[(m^2/s^2)/Hz] % PSD of mass acceleration
H=tfestimate(y,x,hann(nfft),noverlap,nfft,fs);         % FRF (transmissibility)
coh=mscohere(y,x,hann(nfft),noverlap,nfft,fs);         % coherence

%% Plot the results
figure
subplot(2,1,1)                                         % Time domain plots
plot(t,y); axis([0,T,1.1*min(y),1.1*max(y)])           % plot of shaker acceleration
xlabel('time (s)'); ylabel('shak. acc. (m/s^2)');
subplot(2,1,2)
plot(t,x); axis([0,T,1.1*min(x),1.1*max(x)])           % plot of mass acceleration
xlabel('time (s)');
ylabel('mass acc. (m/s^2)');

figure                                                 % Frequency domain plots
semilogx(ff,10*log10(Syy))                             % plot of shaker acc. PSD
axis square; grid; axis([1,200,-50,-20])
xlabel('frequency (Hz)');
ylabel('shak. acc. PSD (dB ref 1 (m^2/s^4)/Hz)');
```

(Continued)

MATLAB Example 7.4   (Continued)

```
figure
semilogx(ff,10*log10(Sxx))                    % plot of mass acc. PSD
axis square; grid; axis([1,200,-80,20])
xlabel('frequency (Hz)');
ylabel('mass acc. PSD (dB ref 1 (m^2/s^4)/Hz)');

figure
semilogx(ff,20*log10(abs(H)))                 % plot of estimated transmissibility
hold on
semilogx(ft,20*log10(abs(Tf)),':')            % plot of theor. transmissibility
axis square; grid; axis([1,200,-60,40])
xlabel('frequency (Hz)');
ylabel('|Transmissibility| (dB ref unity)');

figure
semilogx(ff,180/pi*unwrap(angle(H)))          % plot of phase
hold on
semilogx(ft,180/pi*unwrap(angle(Tf)),':')
axis square; grid; axis([1,200,-200,0])
xlabel('frequency (Hz)');ylabel('phase (degrees)');

figure
semilogx(ff,coh)                              % plot of coherence
axis square; grid; axis([1,200,0,1.1])
xlabel('frequency (Hz)');ylabel('coherence');
```

Results

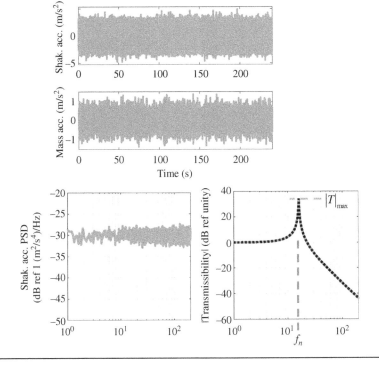

MATLAB Example 7.4   (Continued)

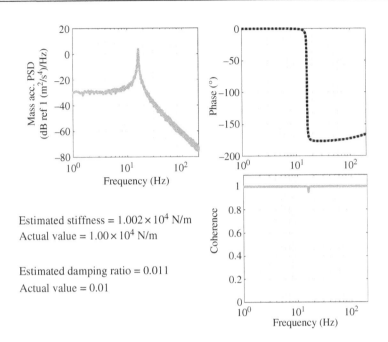

Estimated stiffness = 1.002×10^4 N/m
Actual value = 1.00×10^4 N/m

Estimated damping ratio = 0.011
Actual value = 0.01

Comments:

1. To estimate the stiffness and damping accurately, a very fine frequency resolution is needed. This is particularly important for the damping parameter, especially if the damping is very small. It is particularly challenging to estimate the stiffness and damping accurately when the system has a very low natural frequency and is lightly damped, i.e. if $\zeta f_n \ll 1$.

2. An exercise of the reader is to investigate the effects on the transmissibility, and hence the stiffness and damping when (a) the damping is changed, and (b) when the number of averages is changed.

3. An exercise for the reader is to try changing the excitation signal to a chirp, and draw some conclusions on which is the best for this type of experiment.

7.5.3 Experiment to Investigate the Trade-off Between Decreasing the Response at the Resonance Frequency and Improving Vibration Isolation

Once the stiffness and damping values of an isolator have been estimated, as discussed in the previous subsection, they can be used to predict of the dynamic behaviour of the isolator in any given situation. An illustration of how this may be carried out is described next. To simplify the problem, consider a similar arrangement to that shown in Figure 7.16, but with the isolator connected to ground and the suspended mass is excited by a force which has a constant amplitude that is independent of frequency. The force has similar characteristics to that generated during the start-up of

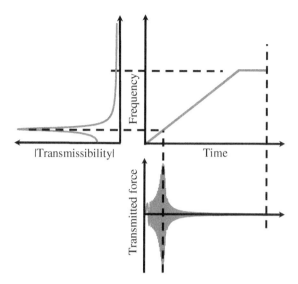

Figure 7.17 Illustration of the vibration of a suspended mass on an isolator as the frequency is increased from 0 Hz to the operating frequency.

a machine from rest to a frequency corresponding to the running speed of the machine (the operational frequency). A graph showing the frequency as a function of time is shown in Figure 7.17. The excitation frequency increases linearly with time, similar to a linear chirp, until it reaches the operational frequency, after which it has a constant value. The force transmissibility for the combined isolator–mass system is shown to the left of the frequency–time graph. Note that this graph has linear axes unlike the previous transmissibility graphs. It can be seen that the force sweeps through the resonance frequency until the operational frequency, which is in the isolation frequency region. The time history of the transmitted force corresponding to the frequency sweep is shown at the bottom of Figure 7.17. It can be seen that the force increases as the excitation frequency passes through the resonance frequency before it reduces rapidly as it passes into the vibration isolation region. For a well-designed vibration isolation system, the operational frequency should be much greater than the natural frequency. The maximum value of the transmitted force is a function of the rate of change of excitation frequency and the damping in the isolator. At the operational speed, the amplitude of the transmitted force is a function of damping in the isolator, and the ratio of the operational frequency and the natural frequency. These relationships are investigated in MATLAB Example 7.5.

MATLAB Example 7.5

In this example, the transmitted force to the rigid foundation supporting a suspended mass is calculated as a function of time. The aim is to illustrate the effect of the rate of change of increasing frequency on the maximum response, and also the effect of damping on the transmitted force.

(Continued)

MATLAB Example 7.5 (Continued)

```
clear all

% Suspended mass
m=1;                                    % [kg]        % suspended mass
% Isolator
k=10000;                                % [N/m]       % stiffness
z=0.01;c=2*z*sqrt(m*k);                 % [Ns/m]      % damping

% Calculation of transmitted force
%slow rate of change of freq.
n=1000;                                              % number of cycles
[fmax,fe,ft,t]=calc(m,k,c,n);                        % function to calculate trans. force
fe1=fe;ft1=ft;t1=t;

%medium rate of change of freq.
n=200;                                               % number of cycles
[fmax,fe,ft,t]=calc(m,k,c,n);
fe2=fe;ft2=ft;t2=t;

%fast rate of change of freq.
n=20;                                                % number of cycles
[fmax,fe,ft,t]=calc(m,k,c,n);
fe3=fe;ft3=ft;t3=t;

%% Plot the results                                  % time domain plots
plot(t1,ft1,'linewidth',3,'Color',[.7 .7 .7])
hold on
plot(t2,ft2,'linewidth',3,'Color',[.5 .5 .5])
hold on
plot(t3,ft3,'linewidth',3,'Color',[.3 .3 .3]); grid
set(gca,'fontsize',24)
axis([0,12,1.1*min(ft1),1.1*max(ft1)])
axis([0,12,-15,15])
xlabel('time (s)');ylabel('transmitted force (N)');

%% Function
function [fmax,fe,ft,t]=calc(m,k,c,n)                % function to calculate trans. force
 fs=2000;dt=1/fs;                       % [Hz,s]      % sampling freq., time resolution
 fmax=100;                              % [Hz]        % operational frequency
 T=n/fmax;t1=0:dt:T;                    % [s]         % time vector
 a=2*pi*fmax/(2*T);                                   % coefficient
 ff1=sin(a*t1.^2);                      % [N]         % force with increasing freq.
 t2=0:dt:15-T;                          % [s]         % time vector
 ff2=sin(n*pi+2*pi*fmax*t2);            % [N]         % steady-state force
 fe=[ff1, ff2];                         % [N]         % total force
 N=length(fe);
 t=0:dt:(N-1)*dt;Tm=max(t);             % [s]         % time vector
 df=1/(N*dt);ff=0:df:fs/2;w=2*pi*ff;ft=ff;           % freq. resolution., freq. vector

 %% Calculation of the transmitted force
 Tf=(k+j*w*c)./(k-w.^2*m+j*w*c);                      % force transmissibility
 TTf=[Tf,fliplr(conj(Tf(1:length(Tf)-1)))];          % double-sided spectrum
 ttf=fs*ifft(TTf);                      % [N/s]       % IDFT of the transmissibility
 ft=conv(real(ttf),fe)/fs;              % [N]         % transmitted force
 ft=ft(1:length(fe));                   % [N]
end
```

(Continued)

MATLAB Example 7.5   (Continued)

Results

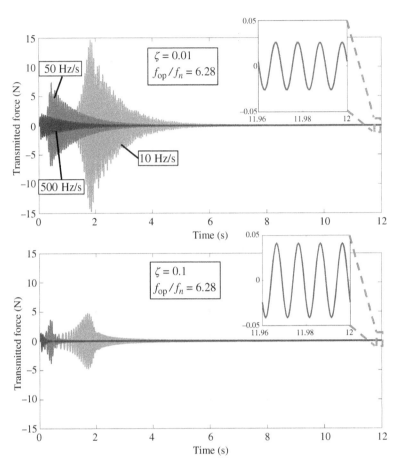

Comments:

1. The results show that if the rate-of-change of excitation frequency is increased, the maximum transmitted force due to the resonance frequency is reduced, but the transmitted force at the operational frequency remains the same. This can be seen in both figures shown above.

2. The trade-off between reducing the response at the resonance frequency by adding damping and increasing the transmitted force at the steady-state operational frequency can be seen. By changing damping, from $\zeta = 0.01$ in the upper graph to $\zeta = 0.1$ in the lower graph, the maximum transmitted force is reduced considerably. However, the penalty for this is an increase in the transmitted force at the operational frequency, as can be seen in the insets shown in the upper and lower graphs.

3. An exercise for the reader is to plot graphs to show the maximum transmitted force due to the resonance frequency changes as the rate-of-change frequency is increased and the damping is changed. To do this you will need to run the simulation several times to calculate the maximum transmitted force for different values of the rate-of-change of frequency and damping. What is the optimum condition?

7.6 Summary

This chapter has shown how to estimate a frequency response function (FRF) from measured input and output data. Two types of input signal have been considered. The first involves transient excitation such as an impact from an instrumented hammer, and chirp excitation using a shaker. The second involves persistent excitation using a shaker driven with random noise. A summary of the method is given in Table 7.2. In all cases, averaging of data is carried out in the frequency domain

Table 7.2 Summary of methods to obtain an FRF.

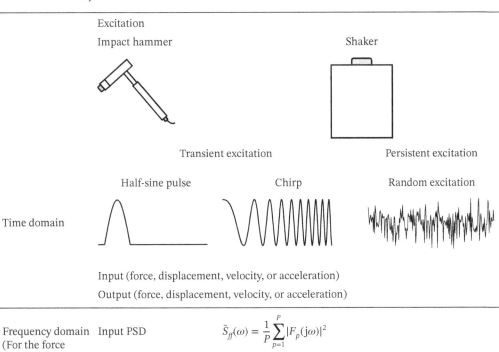

	Excitation	
	Impact hammer	Shaker
	Transient excitation	Persistent excitation
	Half-sine pulse Chirp	Random excitation
Time domain		

Input (force, displacement, velocity, or acceleration)
Output (force, displacement, velocity, or acceleration)

Frequency domain (For the force input and displacement response output case) (*P* averages)		
	Input PSD	$\tilde{S}_{ff}(\omega) = \dfrac{1}{P}\displaystyle\sum_{p=1}^{P} |F_p(j\omega)|^2$
	Output PSD	$\tilde{S}_{xx}(\omega) = \dfrac{1}{P}\displaystyle\sum_{p=1}^{P} |X_p(j\omega)|^2$
	CPSD	$\tilde{S}_{fx}(j\omega) = \dfrac{1}{P}\displaystyle\sum_{p=1}^{P} X_p(j\omega)F_p^*(j\omega) \quad \tilde{S}_{xf}(j\omega) = \tilde{S}_{fx}^*(j\omega)$
		$\tilde{S}_{xf}(j\omega) = \dfrac{1}{P}\displaystyle\sum_{p=1}^{P} X_p^*(j\omega)F_p(j\omega)$
	FRF	$H_1(j\omega) = \dfrac{\tilde{S}_{fx}(j\omega)}{\tilde{S}_{ff}(\omega)}$
		$H_2(j\omega) = \dfrac{\tilde{S}_{xx}(\omega)}{\tilde{S}_{xf}(j\omega)}$
	Coherence	$\gamma_{fx}^2(\omega) = \dfrac{|\tilde{S}_{fx}(j\omega)|^2}{\tilde{S}_{ff}(\omega)\tilde{S}_{xx}(\omega)}$

to remove background/instrumentation noise, and in the case of random excitation to reduce the random error in the frequency domain estimates. The effect of noise on the FRF estimate has been discussed, and it has been shown that it is better to use the power spectral densities (PSDs) and cross power spectral densities (CPSDs) in the estimation of an FRF rather than ratios of the Fourier transforms of the input and output time histories. This helps to remove noise that is uncorrelated between the input and the output, as the cross-spectral density acts as a filter for this noise. Two estimators for the FRF have been discussed, namely the H_1 and the H_2 estimators. The H_1 estimator is insensitive to noise on the output but has a bias caused by noise on the input, which results in the modulus being underestimated. Conversely the H_2 estimator is insensitive to noise on the input, but has a bias, caused by noise on the output, which results in the modulus being overestimated. It is straightforward to calculate both estimators, but in practice the H_1 estimator is more often used, because in many cases it is easier to control the signal-to-noise ratio (SNR) of the input rather than the output. There are other estimators that can be used, but these are not discussed in this book. The interested reader is referred to Shin and Hammond (2008) for further information. Examples have been provided in which three different types of excitation signal are used, and the issue of '*force drop out*' at the resonance frequency of the structure under test when a shaker is used to excite the structure has been illustrated. This occurs because of shaker–structure interaction.

To show how the estimation of an FRF can lead to useful information about the physical properties of a system, a vibration isolator was considered. Using a shaker excited with random noise, the acceleration (or displacement) transmissibility of a mass suspended by an isolator was estimated from a virtual experiment. The stiffness and damping properties of the isolator were estimated from the measured FRF. Using these results, it was shown how the maximum transmitted force to a rigid foundation could be calculated during the run-up of a machine suspended on the isolator.

References

Avitabile, P. (2017). *Modal Testing: A Practitioner's Guide*. Wiley.

Bendat, J.S. and Piersol, A.G. (1980). *Engineering Applications of Correlation and Spectral Analysis*. Wiley.

Bendat, J.S. and Piersol, A.G. (2000). *Random Data: Analysis and Measurement Procedures*, 3rd Edition. Wiley-Interscience.

Brandt, A. (2011). *Noise and Vibration Analysis: Signal Analysis and Experimental Procedures*. Wiley.

ISO 10846. (2008). *Acoustics and vibration — Laboratory measurement of vibro-acoustic transfer properties of resilient elements — Parts 1-5*.

Shin, K. and Hammond, J.K. (2008). *Fundamentals of Signal Processing for Sound and Vibration Engineers*. Wiley.

8

Multi-Degree-of-Freedom (MDOF) Systems: Dynamic Behaviour

8.1 Introduction

In the previous chapters in this book, the vibrating system used in the analysis and simulations was an SDOF system. This was chosen for simplicity and for relatively easy interpretation of the dynamic behaviour. This system has a single resonance frequency, and for an SDOF system attached to ground, it exhibits stiffness-like behaviour below the resonance frequency, damping-like behaviour at frequencies close to the resonance frequency, and mass-like behaviour above the resonance frequency. Although a study of this type of system is useful in terms of understanding some features of a vibrating system, most real systems have several *resonance* frequencies of interest, and frequencies where the response of the system is very small, which are called *anti-resonance* frequencies. Such systems cannot be modelled using a single mass, a single spring, and a single damper. They require several of these components and are called multi-degree-of-freedom (MDOF) systems.

In this chapter two types of MDOF system are described. The first is an extension of the SDOF system. A number of lumped parameter SDOF systems are connected together in a chain-like manner to form an MDOF system, in which the number of degrees-of-freedom (DOF) is equal to the number of masses, i.e. a system with N DOF has N natural frequencies. The second is a continuous system which has distributed rather than lumped parameters, and has an infinite number of DOF. To illustrate the way in which this type of system can be modelled in the time and frequency domains, a rod and a beam are considered.

8.2 Lumped Parameter MDOF System

A simple example of an MDOF system is shown in Figure 8.1. It consists of a chain of N SDOF systems connected in series. Such a system is called a lumped parameter system. The equation of motion is a matrix equation, which has a similar form to the scalar equation of motion for an SDOF system, and is given by

$$\mathbf{M}\ddot{\mathbf{x}} + \mathbf{C}\dot{\mathbf{x}} + \mathbf{K}\mathbf{x} = \mathbf{f}, \tag{8.1}$$

Virtual Experiments in Mechanical Vibrations: Structural Dynamics and Signal Processing,
First Edition. Michael J. Brennan and Bin Tang.
© 2023 John Wiley & Sons Ltd. Published 2023 by John Wiley & Sons Ltd.
Companion website: www.wiley.com/go/brennan/virtualexperimentsinmechanicalvibrations

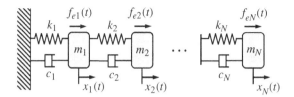

Figure 8.1 Simple lumped-parameter model of an MDOF system.

where the mass, stiffness, and damping matrices are given by

$$\mathbf{M} = \begin{bmatrix} m_1 & 0 & \cdots & 0 \\ 0 & m_2 & \cdots & 0 \\ \vdots & \vdots & \ddots & \vdots \\ 0 & 0 & \cdots & m_N \end{bmatrix}, \quad \mathbf{K} = \begin{bmatrix} k_1 + k_2 & -k_2 & \cdots & 0 \\ -k_2 & k_2 + k_3 & \cdots & 0 \\ \vdots & \vdots & \ddots & \vdots \\ 0 & 0 & \cdots & k_N \end{bmatrix}, \quad \mathbf{C} = \begin{bmatrix} c_1 + c_2 & -c_2 & \cdots & 0 \\ -c_2 & c_2 + c_3 & \cdots & 0 \\ \vdots & \vdots & \ddots & \vdots \\ 0 & 0 & \cdots & c_N \end{bmatrix},$$

and the force and displacement vectors are given by $\mathbf{f} = \{f_{e1}\ f_{e2} \cdots f_{eN}\}^T$ and $\mathbf{x} = \{x_1\ x_2 \cdots x_N\}^T$ in which f_{el} is the force applied to the l-th mass and x_n is the displacement of the n-th mass, respectively. The superscript T denotes the transpose, and as before, the overdots denote differentiation with respect to time.

If harmonic excitation is assumed such that $\mathbf{f} = \bar{\mathbf{f}} e^{j\omega t}$ and $\mathbf{x} = \bar{\mathbf{x}} e^{j\omega t}$, where $\bar{\mathbf{f}} = \{\overline{F}_1\ \overline{F}_2 \cdots \overline{F}_N\}^T$, and $\bar{\mathbf{x}} = \{\overline{X}_1\ \overline{X}_2 \cdots \overline{X}_N\}^T$, in which $\overline{F}_l$ and $\overline{X}_n$ are the complex amplitudes of the force applied to the l-th mass and the complex displacement amplitudes of the n-th mass, respectively (see Chapter 2 to see the meaning of complex amplitude), then Eq. (8.1) becomes

$$[\mathbf{K} - \omega^2 \mathbf{M} + j\omega \mathbf{C}]\bar{\mathbf{x}} = \bar{\mathbf{f}}. \tag{8.2}$$

The vector of complex displacement amplitude responses is then given by

$$\bar{\mathbf{x}} = \mathbf{H}\bar{\mathbf{f}}, \tag{8.3}$$

where $\mathbf{H} = [\mathbf{K} - \omega^2 \mathbf{M} + j\omega \mathbf{C}]^{-1}$ is an $N \times N$ receptance matrix in which the superscript -1 denotes the matrix inverse. The diagonal terms of the receptance matrix are called *point receptances*, and the off-diagonal terms are called the *transfer receptances* (Bishop and Johnson, 1960; Mead, 1999). Both types of FRF contain resonance and anti-resonance frequencies. The resonances occur at the same frequencies in both types of FRF because natural frequencies of a structure are a global phenomenon, but the anti-resonances occur at different frequencies, according to the position measured in the structure. The point receptances have particular properties, which are illustrated along with the transfer receptances using a 3DOF system in Section 8.2.1. Note that the transfer receptance H_{nl}, where the subscripts nl denote the row and column in the matrix $\mathbf{H}$ (displacement of the n-th mass due to the force applied to the l-th mass), is equal to H_{ln} because of reciprocity (Fahy, 2003). This means that that receptance matrix $\mathbf{H}$ is symmetric.

8.2.1 Example – 3DOF System

To illustrate some FRF features of an MDOF system, a 3DOF system is considered. The system is shown at the top of Figure 8.2. The matrix equation of motion that describes the system is Eq. (8.1) with the mass, stiffness, and damping matrices given by

$$\mathbf{M} = \begin{bmatrix} m_1 & 0 & 0 \\ 0 & m_2 & 0 \\ 0 & 0 & m_3 \end{bmatrix}, \quad \mathbf{K} = \begin{bmatrix} k_1 + k_2 & -k_2 & 0 \\ -k_2 & k_2 + k_3 & -k_3 \\ 0 & -k_3 & k_3 \end{bmatrix}, \quad \mathbf{C} = \begin{bmatrix} c_1 + c_2 & -c_2 & 0 \\ -c_2 & c_2 + c_3 & -c_3 \\ 0 & -c_3 & c_3 \end{bmatrix},$$

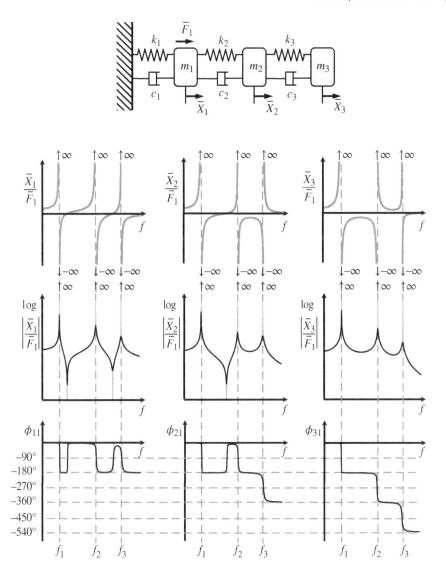

Figure 8.2 Point and transfer receptances of a 3DOF lumped parameter system.

in which $m_1 = m_2 = m_3 = m$, $k_1 = k_2 = k_3 = k$, $c_1 = c_2 = c_3 = c$, and the force and displacement vectors are given by $\bar{\mathbf{f}} = \left\{ \bar{F}_1 \ 0 \ 0 \right\}^T$ and $\bar{\mathbf{x}} = \left\{ \bar{X}_1 \ \bar{X}_2 \ \bar{X}_3 \right\}^T$, respectively. Although this MDOF system appears to be quite specific, its FRFs exhibit some general features that are found in many systems. To illustrate these features, a harmonic force with amplitude $\bar{F}_1$ is applied to mass m_1. The point receptance is then given by $\bar{X}_1 / \bar{F}_1 = H_{11}$ and the two transfer receptances are given by $\bar{X}_2 / \bar{F}_1 = H_{21}$ and $\bar{X}_3 / \bar{F}_1 = H_{31}$. These are plotted in Figure 8.2. In the first row the FRFs are plotted with damping set to zero. The reason for this is that although it gives non-physical results at the resonance frequencies (i.e. infinite displacements), it is helpful in the understanding of why anti-resonances occur. In the remaining two rows of Figure 8.2, the modulus and phase are plotted for the respective FRFs for both the cases when damping is set to zero, and when there is a small amount of damping in the system.

Concerning the FRFs for the undamped case shown in the first row of Figure 8.2, it is important to note that because the masses either move in-phase or out-of-phase with the applied force, it is not necessary to plot the modulus and phase separately. Thus, when an FRF is positive, the displacement of the mass m_n is in-phase with the applied force $\overline{F}_1$, and when an FRF is negative, the phase is $-180°$ (anti-phase). Further, it can be noted that there are three resonance frequencies, f_1, f_2, and f_3 as the system has 3DOF. The way in which these are calculated is discussed in Section 8.2.2. It can be seen that at frequencies below the first resonance frequency, all the masses are vibrating in-phase with each other. At the first resonance frequency, the displacements of all three masses remain in-phase with each other and become infinite, because there is no damping, and then they all undergo a phase change and appear from minus infinity above the resonance frequency.

In between the first and second resonance frequencies, the phase trajectories of m_1 and m_2 differ from the phase trajectory of m_3. Following the first resonance frequency, the displacement of m_1 is initially in anti-phase with the applied force ($\overline{X}_1/\overline{F}_1$ is negative), but then it changes sign at a particular frequency to become in-phase with the force ($\overline{X}_1/\overline{F}_1$ is positive). This means that it passes through zero. The frequency at which this occurs is called an anti-resonance frequency and is an important phenomenon in structural dynamics. In an undamped system, the displacement of m_1 is zero at this frequency, irrespective of magnitude of the applied force. The displacement of m_2 follows a similar trajectory, but the anti-resonance occurs at a different frequency. In general, anti-resonances in FRFs at different measurement positions on a structure occur at different frequencies, unlike a resonance frequency, which occurs at the same frequency in all FRFs, independent of the position. For this reason, resonance frequencies are called a global feature of the structure, whereas anti-resonances are called a local feature. As frequency is increased from just above the first resonance frequency to the second resonance frequency, the displacement of m_3 does not change phase like that of m_1 and m_2. This means that the displacement of m_3 does not pass through zero, and hence there is no anti-resonance. It also means that at the second resonance frequency, the displacement of m_3 is in anti-phase with m_1 and m_2. At the second resonance frequency the displacements of all the masses undergo a phase shift of $180°$ with respect to the applied force as they pass from positive (negative) infinity to negative (positive) infinity.

As frequency increases between the second and third resonance frequencies, a similar pattern to that between the first and second resonance frequencies occurs for m_1 resulting in another anti-resonance frequency for this mass. However, there are no further anti-resonance frequencies for masses m_2 and m_3.

The more conventional way of plotting FRFs is in terms of modulus and phase, and these are shown in the centre and bottom rows of Figure 8.2, respectively. Note that the modulus is plotted on a logarithmic scale, so that the resonances and anti-resonance frequencies can be clearly seen. In fact, for a measurement to determine the FRF which contains both resonances and anti-resonances, the measurement system must be capable of capturing data over a very large dynamic range, especially when the damping is light. This can be a challenging situation. In each graph of the second two rows of Figure 8.2, two plots are shown. One is the FRF for an undamped system, which can be compared directly with the first row in the figure and the other is for a system with the same mass and stiffness, but with a small amount of damping added (black line). It should be further noted that unwrapped phase is plotted, as this gives the actual phase rather than the phase restricted between $-180°$ and $+180°$. The moduli have the classic shapes found in measured data, clearly showing the dynamic behaviour, influenced by the resonance and anti-resonance frequencies. The structure of the point receptance, in terms of alternate resonance and anti-resonance frequencies, is a feature of the point receptance of *all* vibrating systems. However, this is not a feature of a transfer receptance. There may or may not be anti-resonance frequencies depending upon the specific properties of the system. In the example shown in Figure 8.2, there is a single anti-resonance in $\overline{X}_2/\overline{F}_1$, but none in $\overline{X}_3/\overline{F}_1$. The effect of damping on the FRFs can be clearly seen in Figure 8.2. In the modulus plots, it can be seen that damping has the effect of reducing the response at the resonance frequencies, as demonstrated for an SDOF system in Chapter 2. It also has the effect of increasing the response (from zero) at the anti-resonance frequencies. In the phase plots it has the effect of removing the sharp corners evident in the undamped case. Note that in the phase plots, the resonance frequencies occur at phase angles of $-90°$, $-270°$, and $-450°$ which are at an angle of $-90°$ less integer multiples of $180°$, and the anti-resonance frequencies occur at $-90°$. These features are also common to any linear, lightly damped vibrating system.

MATLAB Example 8.1

In this example, the FRFs of a 3DOF system are plotted showing the behaviour of point and transfer receptances.

```
clear all

%% Parameters
m1=1;m2=1;m3=1;                       % [kg]      % masses
k1=1e4;k2=1e4;k3=1e4;k4=0*5e3;        % [N/m]     % stiffnesses
M=[m1 0 0; 0 m2 0; 0 0 m3];                      % mass matrix
K=[k1+k2 -k2 0; -k2 k2+k3 -k3; 0 -k3 k3+k4];     % stiffness matrix
C=1e-4*K;                                         % damping matrix

%% Forced vibration
n=0;
for f=0:0.001:50;                     % frequency vector
 w=2*pi*f;n=n+1;
 An=inv(K-w.^2*M+j*w*C);              % calculate matrix of FRFs
 An11(n)=An(1,1);                     % point receptance
 An21(n)=An(2,1);                     % transfer receptance
 An31(n)=An(3,1);                     % transfer receptance
end

f=0:0.001:50;
figure                                % FRF plots
plot(f,20*log10(abs(An11)))
axis square; grid,axis([1,40,-130,-30])
xlabel('frequency (Hz)');
ylabel('amplitude (dB ref 1 m/N)');
figure
plot(f,180/pi*unwrap(angle(An11)))
axis square; grid; axis([1,40,-600,0])
xlabel('frequency (Hz)');
ylabel('phase (degrees)');
figure
plot(f,20*log10(abs(An21)))
axis square; grid; axis([1,40,-130,-30])
xlabel('frequency (Hz)');
ylabel('amplitude (dB ref 1 m/N)');
figure
plot(f,180/pi*unwrap(angle(An21)))
axis square; grid; axis([1,40,-600,0])
```

(Continued)

MATLAB Example 8.1   (Continued)

```
xlabel('frequency (Hz)');
ylabel('phase (degrees)');
figure
plot(f,20*log10(abs(An31)))
axis square; grid; axis([1,40,-130,-30])
xlabel('frequency (Hz)');
ylabel('amplitude (dB ref 1 m/N)');
figure
plot(f,180/pi*unwrap(angle(An31)))
axis square; grid; axis([1,40,-600,0])
xlabel('frequency (Hz)');
ylabel('phase (degrees)');
```

Results

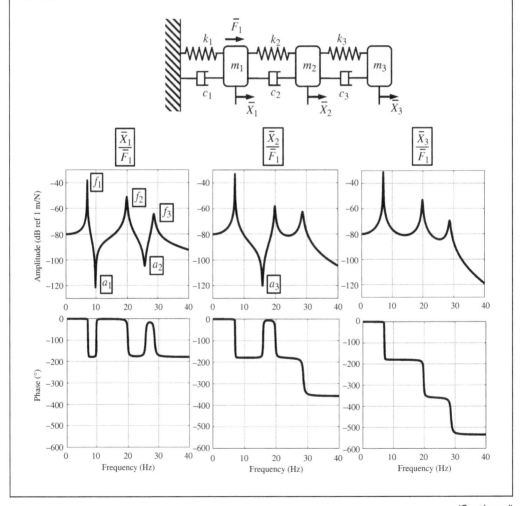

(Continued)

MATLAB Example 8.1   (Continued)

	Resonance frequencies (Hz)			Anti-resonance frequencies (Hz)	
f_1	f_2	f_3	a_1	a_2	a_3
7.1	19.9	28.7	9.8	25.8	15.9

Comments:

1. This example gives similar results to those shown in Figure 8.2.
2. Exercises for the reader are to:
   (a) calculate the point and transfer receptances when the excitation force is applied to masses m_2 and m_3,
   (b) explore the effects of changing the system parameters to see how they affect the resonance and anti-resonance frequencies,
   (c) set the stiffness k_1 and damping c_1 to zero, and to interpret the resulting FRFs, especially at very low frequency,
   (d) plot the real and imaginary parts of the point receptance as a function of frequency, and compare these to that for an SDOF system. Why is the imaginary part always less than or equal to zero?

8.2.2   Free Vibration

To discover the system properties that govern the dynamic behaviour of the MDOF system, free vibration analysis is helpful. This is carried out by assuming an undamped system, as this gives the underlying properties that govern the resonance (or natural) frequencies and the anti-resonance frequencies. In this case, Eq. (8.1) simplifies to

$$\mathbf{M}\ddot{\mathbf{x}} + \mathbf{K}\mathbf{x} = \mathbf{0}. \tag{8.4}$$

Recall in Chapter 2 that the equation of motion for the free vibration of an undamped SDOF system is given by $m\ddot{x} + kx = 0$, which can be written as $\ddot{x} + (k/m)x = 0$, and the square of the angular natural frequency is given by $\omega_n^2 = k/m$. Now, for an N DOF system there are N natural frequencies. They can be determined from Eq. (8.4), which can be written as

$$\ddot{\mathbf{x}} + \mathbf{M}^{-1}\mathbf{K}\mathbf{x} = \mathbf{0}. \tag{8.5}$$

Now, the squares of the N angular natural frequencies are given by the *eigenvalues* of the matrix $\mathbf{M}^{-1}\mathbf{K}$. These are easily calculated in MATLAB using the `eig` function. Accompanying each eigenvalue is an eigenvector, which describes the relative motion of each of the masses as they undergo vibration at a natural frequency. They are also easily calculated in MATLAB using the `eig` function. These are called the *Mode shapes* of the system. Note the words 'shapes', which means that the vector does not have any relationship to the actual amplitudes, rather it gives the relative amplitudes and phases of the masses, so they can be scaled in different ways. In MATLAB Example 8.2, each mode shape is scaled so that the maximum amplitude is set to unity.

MATLAB Example 8.2

In this example, the natural frequencies and mode shapes of the 3DOF system in MATLAB Example 8.1 are calculated.

```
clear all

%% Parameters
m1=1;m2=1;m3=1;                        % [kg]        % See MATLAB example 8.1
k1=1e4;k2=1e4;k3=1e4;k4=0*5e3;         % [N/m]
M=[m1 0 0; 0 m2 0; 0 0 m3];
K=[k1+k2 -k2 0; -k2 k2+k3 -k3; 0 -k3 k3+k4];

%% Undamped natural frequencies
[V W]= eig (inv(M)*K);                 % calculation of eigenvalues
                                       and eigenvectors
R=sqrt(W)/(2*pi);                      % calculation of nat. freqs.
V1=V(:,1)/max(abs(V(:,1)));            % calculation of the normalized
V2=V(:,2)/max(abs(V(:,2)));            mode shapes
V3=V(:,3)/max(abs(V(:,3)));
```

Results

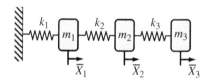

Undamped natural frequencies (Hz)		
f_1	f_2	f_3
7.1	19.9	28.7

Mode shapes		
$\Phi_1 = \begin{Bmatrix} 0.45 \\ 0.80 \\ 1.00 \end{Bmatrix}$	$\Phi_2 = \begin{Bmatrix} 1.00 \\ 0.45 \\ -0.80 \end{Bmatrix}$	$\Phi_3 = \begin{Bmatrix} -0.80 \\ 1.00 \\ -0.45 \end{Bmatrix}$

Comment:

1. If the damping in the system is small, the undamped natural frequencies of the system are almost the same as the resonance frequencies.
2. An exercise for the reader is to determine the natural frequencies and mode shapes for the 3DOF system with different masses and stiffnesses.

The mode shapes for the system determined in MATLAB Example 8.2 are drawn in Figure 8.3. It can be seen that the phases between the masses are different for each natural frequency. For

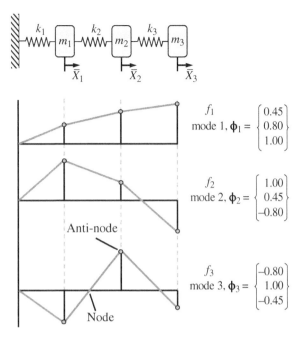

Figure 8.3 Mode shapes of the 3DOF system where the masses are all equal to 1 kg, and the stiffnesses are all equal to 1×10^4 N/m.

example, all the masses move in-phase with each other at the first natural frequency, but at the second and third natural frequencies, two masses move in-phase with each other and the remaining mass moves in anti-phase. The relative displacement amplitudes of each mass are different for each natural frequency, and the motion of each mass at a natural frequency is called an anti-node. If there is a phase difference between the masses at a natural frequency, then there is at least one point on the structure that is motionless at this frequency. This point is called a node. It is evident that there are no nodes for the first natural frequency, a single node for the second natural frequency, and two nodes for the third natural frequency.

8.2.3 Resonance and Anti-resonance Frequencies

As shown in MATLAB Example 8.2, for a lightly damped system, the resonance frequencies can be determined approximately by simply calculating the undamped natural frequencies of the system. The anti-resonance frequencies can be estimated in a similar way. For the chain-like system shown in Figure 8.1, the undamped natural frequencies are calculated for sub-systems of the original system. To determine which sub-system to use, the displacement of the mass, which has the anti-resonance in the FRF is set to zero, i.e. attached to ground. An example of this is shown in Figure 8.4, and in MATLAB Example 8.3. The system in Figure 8.2 is considered. For the point receptance, in which the force is applied to m_1, this position is attached to ground, and the undamped natural frequencies of the system to the right of the excitation point are calculated using the method shown in MATLAB Example 8.2. The undamped natural frequencies of the subsystem are the anti-resonance frequencies of the point receptance of the original undamped system. These are approximately the same as the anti-resonance frequencies of the original damped system, if the damping is small. To determine the anti-resonance frequency in the transfer receptance X_2/F_1, mass m_2 is grounded and the natural frequency of the sub-system to the right of m_2 is calculated. This is illustrated in Figure 8.4 and is calculated in MATLAB Example 8.3.

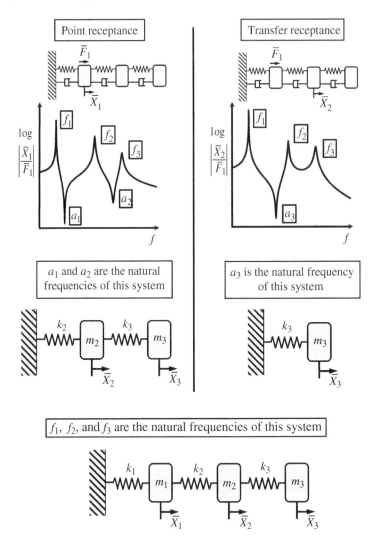

Figure 8.4 Relationship between resonance and anti-resonance frequencies and the free vibration of undamped sub-systems of a 3DOF system.

MATLAB Example 8.3

In this example, the resonance and anti-resonance frequencies of the 3DOF system in MATLAB Example 8.1 are calculated.

```
clear all

%% Parameters
m1=1;m2=1;m3=1;                        % [kg]      % See MATLAB example 8.1
k1=1e4;k2=1e4;k3=1e4;k4=0*5e3;         % [N/m]
M=[m1 0 0; 0 m2 0; 0 0 m3];
K=[k1+k2 -k2 0; -k2 k2+k3 -k3; 0 -k3 k3+k4];
```

(Continued)

MATLAB Example 8.3   (Continued)

```
%% Undamped natural frequencies
[V W]= eig (inv(M)*K);                    % calculation of eigenvalues
f123=sqrt(W)/(2*pi)                       % calculation of nat. freqs.

%% Anti-resonance frequencies X1/Fe1
M=[m2 0; 0 m3];                           % mass matrix
K=[k2+k3 -k3; -k3 k3];                    % stiffness matrix
[V1 W1]= eig (inv(M)*K);                  % calculation of eigenvalues
a12=sqrt(W1)/(2*pi)                       % calculation of anti-resonance frequencies

%% Anti-resonance frequencies X2/Fe1
W2=(k3)/m1;                               % calculation of square of nat. freq.
a3=sqrt(W2)/(2*pi)                        % calculation of anti-resonance frequency
```

Results

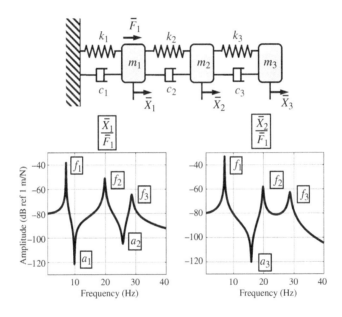

Resonance frequencies (Hz)			Anti-resonance frequencies (Hz)		
f_1	f_2	f_3	a_1	a_2	a_3
7.1	19.9	28.7	9.8	25.8	15.9

Comments:

1. The system is split into the sub-systems shown in Figure 8.4 to calculate the anti-resonance frequencies. The resonance frequencies are also estimated from the undamped natural frequencies of the system.
2. An exercise for the reader is to calculate the anti-resonance frequencies of the point and transfer receptances of the system when the excitation force is applied to masses m_1 or m_2. Try to explain the physics that governs this behaviour.

A particular situation occurs for an FRF of an MDOF system if the measurement position and/or the excitation position coincide with a nodal point. In a lumped parameter system, such as that considered in this chapter, this coincidence occurs at the position of a mass. This situation does not occur for the system shown in Figure 8.2, so an alternative system is considered to illustrate the effect. It is shown at the top of Figure 8.5, and is the same as the previous example but with a spring connected between mass m_3 and the ground. The mode shapes for this system are also shown in Figure 8.5. It should be noted that the system is now symmetric, and the central mass is coincident with a node of the second natural frequency. The behaviour of this system in terms of the modulus of all the FRFs is shown at the bottom of Figure 8.5. The elements of the receptance matrix shown at the top right of Figure 8.5 are plotted. There are several points to be noted from this figure:

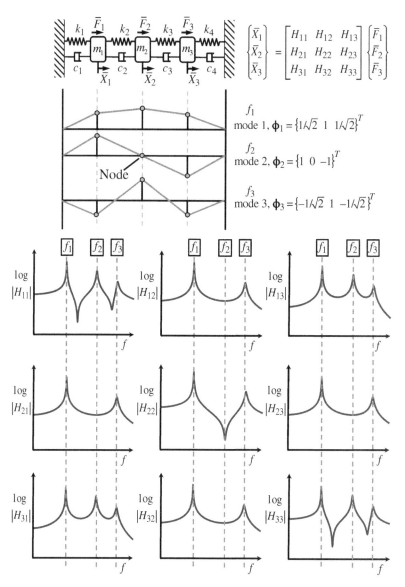

Figure 8.5 Point and transfer receptances of a symmetric 3DOF lumped parameter system.

1. Because the system is symmetric $H_{11} = H_{33}$ and $H_{12} = H_{32}$.
2. Because of reciprocity $H_{12} = H_{21}, H_{13} = H_{31}$, and $H_{23} = H_{32}$.
3. There is a nodal point at the position of the central mass for the second natural frequency. Thus, if an FRF involves either a force being applied to this mass or the displacement of this mass being measured, then the resonance corresponding to the second natural frequency is not evident.
4. If the force and displacement are measured on the central mass, i.e. H_{22}, then an anti-resonance is evident instead of a resonance at the second natural frequency.

8.2.4   Modal Decomposition

The matrix equation of motion for forced vibration of an MDOF system is given by Eq. (8.1). Although this formulation can be used to determine the FRFs of the system as shown above, Eq. (8.1) can be manipulated to obtain greater insight. Furthermore, transformation to the time domain to obtain the IRFs can be achieved analytically for each mode of vibration. First, note that Eq. (8.1) is a vector–matrix formulation of *N-coupled* second-order differential equations. The equations are coupled because the stiffness and damping matrices have off-diagonal terms. It is possible to decouple these equations so that there are P separate equations, one for each mode of vibration. Note that in the MDOF case considered here, the number of modes included in the model is the same as the number of DOF, i.e. $N = P$. However, this is not always the case, for example in distributed rather than lumped parameter systems, as discussed in Section 8.3. Decoupling the equations of motion is possible because undamped vibration modes are *orthogonal*, i.e. they vibrate independently of each other. In general, for a damped system this is not possible, but a similar approach can be applied for a system where the damping matrix is a linear combination of the mass and stiffness matrices, i.e.

$$\mathbf{C} = \alpha\mathbf{M} + \beta\mathbf{K}, \tag{8.6}$$

where α and β are constants. This type of damping is called Rayleigh or proportional damping (Ewins, 2000). It is evident that the damping matrix for the system shown in Figure 8.1 only has the form for Rayleigh damping if $\alpha = 0$, and this is assumed henceforth.

The process of transforming a set of coupled equations to a set of uncoupled equations is a standard linear algebra problem and is extremely useful in structural dynamics. The procedure can be found in many elementary texts, for example (Tse et al., 1978). Eq. (8.1) is transformed using the matrix of eigenvectors (or mode shapes as discussed in Section 8.2.2), given by

$$\mathbf{\Phi} = \begin{bmatrix} \mathbf{\phi}_1 & \mathbf{\phi}_2 & \cdots & \mathbf{\phi}_P \end{bmatrix}. \tag{8.7}$$

The relationship between the physical displacement vector $\mathbf{x}$ and the modal displacements $\mathbf{q}$ is given by

$$\mathbf{x} = \mathbf{\Phi}\mathbf{q}. \tag{8.8}$$

Substituting for $\mathbf{x}$ in Eq. (8.1) results in

$$\mathbf{M}\mathbf{\Phi}\ddot{\mathbf{q}} + \mathbf{C}\mathbf{\Phi}\dot{\mathbf{q}} + \mathbf{K}\mathbf{\Phi}\mathbf{q} = \mathbf{f}. \tag{8.9}$$

Pre-multiplying Eq. (8.9) with the transpose of the matrix of mode shapes, gives

$$\mathbf{\Phi}^T\mathbf{M}\mathbf{\Phi}\ddot{\mathbf{q}} + \mathbf{\Phi}^T\mathbf{C}\mathbf{\Phi}\dot{\mathbf{q}} + \mathbf{\Phi}^T\mathbf{K}\mathbf{\Phi}\mathbf{q} = \mathbf{\Phi}^T\mathbf{f}, \tag{8.10}$$

which can be written as,

$$\tilde{\mathbf{M}}\ddot{\mathbf{q}} + \tilde{\mathbf{C}}\dot{\mathbf{q}} + \tilde{\mathbf{K}}\mathbf{q} = \mathbf{g}, \tag{8.11}$$

where the modal mass, stiffness, and damping matrices are given by

$$\tilde{\mathbf{M}} = \boldsymbol{\Phi}^T \mathbf{M} \boldsymbol{\Phi} = \begin{bmatrix} \tilde{m}_1 & 0 & \cdots & 0 \\ 0 & \tilde{m}_2 & \cdots & 0 \\ \vdots & \vdots & \ddots & \vdots \\ 0 & 0 & \cdots & \tilde{m}_P \end{bmatrix}, \quad \tilde{\mathbf{K}} = \boldsymbol{\Phi}^T \mathbf{K} \boldsymbol{\Phi} = \begin{bmatrix} \tilde{k}_1 & 0 & \cdots & 0 \\ 0 & \tilde{k}_3 & \cdots & 0 \\ \vdots & \vdots & \ddots & \vdots \\ 0 & 0 & \cdots & \tilde{k}_P \end{bmatrix}, \quad \tilde{\mathbf{C}} = \boldsymbol{\Phi}^T \mathbf{C} \boldsymbol{\Phi} = \begin{bmatrix} \tilde{c}_1 & 0 & \cdots & 0 \\ 0 & \tilde{c}_2 & \cdots & 0 \\ \vdots & \vdots & \ddots & \vdots \\ 0 & 0 & \cdots & \tilde{c}_P \end{bmatrix},$$

and the modal force and modal displacement vectors are given by $\mathbf{g} = \boldsymbol{\Phi}^T \mathbf{f} = \{g_1 \; g_2 \; \cdots \; g_P\}^T$ and $\mathbf{q} = \boldsymbol{\Phi}^{-1} \mathbf{x} = \{q_1 \; q_2 \; \cdots \; q_P\}^T$; $\tilde{m}_p, \tilde{k}_p$, and $\tilde{c}_p$ are the modal mass, modal stiffness, and modal damping of the p-th mode, respectively. The damping ratio of the p-th mode is given by $\zeta_p = \tilde{c}_p/(2\sqrt{\tilde{m}_p \tilde{k}_p})$. The fundamental difference between the description of an MDOF system, in terms of its physical coordinates, and the modal description of the system is illustrated in Figure 8.6.

If harmonic excitation is assumed such that $\mathbf{g} = \bar{\mathbf{g}} e^{j\omega t}$ where $\bar{\mathbf{g}} = \{\bar{G}_1 \; \bar{G}_2 \; \cdots \; \bar{G}_P\}^T$, in which $\bar{G}_p = \boldsymbol{\phi}_p^T \bar{\mathbf{f}}$ is the amplitude of the p-th modal force, and $\bar{\mathbf{q}} = \{\bar{Q}_1 \; \bar{Q}_2 \; \cdots \; \bar{Q}_P\}^T$, in which $\bar{Q}_p$ is the modal displacement amplitude of the p-th mode, Eq. (8.11) becomes

$$\bar{\mathbf{q}} = \tilde{\mathbf{H}} \bar{\mathbf{g}}, \tag{8.12}$$

where $\tilde{\mathbf{H}} = [\tilde{\mathbf{K}} - \omega^2 \tilde{\mathbf{M}} + j\omega \tilde{\mathbf{C}}]^{-1} = \mathrm{diag}(\tilde{H}_1, \tilde{H}_2, \dots, \tilde{H}_P)$, in which $\tilde{H}_p = [1/(\tilde{k}_p - \omega^2 \tilde{m}_p + j\omega \tilde{c}_p)]$. Note that $\tilde{\mathbf{H}}$ is a diagonal matrix unlike $\mathbf{H}$ in Eq. (8.3). Now, the relationship between the vector of physical complex displacement amplitudes and the complex modal displacement amplitudes is given by

$$\bar{\mathbf{x}} = \boldsymbol{\Phi} \bar{\mathbf{q}}, \tag{8.13}$$

and the relationship between the complex modal force amplitudes and the physical complex force amplitudes applied to each mass is given by

$$\bar{\mathbf{g}} = \boldsymbol{\Phi}^T \bar{\mathbf{f}}. \tag{8.14}$$

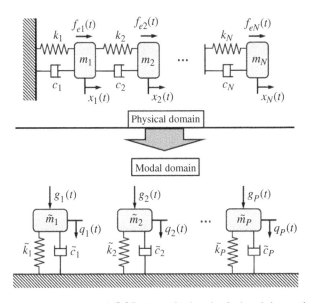

Figure 8.6 Lumped parameter MDOF system in the physical and the modal domain.

Equations (8.12–8.14) are combined to give

$$\bar{\mathbf{x}} = \boldsymbol{\Phi}\tilde{\mathbf{H}}\boldsymbol{\Phi}^T\bar{\mathbf{f}}. \tag{8.15}$$

As $\tilde{\mathbf{H}}$ is a diagonal matrix the receptance corresponding to the displacement of the n-th mass due to a force applied to the l-th mass, can be written as a summation of the modal responses, given by

$$\frac{\overline{X}_n}{\overline{F}_l} = \sum_{p=1}^{P} \frac{\boldsymbol{\phi}_p(l)\boldsymbol{\phi}_p(n)}{\tilde{k}_p - \omega^2\tilde{m}_p + j\omega\tilde{c}_p}, \tag{8.16}$$

where $\boldsymbol{\phi}_p(l)$ is the mode shape of the p-th mode at the l-th mass where the force is applied, and $\boldsymbol{\phi}_p(n)$ is the mode shape of the p-th mode at the n-th mass where the displacement is measured.

To illustrate the modal decomposition of a vibrating system, the 3DOF system shown in Figure 8.5 is considered. Although nine FRFs are shown in that figure, there are only four distinct FRFs due to reciprocity and symmetry of the system. These are shown in Figure 8.7 together with the FRFs in terms of the individual modal responses. Note, that in all the plots, the response at each resonance frequency, marked as f_1, f_2, and f_3, is dominated by a single mode of vibration. This is because the natural frequencies are well-spaced, and the damping is relatively light. Note also that in the FRFs that involve either the force being applied to the central mass, or the response of the central mass is zero, i.e. $|\overline{X}_2/\overline{F}_2|$ and $|\overline{X}_2/\overline{F}_1|$, then the second resonance frequency is not evident. In fact, the second mode response is absent from these plots. This is because the numerator is zero for the second mode, as the corresponding mode shape $\boldsymbol{\phi}_2$ has a value of zero at the position of the central mass (i.e. it is a node), as shown in Figure 8.5. Thus, at the central mass position, this mode of vibration is either not excited or is not observed. To analyse the plots in Figure 8.7 in more detail, Eq. (8.16) is written as

$$\frac{\overline{X}_n}{\overline{F}_l} = \sum_{p=1}^{P} \tilde{H}_p\boldsymbol{\phi}_p(l)\boldsymbol{\phi}_p(n), \tag{8.17}$$

which, for the point receptances simplifies to $\overline{X}_n/\overline{F}_n = \sum_{p=1}^{P} \tilde{H}_p\boldsymbol{\phi}_p^2(n)$. Further, for the purpose of analysis it is assumed that the damping is set to zero. Each modal response is essentially the response of an SDOF system weighted by $\boldsymbol{\phi}_p^2(n)$. Note that $\boldsymbol{\phi}_p^2(n)$ is positive, so for a point receptance, the sign of each modal response is simply governed by the sign of $\tilde{H}_p$. Recall from Chapter 2, that below the resonance frequency, the receptance of an SDOF system is positive and is controlled by the stiffness, and above the resonance frequency it is negative and is controlled by the mass. Examining the moduli of the point receptances $|\overline{X}_1/\overline{F}_1|$ and $|\overline{X}_2/\overline{F}_2|$ in Figure 8.7, it can be seen that the stiffness- and mass-dominated parts of the modal components of the FRF are labelled. For $|\overline{X}_1/\overline{F}_1|$, well below the first resonance frequency, all the modal responses are positive and in-phase with each other, so they simply sum to give the overall FRF. Between the first and second resonance frequencies the anti-resonance marked a_1 occurs. This forms because the FRF of the first mode changes sign (for the undamped case), as it has mass-like behaviour. This means that it is in anti-phase with the combined stiffness characteristics of the second and third modes, so the sum of these modal responses is zero at the anti-resonance frequency (for the undamped case). There is a second anti-resonance frequency between the second and third modes of vibration, marked as a_2. As this occurs above the second resonance frequency, the FRF of the second mode has a mass-like characteristic. The anti-resonance is thus formed when the sum of the combined mass characteristics of the first and second modes is equal to the stiffness characteristic of the third mode. As they

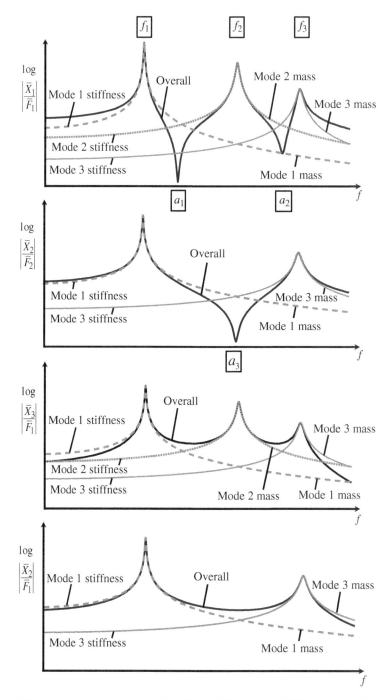

Figure 8.7 Modal decomposition of the different FRFs for the MDOF system in Figure 8.5.

have opposite sign, this is equal to zero in the undamped case, but as seen in Figure 8.7, this is not the case, which is due to damping in the system as discussed previously. Concerning the point receptance $|\overline{X}_2/\overline{F}_2|$ shown in Figure 8.7, it can be seen that it exhibits similar behaviour to $|\overline{X}_1/\overline{F}_1|$, but it appears to be a 2DOF system as the second mode does not feature in the FRF for the reasons given above.

To analyse the behaviour of the transfer receptances $|\overline{X}_3/\overline{F}_1|$ and $|\overline{X}_2/\overline{F}_1|$, shown in the lower part of Figure 8.7, it is helpful to study the mode shape matrix given by

$$\Phi = \begin{matrix} \boldsymbol{\phi}_1 \quad\ \boldsymbol{\phi}_2 \quad\ \boldsymbol{\phi}_3 \\ \begin{bmatrix} \dfrac{1}{2} & -\dfrac{1}{\sqrt{2}} & -\dfrac{1}{2} \\[2mm] \dfrac{1}{\sqrt{2}} & 0 & \dfrac{1}{\sqrt{2}} \\[2mm] \dfrac{1}{2} & \dfrac{1}{\sqrt{2}} & -\dfrac{1}{2} \end{bmatrix} \end{matrix}. \tag{8.18}$$

Note that when the vectors of mode shapes are assembled into a matrix rather than when studying them separately as in Figure 8.5, it is preferable to normalise them so that $\Phi^T M \Phi = I$, where I is the identity matrix. In this case, the mode shapes are generally called the mass-normalised mode shapes. Note that the modal mass is then unity for each mode. Referring to Eq. (8.17), it can be seen that the sign of the FRF for each mode is now also governed by the sign of $\phi_p(l)\phi_p(n)$, which is dependent upon the position and particular mode as can be seen in Eq. (8.16). In the point receptances discussed above, it was noted that the FRFs for each mode were all in-phase at frequencies well below the first mode of vibration, and so summed to give the overall FRF. In the transfer receptances shown in Figure 8.7, this is not the case, which can be seen by examining the form of the mode shapes given in Eq. (8.18). From Figure 8.7, it can be seen that the response of the first mode is greater than the overall response. For $|\overline{X}_3/\overline{F}_1|$, this occurs because the modal response of the first mode in this frequency region (for the undamped case) is positive, the modal response of the second mode is negative, and the modal response of the third mode is positive. This is easily verified by examining Eqs. (8.17) and (8.18). The behaviour for $|\overline{X}_2/\overline{F}_1|$ is similar to that for $|\overline{X}_3/\overline{F}_1|$, but the second mode does not feature in the FRF because it is not observed due to a nodal point for the second mode occurring at the position of the central mass. Note that anti-resonance frequencies do not occur in the transfer receptances, because the characteristics of each modal response do not have the appropriate sign and amplitude required to give a zero in the FRF for an undamped system.

MATLAB Example 8.4

In this example, the FRFs of the 3DOF system in MATLAB Example 8.1 are decomposed into modes.

```
clear all

%% Parameters
% (see MATLAB Example 8.1)

%% Modal matrices
[V W]=eig(inv(M)*K);                   % calculation of eigenvectors
Mm=V'*M*V;Km=V'*K*V;Cm=V'*C*V;         % modal mass, stiff., and damping
zeta=Cm/(2*sqrt(Km*Mm))                % modal damping ratios

%% Forced vibration
n=0;
for f=0:0.001:50                       % frequency vector
  w=2*pi*f;n=n+1;
  An=inv(K-w.^2*M+j*w*C);              % calculate matrix of FRFs
  An11(n)=An(1,1);                     % point receptance
  An21(n)=An(2,1);                     % transfer receptance
  An31(n)=An(3,1);                     % transfer receptance
```

(Continued)

MATLAB Example 8.4   (Continued)

```
% modal responses
Hm=inv(Km-w^2*Mm+j*w*Cm);
% mode 1
F=[1 0 0]';Q=[1 0 0; 0 0 0;0 0 0];
Xa=V*Q*Hm*V'*F;                        % response of first mode in FRFs
% mode 2
Q=[0 0 0; 0 1 0;0 0 0];
Xb=V*Q*Hm*V'*F;                        % response of second mode in FRFs
% mode 3
Q=[0 0 0; 0 0 0;0 0 1];
Xc=V*Q*Hm*V'*F;                        % response of third mode in FRFs
% X1/F1
X11a(n)=Xa(1);X11b(n)=Xb(1);X11c(n)=Xc(1);  % individual modal responses
% X2/F1
X21a(n)=Xa(2);X21b(n)=Xb(2);X21c(n)=Xc(2);  % individual modal responses
% X3/F1
X31a(n)=Xa(3);X31b(n)=Xb(3);X31c(n)=Xc(3);  % individual modal responses
end

f=0:0.001:50;

%% Plot the results
figure                                 % modulus
plot(f,20*log10(abs(An11)));hold on
plot(f,20*log10(abs(X11a)),'-');hold on
plot(f,20*log10(abs(X11b)),':');hold on
plot(f,20*log10(abs(X11c)))
axis square; grid; axis([0,40,-130,-30])
xlabel('frequency (Hz)');
ylabel('amplitude (dB ref 1 m/N)');


figure                                 % phase
plot(f,180/pi*(angle(X11a)),'-'); hold on
plot(f,180/pi*(angle(X11b)),':'); hold on
plot(f,180/pi*(angle(X11c)))
axis square; grid; axis([0,40,-200,200])
xlabel('frequency (Hz)');
ylabel('phase (degrees)');

figure                                 % modulus
plot(f,20*log10(abs(An21)));hold on
plot(f,20*log10(abs(X21a)),'-');hold on
plot(f,20*log10(abs(X21b)),':');hold on
plot(f,20*log10(abs(X21c)))
axis square; grid; axis([0,40,-130,-30])
xlabel('frequency (Hz)');
ylabel('amplitude (dB ref 1 m/N)');

figure                                 % phase
plot(f,180/pi*(angle(X21a)),'-');hold on
plot(f,180/pi*(angle(X21b)),':');hold on
plot(f,180/pi*(angle(X21c)))
axis square; grid; axis([0,40,-200,200])
xlabel('frequency (Hz)');
ylabel('phase (degrees)');

figure                                 % modulus
plot(f,20*log10(abs(An31)));hold on
```

(Continued)

MATLAB Example 8.4   (Continued)

```
plot(f,20*log10(abs(X31a)),'-');hold on
plot(f,20*log10(abs(X31b)),':');hold on
plot(f,20*log10(abs(X31c)))
axis square; grid; axis([0,40,-130,-30])
xlabel('frequency (Hz)');
ylabel('amplitude (dB ref 1 m/N)');

figure                              % phase
plot(f,180/pi*(angle(X31a)),'-'); hold on
plot(f,180/pi*(angle(X31b)),':'); hold on
plot(f,180/pi*(angle(X31c)))
axis square; grid; axis([0,40,-200,200])
xlabel('frequency (Hz)');
ylabel('phase (degrees)');
```

Results

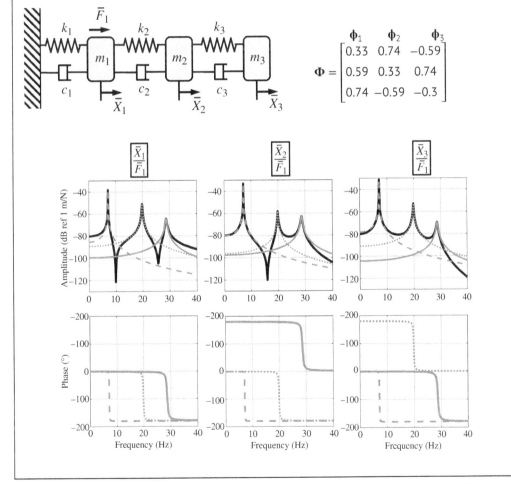

(Continued)

MATLAB Example 8.4 (Continued)

Modal damping ratios		
ζ_1	ζ_2	ζ_3
0.002	0.006	0.009

Comments:

1. This example gives similar results to that shown in Figure 8.2, but with the individual modal contributions shown.
2. Exercises for the reader are to:
   (a) calculate the point and transfer receptances when the excitation force is applied to masses m_1 and m_2,
   (b) explore the effects of changing the system parameters to see how they affect the modal contributions.

8.2.5 Impulse Response Function (IRF)

As with the SDOF system discussed in Chapter 3, the impulse response function (IRF) of a system can be determined by taking the IDFT of the corresponding double-sided FRF. However, there is not a simple analytical formulation unless the FRF is written as a modal summation, as in Eq. (8.16). In this case, because each mode has the form of an SDOF system multiplied by the constant $\phi_p(l)\phi_p(n)$, the IRF of the system can be written as sum of the IRFs of corresponding to each mode. This is given by

$$h(t) = \sum_{p=1}^{P} A_p \sin(\omega_{d,p}t),\tag{8.19}$$

where $A_p = \frac{\phi_p(l)\phi_p(n)}{\tilde{m}_p\omega_{d,p}}e^{-\zeta_p\omega_p t}$, $\omega_{d,p} = \omega_p\sqrt{1-\zeta_p^2}$, $\omega_p = \sqrt{\frac{\tilde{k}_p}{\tilde{m}_p}}$, and $\zeta_p = \frac{\tilde{c}_p}{2\sqrt{\tilde{m}_p\tilde{k}_p}}$.

An example showing the modal components of the FRF of a 3DOF system together with the IRF and its modal components are shown in Figure 8.8. The point receptance $\overline{X}_1/\overline{F}_1$ is considered. From the FRF, it can be seen that second and third modes are relatively highly damped compared to the first mode, and thus the IRFs corresponding to these modes decay away more rapidly compared to the IRF of the first mode. Examining the IRF, $h(t)$ in the upper right graph in Figure 8.8, it can be seen that it has a complicated waveform for the first part of the IRF. This is because it is the sum of the IRFs for the three modes of vibration, and hence has three frequency components. As time increases, the contribution of the second and third modes to the IRF diminishes substantially, leaving only the first mode. Recall that the decay rate of the p-th mode IRF depends on the value of $\zeta_p\omega_p$. Thus, the IRF of a highly damped, higher-frequency mode decays more quickly than that of a lightly damped, lower-frequency mode.

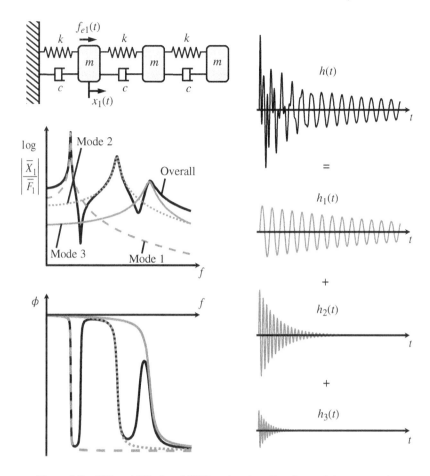

Figure 8.8 FRF and IRF of an MDOF system together its modal components.

MATLAB Example 8.5

In this example, the IRFs of the 3DOF system in MATLAB Example 8.5 are calculated and compared with the theoretical ones.

```
clear all

%% Parameters
% (see Matlab Example 8.1)
C=5e-4*K;                                % different damping values

%% Modal matrices
[V W]=eig(inv(M)*K);                     % calculation of eigenvectors
Mm=V'*M*V;Km=V'*K*V;Cm=V'*C*V;           % modal mass, stiff., and damping
zeta=Cm/(2*sqrt(Km*Mm));                 % modal damping ratios
R=sqrt(W);                               % undamped natural frequencies
```

(Continued)

MATLAB Example 8.5   (Continued)

```
wd=R*sqrt(1-zeta(1,1)^2);                    % damped natural frequencies

%% Modal parameters
Mm1=Mm(1,1);Mm2=Mm(2,2);Mm3=Mm(3,3);          % modal masses
w1=R(1,1);w2=R(2,2);w3=R(3,3);                % undamped natural frequencies
wd1=wd(1,1);wd2=wd(2,2);wd3=wd(3,3);          % damped natural frequencies
z1=zeta(1,1);z2=zeta(2,2);z3=zeta(3,3);       % modal damping ratios

%% Forced vibration
fs=1000;df=0.001;dt=1/fs;n=0;                 % frequency and time parameters
for f=0:0.001:fs/2                            % frequency vector
 w=2*pi*f;n=n+1;
 An=inv(K-w.^2*M+j*w*C);                       % calculate matrix of FRFs
 An11(n)=An(1,1);                              % point receptance
 An21(n)=An(2,1);                              % transfer receptance
 An31(n)=An(3,1);                              % transfer receptance

 % modal responses
 Hm=inv(Km-w^2*Mm+j*w*Cm);
 % mode 1
 F=[1 0 0]';Q=[1 0 0; 0 0 0;0 0 0];
 Xa=V*Q*Hm*V'*F;                               % response of first mode
 % mode 2
 Q=[0 0 0; 0 1 0;0 0 0];
 Xb=V*Q*Hm*V'*F;                               % response of second mode
 % mode 3
 Q=[0 0 0; 0 0 0;0 0 1];
 Xc=V*Q*Hm*V'*F;                               % response of third mode
 % X1/F1
 X11a(n)=Xa(1);X11b(n)=Xb(1);X11c(n)=Xc(1);   % individual modal responses
 % X2/F1
 X21a(n)=Xa(2);X21b(n)=Xb(2);X21c(n)=Xc(2);   % individual modal responses
 % X3/F1
 X31a(n)=Xa(3);X31b(n)=Xb(3);X31c(n)=Xc(3);   % individual modal responses
end

%% Calculation of IRFs
%% IRF corresponding to X1/F1
HH=An11;HH1=X11a;HH2=X11b;HH3=X11c;
[h h1 h2 h3]=IRF(fs,HH,HH1,HH2,HH3);           % function to calculate IRFs
t=0:dt:(length(h)-1)*dt;                       % time vector
h11=h;h11a=h1;h11b=h2;h11c=h3;                 % IRFs - numerical

A1=1/(Mm1*wd1)*exp(-z1*w1*t).*sin(wd1*t);      % part of first mode IRF
A2=1/(Mm2*wd2)*exp(-z2*w2*t).*sin(wd2*t);      % part of second mode IRF
A3=1/(Mm3*wd3)*exp(-z3*w3*t).*sin(wd3*t);      % part of third mode IRF

h11at=V(1,1)^2*A1;                             % first mode IRF - theory
h11bt=V(1,2)^2*A2;                             % second mode IRF - theory
h11ct=V(1,3)^2*A3;                             % third mode IRF - theory
h11t=h11at+h11bt+h11ct;                        % complete IRF - theory

%% IRF corresponding to X2/F1
HH=An21;HH1=X21a;HH2=X21b;HH3=X21c;
[h h1 h2 h3]=IRF(fs,HH,HH1,HH2,HH3);           % function to calculate IRFs
```

(Continued)

MATLAB Example 8.5   (Continued)

```
h21=h;h21a=h1;h21b=h2;h21c=h3;              % IRFs - numerical

h21at=V(1,1)*V(2,1)*A1;                     % first mode IRF - theory
h21bt=V(1,2)*V(2,2)*A2;                     % second mode IRF - theory
h21ct=V(1,3)*V(2,3)*A3;                     % third mode IRF - theory
h21t=h21at+h21bt+h21ct;                     % complete IRF - theory


%% IRF corresponding to X3/F1
HH=An31;HH1=X31a;HH2=X31b;HH3=X31c;
[h h1 h2 h3]=IRF(fs,HH,HH1,HH2,HH3);        % function to calculate IRFs
h31=h;h31a=h1;h31b=h2;h31c=h3;              % IRFs - numerical

h31at=V(1,1)*V(3,1)*A1;                     % first mode IRF - theory
h31bt=V(1,2)*V(3,2)*A2;                     % second mode IRF - theory
h31ct=V(1,3)*V(3,3)*A3;                     % third mode IRF - theory
h31t=h31at+h31bt+h31ct;                     % complete IRF - theory

%% Figures                                  % the figures are plotted
figure                                      for the IRF corresponding
plot(t,h11);hold on                         to X1/F1. To plot the other IRFs
plot(t,h11t,'-k')                           change the code. For example, insert
axis square; grid; axis([0,2,-7e-3,9e-3])   h21 instead.
xlabel('time (s)');ylabel('IRF (m/Ns)')
figure
plot(t,h11a);hold on
plot(t,h11at,'-k')
axis square; grid; axis([0,2,-7e-3,9e-3])
xlabel('time (s)'); ylabel('Mode 1 IRF (m/Ns)')
figure
plot(t,h11b);hold on
plot(t,h11bt,'-k')
axis square; grid; axis([0,2,-7e-3,9e-3])
xlabel('time (s)'); ylabel('Mode 2 IRF (m/Ns)')
figure
plot(t,h11c);hold on
plot(t,h11ct,'-k')
axis square; grid; axis([0,2,-7e-3,9e-3])
xlabel('time (s)'); ylabel('Mode 3 IRF (m/Ns)')

function [h h1 h2 h3]=IRF(fs,HH,HH1,HH2,HH3)
 Hd=[HH fliplr(conj(HH))];
 H=Hd(1:length(Hd)-1);
 h=fs*ifft(H);

 H1d=[HH1 fliplr(conj(HH1))];
 H1=H1d(1:length(H1d)-1);
 h1=fs*ifft(H1);

 H2d=[HH2 fliplr(conj(HH2))];
 H2=H2d(1:length(H2d)-1);
 h2=fs*ifft(H2);

 H3d=[HH3 fliplr(conj(HH3))];
 H3=H3d(1:length(H3d)-1);
 h3=fs*ifft(H3);
end
```

(Continued)

MATLAB Example 8.5   (Continued)

Results

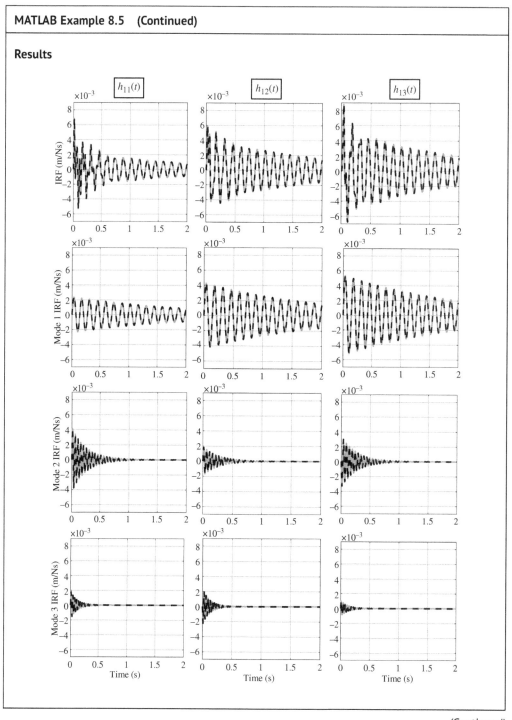

MATLAB Example 8.5   (Continued)

Comments:

1. This example gives the IRFs corresponding to the FRFs shown in MATLAB Example 8.4.
2. Exercises for the reader are to:
   (a) calculate the IRFs when the excitation force is applied to masses m_1 and m_2,
   (b) attach an extra spring to the right-hand mass as in Figure 8.5 and calculate the IRFs of the system,
   (c) calculate the IRFs by numerically integrating the equations of motion using the Runge–Kutta method as described in Appendix D and compare them with the IRFs calculated using the modal approach.

8.3   Continuous Systems

Structures that have distributed mass and stiffness are generally called continuous systems. Because they are not composed of lumped elements, such as mass, stiffness, and damping, they are described by partial differential equations (PDEs), which are functions of space and time, rather than ordinary differential equations (ODEs), that have only been functions of time. It was shown previously in this chapter that the number of natural frequencies and corresponding mode shapes for a lumped parameter system are dependent upon the number of DOF. In a continuous system there are an infinite number of DOF, and hence an infinite number of natural frequencies. However, any model of a continuous system is only valid over a finite frequency range and thus the model will have a finite number of natural frequencies. In this section two continuous systems are considered, which are shown in Figure 8.9. One is a slender rod, in which longitudinal or in-plane waves propagate, and the axial displacement is given by $u(x, t)$. The other is a slender beam in which bending or flexural waves propagate, and the lateral displacement is given by $w(x, t)$. Both structures are homogeneous and uniform, so there is no coupling between longitudinal and bending motion. Of course, there are many other continuous systems, such as strings, shafts, membranes, and plates (Leissa and Qatu, 2011), but they are not covered in this book. However, the treatment applied to the rod and the beam can be applied to the other systems.

8.3.1   Rod

The equation of motion for a force-excited rod is given by Tse et al. (1978)

$$\underbrace{ES\frac{\partial^2 u(x,t)}{\partial x^2}}_{\substack{\text{stiffness force}\\\text{per unit length}}} - \underbrace{\rho S\frac{\partial^2 u(x,t)}{\partial t^2}}_{\substack{\text{inertia force}\\\text{per unit length}}} = \underbrace{-f_e(x,t)}_{\substack{\text{force per}\\\text{unit length}}} \tag{8.20}$$

where E, S, and ρ are Young's modulus, cross-sectional area, and density of the rod, respectively. Note that unlike the ODEs discussed hitherto in this book, Eq. (8.20) is not a balance of forces. Rather, it is a balance of forces per unit length so $f_e(x, t)$ is not a force applied at a point, it is a distributed *axial* force. The *axial* displacement of the rod $u(x, t)$ is a function of both space and time. The solution to Eq. (8.20) can be written as an infinite sum of modal responses, given by

$$u(x, t) = \sum_{p=1}^{\infty} \phi_p(x)q_p(t) \tag{8.21}$$

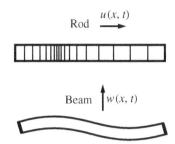

Figure 8.9 Two continuous systems – a rod undergoing axial vibration and a beam undergoing flexural or bending vibration.

where $\phi_p(x)$ is the p-th mode shape, which is only a function of x, and $q_p(t)$ is the p-th modal participation factor, which is a function of time. Note that an infinite sum of modes is required to give the correct displacement solution because of the distributed mass and stiffness. Substituting Eq. (8.21) into (8.20) results in

$$ES\frac{\partial^2}{\partial x^2}\left(\sum_{p=1}^{\infty}\phi_p(x)q_p(t)\right) - \rho S\frac{\partial^2}{\partial t^2}\left(\sum_{p=1}^{\infty}\phi_p(x)q_p(t)\right) = -f_e(x,t) \tag{8.22a}$$

As the time and space dependency are restricted to separate variables, Eq. (8.22a) can be written as an ODE to give

$$ES\sum_{p=1}^{\infty}\left(\frac{d^2\phi_p(x)}{dx^2}q_p(t)\right) - \rho S\sum_{p=1}^{\infty}\left(\phi_p(x)\frac{d^2q_p(t)}{dt^2}\right) = -f_e(x,t) \tag{8.22b}$$

The principle of the orthogonality of modes can be used to simplify Eq. (8.22b). Multiplying each term by the q-th mode shape, $\phi_q(x)$, and integrating over the length l of the rod, results in

$$ES\int_0^l\left[\phi_q(x)\sum_{p=1}^{\infty}\left(\frac{d^2\phi_p(x)}{dx^2}q_p(t)\right)\right]dx - \rho S\int_0^l\left[\phi_q(x)\sum_{p=1}^{\infty}\left(\phi_p(x)\frac{d^2q_p(t)}{dt^2}\right)\right]dx$$

$$= -\int_0^l[\phi_q(x)f_e(x,t)]dx \tag{8.23}$$

Now, the orthogonality conditions are $\int_0^l\phi_q(x)\phi_p(x)dx = 0$ and $\int_0^l\frac{d^2\phi_q(x)}{dx^2}\phi_p(x)dx = 0$, so Eq. (8.23) becomes

$$\tilde{m}_p\ddot{q}_p(t) + \tilde{k}_pq_p(t) = g_p(t) \tag{8.24}$$

where the overdot represent differentiation with the respect to time, and

$$\tilde{m}_p = \rho S\int_0^l\phi_p^2(x)dx$$

is the modal mass,

$$\tilde{k}_p = -ES\int_0^l\phi_p(x)\frac{d^2\phi_p(x)}{dx^2}dx$$

is the modal stiffness,

$$g_p(t) = \int_0^l\phi_p(x)f_e(x,t)dx$$

is the modal force.

Damping can be added to each mode by including a damping term to Eq. (8.24). Dividing Eq. (8.24) by the modal mass and adding damping in the form of a modal damping ratio ζ_p results in

$$\ddot{q}_p(t) + 2\zeta_p\omega_p\dot{q}_p + \omega_p^2 q_p(t) = \frac{g_p(t)}{\tilde{m}_p} \tag{8.25}$$

If a point excitation force is applied at x_1 then $f_e(x_1, t) = f_e(t)\delta(x - x_1)$, where $f_e(t)$ is the time history of the excitation force, which has units of N, and $\delta(x - x_1)$ is a delta function at $x = x_1$ and has units of $1/m$, then

$$g_p(t) = \int_0^l \phi_p(x)f_e(t)\delta(x - x_1)\mathrm{d}x. \tag{8.26}$$

Using the sifting property of the delta function described in Appendix E, Eq. (8.26) becomes $g_p(t) = \phi_p(x_1)f_e(t)$, so Eq. (8.25) becomes

$$\ddot{q}_p(t) + 2\zeta_p\omega_p\dot{q}_p + \omega_p^2 q_p(t) = \frac{\phi_p(x_1)f_e(t)}{\tilde{m}_p} \tag{8.27}$$

The modal participation factor $q_p(t)$ can thus be determined from Eq. (8.27), which can then be substituted into Eq. (8.21) to give the displacement response $u(x, t)$. If the force applied at $x = x_1$ is harmonic with the form $f_e(t) = \overline{F}(x_1)e^{\mathrm{j}\omega t}$ and the displacement response $u(t) = \overline{U}(x_2)e^{\mathrm{j}\omega t}$ is measured at $x = x_2$, the receptance (displacement response per unit input force) is given by

$$\frac{\overline{U}(x_2)}{\overline{F}(x_1)} = \sum_{p=1}^{\infty} \frac{\phi_p(x_1)\phi_p(x_2)}{\tilde{m}_p\left(\omega_p^2 - \omega^2 + \mathrm{j}2\zeta_p\omega\omega_p\right)}, \tag{8.28}$$

Note the similarity between the receptance for a continuous system and the receptance for a lumped parameter system. The key difference is that the number of modes for a continuous system is infinite. Note also, the modes shapes are given by continuous functions and can thus be evaluated at any position in the rod. As with the mass and stiffness, the damping is assumed to be distributed throughout the structure and it is convenient to add damping to each mode as discussed above. For the receptance of the free–free rod, there is no grounded spring to determine its equilibrium position. This means that if the free–free rod is impacted it will not return to its previous at-rest position. To account for this, an additional term needs to be added to Eq. (8.28), which is called a rigid-body mode. It occurs at a natural frequency of zero, i.e. $\omega_p = 0$, and is given by $\overline{U}(x_2)/\overline{F}(x_1)\big|_{\omega_p=0} = 1/(-\tilde{m}_p\omega^2)$. The way in which the natural frequencies and mode shapes are determined is discussed in the next subsection.

8.3.1.1 Natural Frequencies and Mode Shapes

As with the MDOF system, the natural frequencies and mode shapes are determined by considering free vibration. Thus, Eq. (8.20) becomes

$$ES\frac{\partial^2 u(x, t)}{\partial x^2} - \rho S\frac{\partial^2 u(x, t)}{\partial t^2} = 0 \tag{8.29}$$

Assuming a harmonic displacement of the form $u(x, t) = \overline{U}(x)e^{\mathrm{j}\omega t}$ results in

$$\overline{U}''(x) + \beta_R^2\overline{U}(x) = 0 \tag{8.30}$$

where the dash denotes differentiation with respect to x, and $\beta_R = \omega/c_R$ is the longitudinal wavenumber in which $c_R = \sqrt{E/\rho}$ is the wave speed. The displacement at any point in the rod

Table 8.1 Natural frequencies and mode shapes for some rod configurations.

Configuration	(a) Frequency equation (b) Mode shapes
$\overline{U}(0) = 0$ $\overline{U}(l) = 0$	(a) $\sin(\beta_{R,p}l) = 0 \Rightarrow \beta_{R,p}l = p\pi$ (b) $\phi_p(x) = \sqrt{2}\sin(p\pi x/l)$
$\overline{U}'(0) = 0$ $\overline{U}'(l) = 0$	(a) $\sin(\beta_{R,p}l) = 0 \Rightarrow \beta_{R,p}l = p\pi$ (b) $\phi_p(x) = \sqrt{2}\cos(p\pi x/l)$ rigid body mode, $\phi_0(x) = 1$
$\overline{U}(0) = 0$ $\overline{U}'(l) = 0$	(a) $\cos(\beta_{R,p}l) = 0 \Rightarrow \beta_{R,p}l = (p-1/2)\pi$ (b) $\phi_p(x) = \sqrt{2}\sin((p-1/2)\pi x/l)$

Natural frequency $\omega_p = \beta_{R,p}l\left(\dfrac{E}{\rho l^2}\right)^{\frac{1}{2}}$; mode shape normalisation $\displaystyle\int_0^l \phi_p^2(x)\mathrm{d}x = l$.

consists of a left-going and a right-going wave, and is given by

$$\overline{U}(x) = \overline{A}e^{\mathrm{j}\beta_R x} + \overline{B}e^{-\mathrm{j}\beta_R x} \tag{8.31}$$

where $\overline{A}$ and $\overline{B}$ are complex amplitudes of left- and right-going propagating waves, respectively, which depend upon the boundary conditions. Three configurations are considered, which are shown in Table 8.1. For a fixed boundary $\overline{U}(x) = 0$ and for a free boundary $\overline{U}'(x) = 0$. An example of the method to determine the natural frequency and mode shape is illustrated below for a fixed–fixed rod. It is left as an exercise for the reader to derive expressions for the natural frequencies and mode shapes for the other two configurations.

Example – Fixed–Fixed Rod For a fixed left-hand boundary, $\overline{U}(0) = 0$, so from Eq. (8.31), $\overline{A} + \overline{B} = 0$ or $\overline{B} = -\overline{A}$, thus $\overline{U}(x) = \overline{A}\left(e^{\mathrm{j}\beta_R x} - e^{-\mathrm{j}\beta_R x}\right)$. For a fixed right-hand boundary $\overline{U}(l) = 0$, so that $e^{\mathrm{j}\beta_R l} + e^{-\mathrm{j}\beta_R l} = 0$, or $\sin(\beta_R l) = 0$. This means that $\beta_{R,p}l = p\pi$. Noting that $\beta_{R,p} = \omega_p/c_R$ results in the expression for the p-th natural frequency, which is given by

$$\omega_p = \frac{p\pi}{l}\sqrt{\frac{E}{\rho}} \tag{8.32}$$

At the p-th natural frequency, the displacement of the rod is given by $\overline{U}_p(x) = |\overline{C}_p|\sin(p\pi x/l)$ where $|\overline{C}_p|$ is a constant, the value of which is dependent upon a normalisation factor. If the modal mass is set to be the mass of the rod, i.e. $\tilde{m}_p = \rho Sl$, then the mode shape is normalised such that

$\int_0^l \phi_p^2(x)\mathrm{d}x = l$. Now $\phi_p(x) = \overline{U}_p(x)$, so $|\overline{C}_p| = \sqrt{2}$; therefore, for a fixed–fixed rod the mode shape is given by

$$\phi_p(x) = \sqrt{2}\sin(p\pi x/l). \tag{8.33}$$

The natural frequencies and modes shapes for a rod with three different boundary conditions are given in Table 8.1.

8.3.1.2 Impulse Response Function (IRF)

The IRF of a continuous system can be calculated in the same way as for a lumped parameter MDOF system. Once the modal parameters have been determined, the IRFs for each mode of vibration are calculated, which are then summed to give the overall IRF as in Eq. (8.19). An example is shown for the **fixed–free** rod shown in Table 8.1, for the force and displacement response at the free end. The modal response is compared with the exact solution, which is given by Kinsler et al. (1982) as

$$H_{\mathrm{exact}}(\mathrm{j}\omega) = \frac{\tan(\beta_{\mathrm{R}}l)}{\omega S\sqrt{\rho E}}. \tag{8.34}$$

The modal response is given by

$$H(\mathrm{j}\omega) = \frac{\overline{U}(l)}{\overline{F}(l)} = \sum_{p=1}^{\infty} \frac{\phi_p^2(l)}{\tilde{m}_p\left(\omega_p^2 - \omega^2 + \mathrm{j}2\zeta_p\omega\omega_p\right)}, \tag{8.35}$$

where $\tilde{m}_p = \rho S l$, $\phi_p(l) = \sqrt{2}$, and $\omega_p = \frac{(p-1/2)\pi}{l}\sqrt{\frac{E}{\rho}}$. The exact and approximate FRFs are shown in the top part of Figure 8.10 for a wide frequency range. Note that to plot the exact FRF given in Eq. (8.34) some damping must be included. This is done by using a material loss factor (structural damping) η, so that Young's modulus becomes complex, and is given by $E(1+\mathrm{j}\eta)$, (Nashif et al., 1985). This means that the damping force is proportional to displacement, but is in-phase with the velocity, whereas a viscous damping force is proportional to, and is in-phase with the velocity. The relationship between the loss factor and the damping ratio that is commonly used is $\eta = 2\zeta$, which is derived by assuming the same response at the resonance frequency for two SDOF systems, one of which has viscous damping and the other has structural damping (Brennan and Ferguson, 2004). Although the structural damping model is convenient to use in continuous structures, a minor problem is that it causes a small amount of acausality (the system responds before it is impacted) in the IRF. The modulus of the FRF and the IRF for the fixed–free rod is shown in Figure 8.10 for two levels of damping. The graph in the upper left part of Figure 8.10 shows the FRF for a rod with a modal damping ratio of $\zeta_p = 0.01$. It is calculated using Eqs. (8.34) and (8.35). Note that both the modulus and frequency axes are logarithmic because the dynamic range of the FRF is very large as it has both resonance and anti-resonance frequencies, and also the frequency range is very wide. It can be seen that the FRF from the modal model gives a result that is very similar to that from the exact model, provided that enough modes are included in the response. The corresponding IRFs are shown in the lower left part of Figure 8.10. Again, it can be seen that the IRFs from both models are similar, demonstrating that the acausality due to structural damping is small, at least for $\zeta_p = 0.01$. There are some additional acausality issues for both models, however, due to the rectangular window applied in the frequency domain. This is evident at the beginning of the IRF which was discussed in detail in Chapter 4. It is unavoidable when using the IDFT to obtain an IRF from an analytical FRF, but it can be minimised by using a high sampling frequency. Examination of the IRF for the rod shows that its shape for the fixed–free rod is different to that for the 3DOF lumped parameter system shown in Figure 8.8. This is because the natural frequencies for the rod

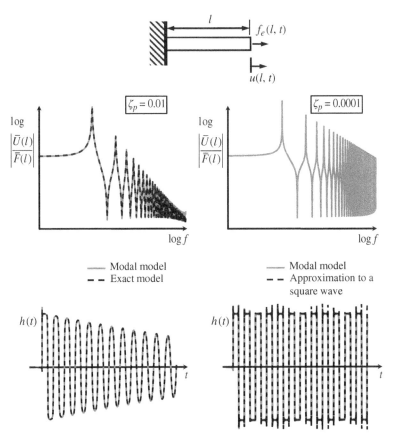

Figure 8.10 FRF and IRF of a fixed–free rod.

are harmonically related, which is not the case for the 3DOF system. Closer examination of the IRF for the fixed–free rod shows that at the beginning of the time history, the waveform tends to a square-like wave, but as time increases it tends to a decaying sine wave, suggesting that the responses due to the higher-order modes decay away rapidly leaving the first mode to dominate the IRF towards the end of the time history. This also occurs in the IRF for the 3DOF system in Figure 8.8, and is a general feature of most IRFs of structural systems.

To further investigate the shape of the IRF, a fixed–free rod with a modal damping ratio of $\zeta_p = 0.0001$ is considered. The FRF for this structure, calculated using only the modal model, is shown in the upper right part of Figure 8.10. Note how the reduced damping affects the FRF at the resonance and anti-resonance frequencies. The corresponding IRF for an *undamped* fixed–free rod is given by

$$h(t) = \sum_{p=1}^{\infty} \frac{\phi_p^2(l)}{\tilde{m}_p \omega_p} \sin(\omega_p t). \tag{8.36a}$$

Substituting for $\tilde{m}_p$, $\phi_p^2(l)$, and ω_p, and noting that $\omega_1 = \frac{\pi}{2l}\sqrt{\frac{E}{\rho}}$, Eq. (8.36a) can be written as

$$h(t) = \frac{4}{\pi} \frac{1}{S\sqrt{\rho E}} \sum_{p=1}^{\infty} \frac{1}{(2p-1)} \sin((2p-1)\omega_1 t), \tag{8.36b}$$

which is the Fourier series representation of a square wave with amplitude $1/(S\sqrt{\rho E})$. Thus, the IRF for an undamped fixed–free rod is a square wave. This is because the natural frequencies occur at odd multiples of the fundamental frequency i.e. ω_1, ω_3, ω_5, ..., and the amplitudes of the corresponding modal IRFs have normalised amplitudes of 1, 1/3, 1/5, ... The IRF for a very lightly damped fixed–free rod is compared with an approximation to a square wave in the lower right part of Figure 8.10. It can be seen there is very good agreement between the two plots. This explains the shape of the IRF directly after the impact for the more heavily damped rod shown in the lower left part of Figure 8.10.

MATLAB Example 8.6

In this example, the transfer receptance FRF and corresponding IRF of a fixed–free rod are calculated using the modal approach. The effects of changing the number of modes used in the calculation are investigated.

```
clear all

%% Parameters
E=69e9;                              % [N/m^2]    % Youngs modulus of aluminium
rho=2700;                            % [kg/m^3]   % density of aluminium
l=10;b=0.02;d=0.01;S=b*d;            % [m, m^2]   % geometrical parameters
z=0.01;n=2*z;                                     % damping ratio and loss factor
Ed=E*(1+j*n);                        % [N/m^2]    % complex Young's modulus
m=rho*S*l;                           % [kg]       % mass of the rod

%% Modal solution
fs=20000;df=0.001;dt=1/fs;                        % frequency parameters
f=0.001:0.01:fs/2;                                % frequency vector
w=2*pi*f;
for n=1:40                                        % 40 modes (can change this number)
  x=l;                               % [m]        % force position
  phi1=sqrt(2)*sin((n-1/2)*pi*x/l);              % mode shape at force position
  x=0.5*l;                           % [m]        % response position
  phi2=sqrt(2)*sin((n-1/2)*pi*x/l);              % mode shape at response position
  wn=(n-1/2)*pi/l*sqrt(E/rho);       % [rad/s]    % natural frequencies
  Ht(n,:)=phi1*phi2./(m*(wn^2-w.^2+j*2*w*wn*z));  % FRF for each mode
end
Htt=sum(Ht);                         % [m/N]      % overall FRF

%% IRF
Htd=[Htt fliplr(conj(Htt))];                      % form the double-sided spectrum
Hm=Htd(1:length(Htd)-1);                          % set the length of the FRF
h=fs*ifft(Hm);                       % [m/Ns]     % calculation of the IRF
h=circshift(h,100);                               % circular shift of the IRF
t=0:dt:(length(h)-1)*dt;             % [s]        % time vector

%% Plot the results
figure
semilogx(f,20*log10(abs(Htt)),'linewidth',3)      % FRF
axis square; grid; axis([10,10000,-180,-90])
xlabel('frequency (Hz)')
ylabel('|FRF| (dB ref 1m/N)')

figure
plot(t,h,'linewidth',3)                           % IRF
axis square; grid; axis([0,0.1,-4e-4,4e-4])
xlabel('time (s)'); ylabel('IRF (m/Ns)')
```

(Continued)

MATLAB Example 8.6   (Continued)

Results

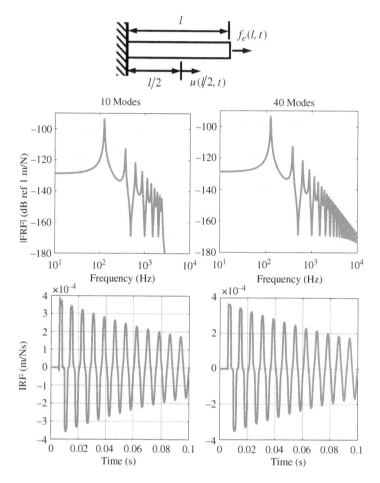

Comments:

1. Compare the difference between the shape of waveform for the IRF corresponding to the transfer receptance calculated in this example with the IRF corresponding to the point receptance given in Figure 8.10. Try to think of a reason why this difference occurs.
2. Note the way in which the reduction in the number of modes changes the shape of the waveform from being rectangular to triangular at the beginning of the IRF. Note also that it does not appreciably change the shape of the waveform as time increases.
3. An exercise for the reader is to investigate what happens to the IRF, when the:
   (a) geometrical parameters are changed,
   (b) damping is changed,
   (c) number of modes is reduced,
   (d) sampling frequency (the frequency range of the FRF) is changed.

8.3.2 Beam

The equation of motion for a force-excited beam is also a PDE because of its distributed mass and stiffness. However, its stiffness force has a different form to that of rod, as it is based on bending stiffness, which is related to the bending moment rather than the shear force, and this results in a fourth-order, instead of a second-order differential equation. An Euler–Bernoulli beam is considered, in which the shear stiffness is assumed to be infinite and rotational inertia is neglected. For a rectangular beam, the equation is valid in the frequency range where the bending wavelength is greater than about 10 times the thickness of the beam. The equation of motion is given by Tse et al. (1978)

$$EI\frac{\partial^4 w(x,t)}{\partial x^4} + \rho S\frac{\partial^2 w(x,t)}{\partial t^2} = f_e(x,t) \tag{8.37}$$

$$\underbrace{\phantom{EI\frac{\partial^4 w(x,t)}{\partial x^4}}}_{\substack{\text{stiffness force} \\ \text{per unit length}}} \quad \underbrace{\phantom{\rho S\frac{\partial^2 w(x,t)}{\partial t^2}}}_{\substack{\text{inertia force} \\ \text{per unit length}}} \quad \underbrace{}_{\substack{\text{force per} \\ \text{unit length}}}$$

where E, I, ρ, and S are Young's modulus, second moment of area, density, and cross-sectional area of the beam, respectively. Note that as with the beam, Eq. (8.37) is a balance of forces per unit length so that $f_e(x, t)$ is not a force applied at a point, it is a distributed *lateral* force. The *lateral* displacement of the beam $w(x, t)$ is a function of both space and time. The transformation of Eq. (8.37) to an ODE follows the same procedure as for the rod. The lateral displacement can be written as an infinite sum of modal responses, given by

$$w(x,t) = \sum_{p=1}^{\infty} \phi_p(x)q_p(t) \tag{8.38}$$

where $\phi_p(x)$ is the p-th bending mode shape and $q_p(t)$ is the p-th modal participation factor. Substituting Eq. (8.38) into (8.37) results in

$$EI\frac{\partial^4}{\partial x^4}\left(\sum_{p=1}^{\infty}\phi_p(x)q_p(t)\right) + \rho S\frac{\partial^2}{\partial t^2}\left(\sum_{p=1}^{\infty}\phi_p(x)q_p(t)\right) = f_e(x,t) \tag{8.39a}$$

As the time and space variables are separable, Eq. (8.39a) can be written as an ODE to give

$$EI\sum_{p=1}^{\infty}\left(\frac{d^4\phi_p(x)}{dx^4}q_p(t)\right) + \rho S\sum_{p=1}^{\infty}\left(\phi_p(x)\frac{d^2q_p(t)}{dt^2}\right) = f_e(x,t) \tag{8.39b}$$

Multiplying each term by the q-th mode shape, $\phi_q(x)$, and integrating over a point length l of the beam, results in

$$EI\int_0^l\left[\phi_q(x)\sum_{p=1}^{\infty}\left(\frac{d^4\phi_p(x)}{dx^4}q_p(t)\right)\right]dx + \rho S\int_0^l\left[\phi_q(x)\sum_{p=1}^{\infty}\left(\phi_p(x)\frac{d^2q_p(t)}{dt^2}\right)\right]dx$$

$$= \int_0^l[\phi_q(x)f_e(x,t)]dx \tag{8.40}$$

Now, the orthogonality conditions are $\int_0^l \phi_q(x)\phi_p(x)dx = 0$ and $\int_0^l \frac{d^4\phi_q(x)}{dx^4}\phi_p(x)dx = 0$, so Eq. (8.40) becomes

$$\tilde{m}_p\ddot{q}_p(t) + \tilde{k}_p q_p(t) = g_p(t) \tag{8.41}$$

where

$$\tilde{m}_p = \rho S\int_0^l \phi_p^2(x)dx$$

is the modal mass,

$$\tilde{k}_p = EI \int_0^l \phi_p(x) \frac{\mathrm{d}^4 \phi_p(x)}{\mathrm{d}x^4} \mathrm{d}x$$

is the modal stiffness,

$$g_p(t) = \int_0^l [\phi_p(x) f_e(x, t)] \mathrm{d}x$$

is the modal force.

As with the rod, damping can be added to each mode. Moreover, if a point lateral excitation force is applied at x_1 then $g_p(t) = \phi_p(x_1) f_e(t)$, where $f_e(t)$ is the time history of the point lateral force applied, so Eq. (8.41) becomes

$$\ddot{q}_p(t) + 2\zeta_p \dot{q}_p + \omega_p^2 q_p(t) = \frac{\phi_p(x_1) f_e(t)}{\tilde{m}_p} \tag{8.42}$$

From this, the modal participation factor $q_p(t)$ can be determined, which can then be substituted into Eq. (8.38) to give the displacement response $w(x, t)$. If the lateral force applied at $x = x_1$ is harmonic with the form $f_e(t) = \overline{F} e^{\mathrm{j}\omega t}$ and the displacement response $w(t) = \overline{W}(x_2) e^{\mathrm{j}\omega t}$ is measured at $x = x_2$, the receptance is given by

$$\frac{\overline{W}(x_2)}{\overline{F}(x_1)} = \sum_{p=1}^{\infty} \frac{\phi_p(x_1) \phi_p(x_2)}{\tilde{m}_p \left(\omega_p^2 - \omega^2 + \mathrm{j}2\zeta_p \omega \omega_p \right)}. \tag{8.43a}$$

Note the similarity between the receptance for a beam and a rod. The form of the equations is exactly the same, with the differences being in the modal parameters of natural frequencies, mode shapes, modal mass, and modal damping, which are discussed in the next subsection. Unlike a rod, which has only axial forces and axial displacements, a beam can be excited by a moment with complex amplitude $\overline{M}$ and also a rotation with amplitude $\overline{W}'$ can be measured. The receptances in terms of these quantities can be determined in a very simply way by considering the spatial derivative of the mode shapes, which is denoted by $(\bullet)' = \mathrm{d}(\bullet)/\mathrm{d}x$. The results are given by

$$\frac{\overline{W}'(x_2)}{\overline{F}(x_1)} = \sum_{p=1}^{\infty} \frac{\phi_p(x_1) \phi_p'(x_2)}{\tilde{m}_p \left(\omega^2 - \omega_p^2 + \mathrm{j}2\zeta_p \omega \omega_p \right)}, \tag{8.43b}$$

$$\frac{\overline{W}(x_2)}{\overline{M}(x_1)} = \sum_{p=1}^{\infty} \frac{\phi_p'(x_1) \phi_p(x_2)}{\tilde{m}_p \left(\omega^2 - \omega_p^2 + \mathrm{j}2\zeta_p \omega \omega_p \right)}, \tag{8.43c}$$

$$\frac{\overline{W}'(x_2)}{\overline{M}(x_1)} = \sum_{p=1}^{\infty} \frac{\phi_p'(x_1) \phi_p'(x_2)}{\tilde{m}_p \left(\omega^2 - \omega_p^2 + \mathrm{j}2\zeta_p \omega \omega_p \right)}. \tag{8.43d}$$

8.3.2.1 Natural Frequencies and Mode Shapes

The natural frequencies and mode shapes are again determined by considering free vibration. In this case the equation of motion is

$$EI \frac{\partial^4 w(x, t)}{\partial x^4} + \rho S \frac{\partial^2 w(x, t)}{\partial t^2} = 0 \tag{8.44}$$

Assuming a harmonic displacement of the form $w(x, t) = \overline{W}(x) e^{\mathrm{j}\omega t}$ results in

$$\overline{W}''''(x) - \beta_B^4 \overline{W}(x) = 0 \tag{8.45}$$

where $\beta_B = \omega/c_B$ is the bending wavenumber in which $c_B = (EI/\rho S)^{1/4}\omega^{1/2}$ is the bending wave speed. Note that for bending vibration the wave speed is a function of frequency, which is not the case for the axial vibration of a rod. This means that high-frequency waves travel faster than low-frequency waves. Such a system is called *dispersive*, because the envelope of a wave packet containing a range of frequencies changes shape as it propagates along the beam (Cremer et al., 2005). A physical explanation of this phenomenon can be found in (Brennan et al., 2016). The solution to Eq. (8.45) is given by

$$\overline{W}(x) = \overline{A}_1 e^{\beta_B x} + \overline{A}_2 e^{-\beta_B x} + \overline{A}_3 e^{j\beta_B x} + \overline{A}_4 e^{-j\beta_B x} \tag{8.46a}$$

where $\overline{A}_1$ and $\overline{A}_2$ are the complex amplitudes of evanescent waves that are confined to within about one wavelength of a discontinuity, and $\overline{A}_3$ and $\overline{A}_4$ are complex amplitudes of left- and right-going propagating waves, respectively. By collecting the wave types, Eq. (8.46a) can be written in terms of hyperbolic and trigonometric functions as

$$\overline{W}(x) = \overline{A}\sinh(\beta_B x) + \overline{B}\cosh(\beta_B x) + \overline{C}\sin(\beta_B x) + \overline{D}\cos(\beta_B x) \tag{8.46b}$$

where $\overline{A}, \overline{B}, \overline{C}$, and $\overline{A}$ are coefficients (which in this case are real numbers).

Four configurations are considered, which are shown in Table 8.2. For a pinned boundary at $x = x_b$ the displacement and bending moment of the beam are zero, i.e. $\overline{W}(x_b) = 0$ and $\overline{W}''(x_b) = 0$. For a fixed boundary the displacement and the slope of the beam are zero, i.e. $\overline{W}(x_b) = 0$ and $\overline{W}'(x_b) = 0$. For a free boundary the bending moment and shear force are zero, i.e. $\overline{W}''(x_b) = 0$ and $\overline{W}'''(x_b) = 0$. An example of the method to determine the natural frequency and mode shape is illustrated below for a pinned–pinned beam. It is left as an exercise for the reader to derive expressions for the natural frequencies and mode shapes for the other configurations. Note, however, that at high frequencies there can be issues in the numerical evaluation of the hyperbolic functions when calculating the mode shapes. This can be addressed by reformulating the equation describing the beam displacement in an appropriate manner (Gonçalves et al., 2018).

Example – Pinned–Pinned Beam For a pinned left-hand boundary, $\overline{W}(0) = 0$ and $\overline{W}''(0) = 0$, so from Eq. (8.46b), $\overline{B} = \overline{D} = 0$. For a pinned right-hand boundary, $\overline{W}(l) = 0$ and $\overline{W}''(l) = 0$, which results in $\overline{A} = 0$, so $\sin(\beta_B l) = 0$. At the p-th natural frequency, the displacement of the beam is given by $\overline{W}_p(x) = \overline{C}_p \sin(p\pi x/l)$, where the value of $\overline{C}_p$ is dependent upon a normalisation factor. If the modal mass is set to be the mass of the beam, i.e. $\tilde{m}_p = \rho Sl$, then the mode shape is normalised such that $\int_0^l \phi_p^2(x)\mathrm{d}x = l$ in the same way that it was for the rod, so the mode shape is given by $\phi_p(x) = \sqrt{2}\sin(p\pi x/l)$. The natural frequencies and modes shapes for a beam with four different boundary conditions are given in Table 8.2. Note that the simply supported (pinned) beam is the simplest case as the beam deflection can be described in terms of a trigonometric function. For all the other boundary conditions both hyperbolic and trigonometric functions are needed. The first three beam mode shapes for the configurations given in Table 8.2 are shown in Figure 8.11.

8.3.2.2 Impulse Response Function (IRF)

The IRF of the beam can be calculated in the same way as for the rod. The modal parameters must be determined first and then the receptances for each mode are calculated, which are then summed to give the overall FRF as in Eq. (8.43a). The overall IRF can be written as a sum of the IRFs corresponding to each mode, similar to the way shown in Eq. (8.36a). An example is given for the **fixed–free** (or cantilever) beam shown in Table 8.2, for the force and displacement response at the free end. The modal response is compared with the exact solution as with the rod. However, this

Table 8.2 Natural frequencies and mode shapes for some beam configurations.

Configuration	(a) Frequency equation (b) Mode shapes	
$\bar{W}(0) = 0 \quad \bar{W}(l) = 0$ $\bar{W}''(0) = 0 \quad \bar{W}''(l) = 0$ (beam with length l)	(a) $\sin(\beta_{\mathrm{B},p}l) = 0$ (b) $\phi_p(x) = \sqrt{2}\sin(p\pi x/l)$	$\beta_{\mathrm{B},p}l = p\pi$
$\bar{W}(0) = 0 \quad \bar{W}(l) = 0$ $\bar{W}'(0) = 0 \quad \bar{W}'(l) = 0$ (beam with length l)	(a) $\cos(\beta_{\mathrm{B},p}l)\cosh(\beta_{\mathrm{B},p}l) = 1$ (b) $\phi_p(x) = \cosh(\beta_{\mathrm{B},p}x) - \cos(\beta_{\mathrm{B},p}x) - \sigma_p(\sinh(\beta_{\mathrm{B},p}x) - \sin(\beta_{\mathrm{B},p}x))$, where $\sigma_p = \dfrac{(\cosh(\beta_{\mathrm{B},p}l) - \cos(\beta_{\mathrm{B},p}l))}{\sinh(\beta_{\mathrm{B},p}l) - \sin(\beta_{\mathrm{B},p}l)}$	$\beta_{\mathrm{B},1}l = 4.73004$ $\beta_{\mathrm{B},2}l = 7.85320$ $\beta_{\mathrm{B},3}l = 10.9956$ $\beta_{\mathrm{B},4}l = 14.1372$ $\beta_{\mathrm{B},5}l = 17.2788$ $\beta_{\mathrm{B},p}l \approx (2n+1)\pi/2,\, p \geq 6$

$\bar{W}''(0) = 0$ $\quad\quad \bar{W}''(l) = 0$
$\bar{W}'''(0) = 0$ $\quad\quad \bar{W}'''(l) = 0$

(a) $\cos(\beta_{B,p}l)\cosh(\beta_{B,p}l) = 1$

(b) $\phi_p(x) = \cosh(\beta_{B,p}x) + \cos(\beta_{B,p}x) - \sigma_p(\sinh(\beta_{B,p}x) + \sin(\beta_{B,p}x))$,

where $\sigma_p = \dfrac{(\cosh(\beta_{B,p}l) - \cos(\beta_{B,p}l)}{\sinh(\beta_{B,p}l) - \sin(\beta_{B,p}l)}$

rigid body modes

$\phi_{01}(x) = 1$, $\phi_{02}(x) = \sqrt{3}(1 - 2x/l)$

$\beta_{B,1}l = 4.73004$
$\beta_{B,2}l = 7.85320$
$\beta_{B,3}l = 10.9956$
$\beta_{B,4}l = 14.1372$
$\beta_{B,5}l = 17.2788$
$\beta_{B,p}l \approx (2n+1)\pi/2, p \geq 6$

$\bar{W}(0) = 0$ $\quad\quad \bar{W}''(l) = 0$
$\bar{W}'(0) = 0$ $\quad\quad \bar{W}'''(l) = 0$

(a) $\cos(\beta_{B,p}l)\cosh(\beta_{B,p}l) = -1$

(b) $\phi_p(x) = \cosh(\beta_{B,p}x) - \cos(\beta_{B,p}x) - \sigma_p(\sinh(\beta_{B,p}x) - \sin(\beta_{B,p}x))$,

where $\sigma_p = \dfrac{(\sinh(\beta_{B,p}l) - \sin(\beta_{B,p}l)}{\cosh(\beta_{B,p}l) + \cos(\beta_{B,p}l)}$

$\beta_{B,1}l = 1.87510$
$\beta_{B,2}l = 4.69409$
$\beta_{B,3}l = 7.85476$
$\beta_{B,4}l = 10.9955$
$\beta_{B,5}l = 14.1372$
$\beta_{B,p}l \approx (2n-1)\pi/2, p \geq 6$

Natural frequency $\omega_p = (\beta_{B,p}l)^2 \left(\dfrac{EI}{\rho Sl^4}\right)^{\frac{1}{2}}$; mode shape normalisation $\displaystyle\int_0^l \phi_p^2(x)dx = l$.

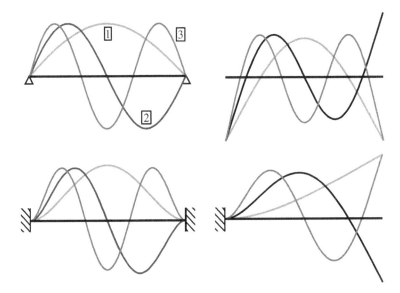

Figure 8.11 First three mode shapes of a pinned–pinned beam, a fixed–fixed beam, a free–free beam, and a fixed–free beam.

is slightly more complicated because displacement and rotation of the beam have to be accounted for, and additional hyperbolic functions are required to describe the beam vibration. To determine the exact FRF at the end of the cantilever beam, the dynamic stiffness of the beam is first needed. It is given by Gardonio and Brennan (2004)

$$
\begin{Bmatrix} \overline{F}(0) \\ \overline{M}(0) \\ \overline{F}(l) \\ \overline{M}(l) \end{Bmatrix} = \frac{EI\beta_B^3}{N} \begin{bmatrix} -K_{11} & -P & K_{12} & V \\ -P & Q_{11} & -V & Q_{12} \\ K_{12} & -V & -K_{11} & P \\ V & Q_{12} & P & Q_{11} \end{bmatrix} \begin{Bmatrix} \overline{W}(0) \\ \overline{W}'(0) \\ \overline{W}(l) \\ \overline{W}'(l) \end{Bmatrix}
\tag{8.47}
$$

where $\overline{F}(0)$ and $\overline{M}(0)$, and $\overline{F}(l)$ and $\overline{M}(l)$ are the forces and moments at the left- and right-hand of the beam, respectively, and the elements of the matrix are given by

$$
\begin{aligned}
K_{11} &= \cos(\beta_B l)\sinh(\beta_B l) + \sin(\beta_B l)\cosh(\beta_B l) \\
K_{12} &= \sin(\beta_B l) + \sinh(\beta_B l) \\
P &= (\sin(\beta_B l)\sinh(\beta_B l))/\beta_B \\
V &= (\cos(\beta_B l) - \cosh(\beta_B l))/\beta_B
\end{aligned}
\qquad
\begin{aligned}
Q_{11} &= (\cos(\beta_B l)\sinh(\beta_B l) - \sin(\beta_B l)\cosh(\beta_B l))/\beta_B^2 \\
Q_{12} &= (\sin(\beta_B l) - \sinh(\beta_B l))/\beta_B^2 \\
N &= \cos(\beta_B l)\cosh(\beta_B l) - 1
\end{aligned}
$$

Now the boundary conditions at $x = 0$ (the left-hand end of the beam) are $\overline{W}(0) = 0$ and $\overline{W}'(0) = 0$ as shown in Table 8.2. Thus Eq. (8.47) can be reduced to

$$
\begin{Bmatrix} \overline{F}(l) \\ \overline{M}(l) \end{Bmatrix} = \frac{EI\beta_B^3}{N} \begin{bmatrix} -K_{11} & P \\ P & Q_{11} \end{bmatrix} \begin{Bmatrix} \overline{W}(l) \\ \overline{W}'(l) \end{Bmatrix}
\tag{8.48}
$$

which can be further rearranged to give

$$
\begin{Bmatrix} \overline{W}(l) \\ \overline{W}'(l) \end{Bmatrix} = \frac{N}{EI\beta_B^3} \begin{bmatrix} -K_{11} & P \\ P & Q_{11} \end{bmatrix}^{-1} \begin{Bmatrix} \overline{F}(l) \\ \overline{M}(l) \end{Bmatrix}
\tag{8.49}
$$

As $\overline{M}(l) = 0$, the point receptance at $x = l$ is given by

$$
\frac{\overline{W}(l)}{\overline{F}(l)} = \frac{-N}{EI\beta_B^3} \left(\frac{Q_{11}}{Q_{11}K_{11} + P^2} \right)
\tag{8.50}
$$

The modal response is given by

$$H(j\omega) = \frac{\overline{W}(l)}{\overline{F}(l)} = \sum_{p=1}^{\infty} \frac{\phi_p^2(l)}{\tilde{m}_p\left(\omega^2 - \omega_p^2 + j2\zeta_p\omega\omega_p\right)}, \tag{8.51}$$

where $\tilde{m}_p = \rho Sl$, $\phi_p(l) = 2$, and $\omega_p = (\beta_{B,p}l)^2\left(\frac{EI}{\rho Sl^4}\right)^{\frac{1}{2}}$ in which $\beta_{B,p}l$ for each natural frequency is given in Table 8.2. The exact and approximate FRFs are shown in Figure 8.12 for a wide range of frequencies. Note that both the modulus and frequency axes are logarithmic. Note also that, as with the rod, damping has been included in the exact model using a material loss factor η, so that Young's modulus becomes complex and is given by $E(1 + j\eta)$. The relationship between the loss factor and the modal damping ratio is $\eta = 2\zeta_p$. It can be seen that the FRF from the modal model gives a result which is very similar to that from the exact model, provided that enough modes are included in the response. Comparing the FRF for the rod in Figure 8.10 and the FRF for the beam in Figure 8.12, it can be seen that although they have similar structure (an anti-resonance between each resonance frequency, which occurs because they are both point receptances), the natural frequencies for the beam are not harmonically related, unlike for the rod. This is because the wave speed in the beam is a function of frequency, so that $\omega_p \propto \beta_B^2$ for the beam whereas $\omega_p \propto \beta_R$ for the rod. This means that the IRF for the beam will have a different shape compared to that for the rod. The IRF for the fixed–free beam is shown in the lower right part of Figure 8.12. It can be seen that the IRF for the exact model and for the modal model are similar, which means that the acausality due to structural damping is small for $\zeta_p = 0.01$. As with the rod, there are some acausality issues for both models due to the rectangular window applied in the frequency domain, which was discussed in Chapter 4. Examination of the IRF in Figure 8.12 shows that the higher-frequency modal responses are evident at the beginning of the time history, but as time increases it tends to a decaying sine wave as in the other cases studied in this book.

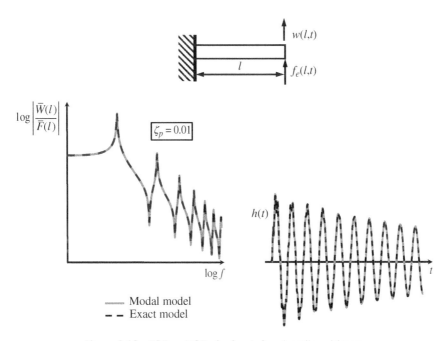

Figure 8.12 FRF and IRF of a fixed–free (cantilever) beam.

MATLAB Example 8.7

In this example, the transfer receptance FRF and corresponding IRF of a fixed–free beam is calculated using the modal approach. The effects of changing the number of modes used in the calculation are investigated.

```
clear all

%% Parameters
E=69e9;                                      % [N/m^2]    % Youngs modulus of aluminium
rho=2700;                                    % [kg/m^3]   % density of aluminium
l=1;b=0.02;d=0.01;S=b*d;I=b*d^3/12;                      % geometrical parameters
z=0.01;n=2*z;                                            % damping ratio and loss factor
Ed=E*(1+j*n);                                % [N/m^2]    % complex Young's modulus
m=rho*S*l;                                   % [kg]       % mass of the beam

%% Modal solution
fs=2000;df=0.001;dt=1/fs;                                % frequency parameters
f=0.001:0.01:fs/2;                                       % frequency vector
w=2*pi*f;
k=w.^0.5*(rho*S/(Ed*I))^.25;                             % wavenumber

nmax=10;                                                 % number of modes
kl(1)=1.87510;kl(2)=4.69409;kl(3)=7.85476;              % kl values 1-3
kl(4)=10.9956;kl(5)=14.1372;                             % kl values 4,5
n=6:nmax;
kl(n)=(2*n-1)*pi/2;                                      % kl values > 5

for n=1:nmax
 A=(sinh(kl(n))-sin(kl(n)))./(cosh(kl(n))+cos(kl(n)));
 x=0.2;                                                  % force position
 phi1=cosh(kl(n)*x/l)-cos(kl(n)*x/l)-...                 % mode shape at force position
 A.*(sinh(kl(n)*x/l)-sin(kl(n)*x/l));
 x=l;                                                    % displacement position
 phi2=cosh(kl(n)*x/l)-cos(kl(n)*x/l)-...                 % mode shape at response position
 A.*(sinh(kl(n)*x/l)-sin(kl(n)*x/l));
 wn=sqrt((E*I)./(rho*S))*(kl(n)).^2;                     % natural frequency
 Ht(n,:)=phi1*phi2./(m*(wn^2-w.^2+j*2*w*wn*z));          % FRF of each mode
end
Htt=sum(Ht);                                             % overall FRF

%% IRF
Htd=[Htt fliplr(conj(Htt))];                             % form the double-sided spectrum
Hm=Htd(1:length(Htd)-1);                                 % set the length of the FRF
h=fs*ifft(Hm);                                           % calculation of the IRF
h=circshift(h,10);                                       % shift the end of the IRF to
                                                         the beginning
t=0:dt:(length(h)-1)*dt;                                 % time vector

%% Plot the results
figure
semilogx(f,20*log10(abs(Htt)))                           % FRF
axis square; grid; axis([1,1010,-150,-30])
xlabel('frequency (Hz)');ylabel('|FRF| (dB ref 1m/N)')

figure
plot(t,h)                                                % IRF
axis square; grid; axis([0,1,-0.02,0.02])
xlabel('time (s)');ylabel('IRF (m/Ns)')
```

(Continued)

MATLAB Example 8.7   (Continued)

Results

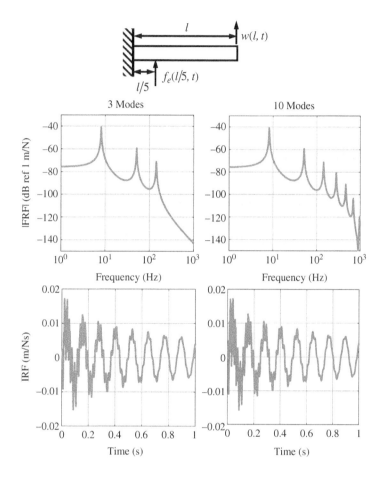

Comments:

1. Compare the difference between the shape of waveform for the IRF corresponding to the transfer receptance calculated in this example with the IRF corresponding to the point receptance given in Figure 8.12. Try to think of a reason why this difference occurs.

2. Note the way in which the reduction in the number of modes does not appreciably change the shape of the waveform as time increases.

3. An exercise for the reader is to investigate what happens to the IRF, when the:
   (a) geometrical parameters are changed,
   (b) damping is changed,
   (c) number of modes is reduced,
   (d) sampling frequency (the frequency range of the FRF) is changed.

8.4 Summary

This chapter has described how to model two classes of MDOF system. One of these is a lumped parameter system comprising point masses, massless springs, and viscous dampers. It is effectively an extension of the SDOF system considered at length in previous chapters. A chain-like system has been discussed, introducing the concept of an anti-resonance, whose frequency depends on the location of the force and the measured response. A description of the receptance FRF in terms of the modal parameters has been derived for a system in which the damping is of the Rayleigh type. This enables an analytical expression for the corresponding IRF to be derived in terms of a

Table 8.3 Summary of the expression for the FRF and IRF of a lumped parameter MDOF system.

Lumped parameter system
The equation of motion is given by $\mathbf{M\ddot{x} + C\dot{x} + Kx = f}$,
where $\mathbf{M}$, $\mathbf{C}$, and $\mathbf{K}$ are the mass, damping, and stiffness matrices, and $\mathbf{f}$ is the vector of applied forces.
The undamped natural frequencies diag$\{\omega_p\}$ are given by the square root of the *eigenvalues* of the matrix $\mathbf{M^{-1}K}$.
The corresponding mode shapes are given by the eigenvectors of the matrix $\mathbf{M^{-1}K}$.

The receptance FRF is given by

$$\frac{\overline{X}_n}{\overline{F}_l} = \sum_{p=1}^{P} \frac{\phi_p(l)\phi_p(n)}{\tilde{m}_p\left(\omega_p^2 - \omega^2 + \mathrm{j}2\zeta_p\omega\omega_p\right)}$$

where $\phi_p(l)$ is the p-th mode shape at the l-th mass where the force is applied, and $\phi_p(n)$ is the p-th mode shape at the n-th mass where the displacement is measured.

The mode shapes are normalised such that $\mathbf{\Phi}^T\mathbf{M}\mathbf{\Phi} = \mathbf{I}$, where $\mathbf{\Phi} = [\phi_1\phi_2\cdots\phi_P]$ is the matrix of mode shape vectors.

$\tilde{m}_p = 1$ is the modal mass, and $\zeta_p = \tilde{c}_p/(2\tilde{m}_p\omega_p)$ is the modal damping ratio, in which $\tilde{c}_p$ is the modal damping coefficient (for proportional damping) which is an element in the diagonal matrix $\tilde{\mathbf{C}} = \mathbf{\Phi}^T\mathbf{C}\mathbf{\Phi}$.

The IRF is given by the sum of the IRFs for each mode

$$h(t) = \sum_{p=1}^{P} A_p \sin(\omega_{d,p}t), \text{ where } A_p = \frac{\phi_p(l)\phi_p(n)}{\tilde{m}_p\omega_{d,p}}e^{-\zeta_p\omega_p t} \text{ and } \omega_{d,p} = \omega_p\sqrt{1 - \zeta_p^2}.$$

finite number of modes of the system. A summary of the important points for a lumped parameter MDOF system is given in Table 8.3.

The second type of MDOF system considered in this chapter is a distributed parameter system, which is classified as a continuous system. Unlike a lumped parameter MDOF system which can be described in terms of an ODE which is a function of time, this type of structure is described in terms of a PDE, which is a function of both space and time. Because of the distributed nature of the mass and stiffness, this model has an infinite number of DOF, which means that, in principle at least, it also has an infinite number of natural frequencies. The models used for continuous structures generally have an upper frequency limit in terms of their validity, however, and so are limited

Table 8.4 Summary of the expressions for the FRF and IRF of a rod.

<table>
<tr><td colspan="1" align="center">Distributed parameter system: Rod</td></tr>
</table>

The equation of motion is given by $ES\dfrac{\partial^2 u(x,t)}{\partial x^2} - \rho S\dfrac{\partial^2 u(x,t)}{\partial t^2} = -f_e(x,t)$

where E, S, and ρ are the Young's modulus, cross-sectional area, and density of the rod, respectively, and $f_e(x, t)$ is the applied axial force per unit length.

The undamped natural frequencies are given by $\omega_p = \beta_{R,p} l \left(\dfrac{E}{\rho l^2}\right)^{\frac{1}{2}}$, where $\beta_{R,p} l$ is a function of the boundary conditions and can be determined using Table 8.1.

The corresponding mode shapes $\phi_p(x)$ are dependent upon the boundary conditions and are listed in Table 8.1.

The receptance FRF is given by

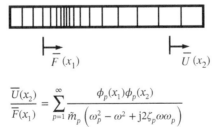

$$\frac{\overline{U}(x_2)}{\overline{F}(x_1)} = \sum_{p=1}^{\infty} \frac{\phi_p(x_1)\phi_p(x_2)}{\tilde{m}_p\left(\omega_p^2 - \omega^2 + j2\zeta_p\omega\omega_p\right)}$$

where $\phi_p(x_1)$ is the p-th mode shape at the position where the force is applied, and $\phi_p(x_2)$ is the p-th mode shape at the position where the displacement is measured.

The mode shapes are normalised so that $\displaystyle\int_0^l \phi_p^2(x)\mathrm{d}x = l$.

The modal mass $\tilde{m}_p = \rho S l$ and the modal damping ratio $\zeta_p = \eta/2$, where η is the material loss factor of the rod.

The IRF is given by the sum of the IRFs for each mode

$$h(t) = \sum_{p=1}^{\infty} A_p \sin(\omega_{d,p}t), \text{ where } A_p = \frac{\phi_p(x_1)\phi_p(x_2)}{\tilde{m}_p\omega_{d,p}}e^{-\zeta_p\omega_p t} \text{ and } \omega_{d,p} = \omega_p\sqrt{1 - \zeta_p^2}.$$

Table 8.5 Summary of the expressions for the FRF and IRF of a beam.

Distributed parameter system: Beam

The equation of motion is given by $EI\dfrac{\partial^4 w(x,t)}{\partial x^4} + \rho S\dfrac{\partial^2 w(x,t)}{\partial t^2} = f_e(x,t)$

where E, I, S, and ρ are Young's modulus, second moment of area, cross-sectional area, and density of the beam, respectively, and $f_e(x,t)$ is the applied lateral force per unit length.

The undamped natural frequencies are given by $\omega_p = (\beta_{B,p}l)^2\left(\dfrac{EI}{\rho S l^4}\right)^{\frac{1}{2}}$, where $\beta_{B,p}l$ is a function of the boundary conditions and can be determined using Table 8.2.

The corresponding mode shapes $\phi_p(x)$ are dependent upon the boundary conditions and are listed in Table 8.2.

The receptance FRF is given by

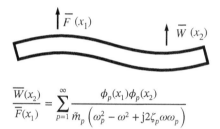

$$\frac{\overline{W}(x_2)}{\overline{F}(x_1)} = \sum_{p=1}^{\infty} \frac{\phi_p(x_1)\phi_p(x_2)}{\tilde{m}_p\left(\omega_p^2 - \omega^2 + j2\zeta_p\omega\omega_p\right)}$$

where $\phi_p(x_1)$ is the p-th mode shape at the position where the force is applied, and $\phi_p(x_2)$ is the p-th mode shape at the position where the displacement is measured.

The mode shapes are normalised so that $\displaystyle\int_0^l \phi_p^2(x)\mathrm{d}x = l$.

The modal mass $\tilde{m}_p = \rho S l$ and the modal damping ratio $\zeta_p = \eta/2$, where η is the material loss factor of the beam.

The IRF is given by the sum of the IRFs for each mode

$$h(t) = \sum_{p=1}^{\infty} A_p \sin(\omega_{d,p}t), \text{ where } A_p = \frac{\phi_p(x_1)\phi_p(x_2)}{\tilde{m}_p\omega_{d,p}} e^{-\zeta_p\omega_p t} \text{ and } \omega_{d,p} = \omega_p\sqrt{1-\zeta_p^2}.$$

---

to a finite number of DOF in practice. Two types of one-dimensional structure were discussed in this chapter, a rod, which supports axial motion, and a beam, which supports lateral motion. There are two main differences between these two structures. One is that the rod has a wave speed that is independent of frequency, and the beam has a wave speed that is proportional to the square root of frequency. This influences the spacing between the natural frequencies for each structure. The second is that the behaviour of the beam is a function of the shear force and bending moment, and the motion is described in terms of a lateral displacement and a rotation, whereas the behaviour of the rod is only governed by the axial force and its motion is simply described by the axial displacement. Both the rod and the beam can be described in terms of modal parameters in the same way as a lumped parameter structure, which enables analytical expressions for their IRFs to be written down. A summary of the important points for the vibration of a rod and a beam is given in Tables 8.4 and 8.5, respectively.

References

Bishop, R.E.D. and Johnson, D.C. (1960). *The Mechanics of Vibration.* Cambridge University Press.

Brennan, M.J. and Ferguson, N.S. (2004). *Vibration Control, Chapter 12 in Advanced Applications in Acoustics, Noise and Vibration*, (eds. F.J. Fahy and J. Walker). Spon Press.

Brennan, M.J., Tang, B., and Almeida, F.C.L. (2016). *Wave Motion in Elastic Structures, in Dynamics of Smart Systems and Structures: Concepts and Applications* (eds. V. Lopes Jr., V. Steffen Jr. and M.A. Savi). Springer.

Cremer, L., Heckl, M., and Petersson, B.A.T. (2005). *Structure-Borne Sound: Structural Vibrations and Sound Radiation at Audio Frequencies*, 3rd Edition. Springer.

Ewins, D.J. (2000). *Modal Testing: Theory, Practice and Application*, 2nd Edition. Research Studies Press.

Fahy, F.J. (2003). Some applications of the reciprocity principle in experimental vibroacoustics. *Acoustical Physics*, 49(2), 217–229. https://doi.org/10.1134/1.1560385.

Gardonio, P. and Brennan, M.J. (2004). *Mobility and Impedance Methods in Structural Dynamics, Chapter 9 in Advanced Applications in Acoustics, Noise and Vibration*, (eds. F.J. Fahy and J. Walker). Spon Press.

Gonçalves, P.J.P., Peplow, A. and Brennan, M.J. (2018). Exact expressions for numerical evaluation of high order modes of vibration in uniform Euler-Bernoulli beams, *Applied Acoustics*, 141, 371–373. https://doi.org/10.1016/j.apacoust.2018.05.014

Kinsler, L.E., Frey, A.R., Coppens, A.B. and Sanders, J.V. (1982). *Fundamentals of Acoustics*, 3rd Edition. Wiley.

Leissa, A.W. and Qatu, M.S. (2011). *Vibrations of Continuous Systems.* McGraw-Hill.

Mead, D.J. (1999). *Passive Vibration Control.* Wiley.

Nashif, A.D., Jones, D.I.G., and Henderson, J.P. (1985). *Vibration Damping.* Wiley.

Tse, F.S., Morse, I.E., and Hinkle, R.T. (1978). *Mechanical Vibrations - Theory and Applications*, 2nd Edition. Ally and Bacon, Inc.

9

Multi-Degree-of-Freedom (MDOF) Systems: Virtual Experiments

9.1 Introduction

This chapter uses many of the topics discussed in previous chapters to conduct some virtual experiments with multi-degree-of-freedom (MDOF) systems. Two structures are investigated. One is a 2 DOF lumped parameter system that exhibits both resonance and anti-resonance frequencies, and the other is a cantilever beam comprising distributed mass and stiffness. In both cases the structures are excited using a shaker driven with random noise, and their modal properties are determined to develop modal models. This type of model can be used to gain insight into the dynamic behaviour of the structure under test, or to facilitate predictions which can be used for structural design changes or to help solve vibration control problems.

The concept of a virtual experiment was outlined in Chapter 1 and is further discussed here for convenience. The arrangement is shown in Figure 9.1. In many real vibration experiments the aim is to determine the dynamic properties of the structure under test from input and output measurements. Time histories of the excitation force and the resulting responses at several locations are measured. In many experiments, accelerometers are used to measure the responses, as they provide absolute rather than relative measurements, and are small so do not adversely affect the structure under test. In the virtual experiments described in this chapter, the dynamic properties of the structure under test are also determined from force and acceleration time histories, but these time histories are from simulations rather than measurements. Thus, the first step in a virtual experiment is to determine the time histories as indicated in Figure 9.1, and the ways this can be achieved are discussed in Chapters 5 and 6. Once the force and acceleration time histories have been determined, they can be processed, by transformation to the frequency domain as discussed in Chapter 7 to determine the modal and dynamic properties of the system. This is step 2 in Figure 9.1.

Following the virtual experiments on the 2DOF system and the cantilever beam, the use of a vibration absorber to control vibration is discussed. Its performance is illustrated using an SDOF host structure and is then applied to the cantilever beam. Virtual experiments show how the performance of a vibration absorber on a multi-modal distributed parameter system can be predicted both in the time and frequency domains.

9.2 Two Degree-of-Freedom System: FRF Estimation

The virtual experiment with the 2DOF system is shown in Figure 9.2. A shaker is used to excite the structure on the left-hand mass. Random noise, which has a constant amplitude up to half the

Virtual Experiments in Mechanical Vibrations: Structural Dynamics and Signal Processing,
First Edition. Michael J. Brennan and Bin Tang.
© 2023 John Wiley & Sons Ltd. Published 2023 by John Wiley & Sons Ltd.
Companion website: www.wiley.com/go/brennan/virtualexperimentsinmechanicalvibrations

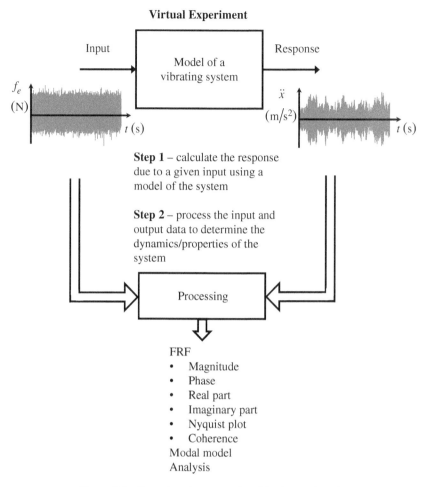

Figure 9.1 Procedure to carry out a virtual experiment.

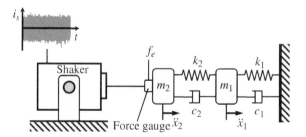

Figure 9.2 Measurement of the accelerance of a 2DOF system.

sampling frequency and a standard deviation of unity, is supplied to the shaker. As discussed in Chapter 7, the force generated by the shaker is proportional to the current supplied and is divided into two parts. One part drives the shaker itself. The other part f_e is applied to the structure and is measured by the force gauge. The resulting acceleration signals of the two masses are calculated using the 2DOF model. In this chapter, this is achieved by convolving the applied force f_e with the impulse response function (IRF) of the 2DOF system to give the acceleration responses, i.e.

$$\ddot{x}_{1,2}(t) = f_e(t) * \ddot{h}_{1,2}(t), \tag{9.1}$$

where $\ddot{h}_{1,2}(t) = \mathcal{F}^{-1}(-\omega^2 H_{1,2})$ in which $\mathcal{F}^{-1}$ denotes the inverse Fourier transform, and $H_1 = \mathbf{H}(1,2)$ and $H_2 = \mathbf{H}(2,2)$, where (i, j) are the elements in the receptance matrix $\mathbf{H} = [\mathbf{K} - \omega^2 \mathbf{M} + j\omega \mathbf{C}]^{-1}$, in which $\mathbf{M}$, $\mathbf{K}$, and $\mathbf{C}$ are the mass, stiffness, and damping matrices, respectively, given by

$$\mathbf{M} = \begin{bmatrix} m_1 & 0 \\ 0 & m_2 \end{bmatrix}, \quad \mathbf{K} = \begin{bmatrix} k_1 + k_2 & -k_2 \\ -k_2 & k_2 \end{bmatrix}, \quad \mathbf{C} = \begin{bmatrix} c_1 + c_2 & -c_2 \\ -c_2 & c_2 \end{bmatrix}.$$

Once the acceleration time histories have been determined, the second step of the process can be carried out, using these time histories together with the force time history, to determine the accelerances FRFs of the system. This is illustrated in MATLAB Example 9.1 for H_2.

MATLAB Example 9.1

In this example, the point acceleration FRF of the 2DOF system shown in Figure 9.2 is estimated when excited by a shaker with random noise.

```
clear all

%% Parameters
% structure
m1=1;m2=1;k1=1e4;k2=1e4;         % [kg, N/m]      % masses and stiffnesses
M=[m1 0; 0 m2];                                   % mass matrix
K=[k1+k2 -k2; -k2 k2];                            % stiffness matrix
C=2e-4*K;                         % [Ns/m]        % damping matrix

% shaker
ms=0.1;                           % [kg]          % mass
ws=2*pi*10;                       % [(rad/s)]     % natural frequency
ks=ws^2*ms;                       % [N/m]         % stiffness
zs=0.1;cs=2*zs*sqrt(ms*ks);       % [ ,Ns/m]      % damping ratio, coefficient

%% Direct FRF calculation
fs=1000;df=0.01;dt=1/fs;          % [Hz, s]       % frequency and time parameters
n=0;
for f=0:df:fs/2                   % [Hz]          % frequency loop
  w=2*pi*f;                       % [rad/s]
  n=n+1;
  A=inv(K-w.^2*M+j*w*C);          % [m/N]         % receptance matrix
  Acc=-w^2*A;                     % [m/Ns²]       % accelerance matrix
  R(n)=A(2,2);                    % [m/N]         % point receptance
  H(n)=Acc(2,2);                  % [m/Ns²]       % point accelerance
end

%% IRF
Hd=[H fliplr(conj(H))];           % [m/Ns²]       % form the double-sided spectrum
Ht=Hd(1:length(Hd)-1);            % [m/Ns²]       % set the length of the FRF
h=fs*ifft(Ht);                    % [m/Ns³]       % calculation of the IRF

%% Input and output
T=250;t=0:dt:T;                   % [s]           % signal duration; time vector
```

(Continued)

MATLAB Example 9.1   (Continued)

```
fe = randn(1,length(t));fe=fe-mean(fe);          % random force signal
f=0:df:fs/2; w=2*pi*f;            % [Hz, rad/s]   % frequency vector
K=1./R;                           % [N/m]         % dynamic stiffness of structure
Ks=ks-w.^2*ms+j*w*cs;             % [N/m]         % dynamic stiffness of shaker
Fe=K./(K+Ks);                                     % FRF for structural force
G=[Fe,fliplr(conj(Fe(1:length(Fe)-1)))];          % double-sided spectrum
g=fs*ifft(G);                     % [1/s]         % IRF for structural force
fes=conv(real(g),fe)/fs;fes=fes(1:length(fe));    % force applied to structure
N=length(fes);
x=conv(h,fes)/fs;x=x(1:length(fe));               % acceleration response

%% Frequency domain calculations
Na = 8;                                           % number of averages
nfft=round(N/Na);                                 % number of points in the DFT
noverlap=round(nfft/2);                           % number of points in the overlap
Sff=cpsd(fes,fes,hann(nfft),noverlap,nfft,fs);    % PSD of force applied
Sxx=cpsd(x,x,hann(nfft),noverlap,nfft,fs);        % PSD of acceleration response
He=tfestimate(fes,x,hann(nfft),noverlap,nfft,fs); % accelerance FRF
CoHe=mscohere(fes,x,hann(nfft),noverlap,nfft,fs); % coherence

dff=1/(nfft*dt);ff=0:dff:fs/2;                    % frequency resolution and vector

%% Plot the results
figure
plot(t,fe)                                        % force time history
axis square; axis([0,250,-6,6]); grid
xlabel('time (s)');ylabel('force (N)');

figure
plot(t,x)                                         % acceleration time history
axis square; axis([0,250,-8,8]); grid
xlabel('time (s)');
ylabel('acceleration (m/s^2)');

figure
plot(ff,10*log10(abs(Sff)))                       % force PSD
axis square; axis([0,40,-50,-10]); grid
xlabel('frequency (Hz)');
ylabel('force PSD (dB ref 1 N^2/Hz)');

figure
plot(ff,10*log10(abs(Sxx)))                       % acceleration PSD
axis square; axis([0,40,-80,20]); grid
xlabel('frequency (Hz)');
ylabel('acc. PSD (dB ref 1 m^2/s^4Hz)');

figure
plot(ff,20*log10(abs(He))); hold on               % FRF modulus
plot(f,20*log10(abs(H)),'-k')
axis square; axis([0,40,-40,40]); grid
```

(Continued)

MATLAB Example 9.1   (Continued)

```
xlabel('frequency (Hz)');
ylabel('|accelerance| (dB ref 1 m/Ns^2)');

figure
plot(ff,180/pi*unwrap(angle(He))); hold on          % phase
plot(f,180/pi*unwrap(angle(H)),'--k')
axis square; axis([0,40,0,200]); grid
xlabel('frequency (Hz)');
ylabel('phase (degrees)');

figure
plot(ff,CoHe)                                       % coherence
axis square; axis([0,40,0,1]); grid
xlabel('frequency (Hz)');
ylabel('coherence');

figure
plot(real(He),imag(He)); hold on                    % Nyquist plot
plot(real(H),imag(H),'--k')
axis square; axis([-40,40,0,80]); grid
xlabel('real \{accelerance\} (m/Ns^2)');
ylabel('imag \{accelerance\} (m/Ns^2)');
```

Results

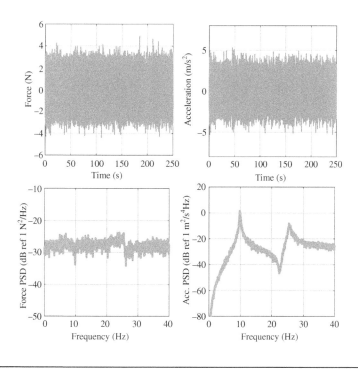

(Continued)

MATLAB Example 9.1   (Continued)

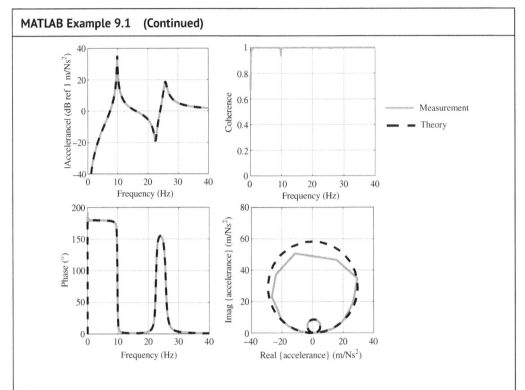

Comments:

1. This example gives the accelerance of a 2DOF system using random noise excitation. An exercise for the reader is to estimate the accelerance FRF using chirp excitation with the shaker and also to use an impact hammer. Compare the results.
2. In this example the output acceleration is determined by convolving the input force with IRF of the system. An exercise for the reader is to determine the acceleration response by using the Runge–Kutta method to numerically integrate the equations of motion and differentiating the velocity time history in the time domain to give acceleration.
3. Repeat the exercise by estimating the transfer accelerance with the shaker in its current position. Also change the position of the shaker so that it is attached to m_1 and then estimate the point and transfer accelerances. Check that reciprocity holds by overlaying the two transfer accelerances.
4. An exercise for the reader is to investigate what happens when the mass and natural frequency of the shaker are changed. Also investigate the effects of changing damping in the structure.

9.2.1   Determination of a Modal Model

It was shown in Chapter 8 that a lumped parameter model can be described in terms of its modal parameters. A modal model can also be determined from a measured FRF by extracting the modal parameters from the FRF, and this is illustrated in this section using the accelerance FRF estimated in MATLAB Example 9.1. In keeping with the spirit of this book, this process is described in an elementary way, but there are advanced techniques to achieve this, and the interested reader is referred to more specialist texts for details, for example Avitabile (2017) and Ewins (2000). In the simple approach described here, it is assumed that the modes are well separated, such that at frequencies close to a resonance frequency, the FRF can be approximated by a single mode or SDOF system. Now, the receptance given in Eq. (8.16), can be written as

$$\frac{\overline{X}_n}{\overline{F}_l} = \sum_{p=1}^{P} \frac{(A_{n,l})_p}{\omega_p^2 - \omega^2 + j2\zeta_p\omega\omega_p},$$ (9.2)

where the subscripts n and l refer to masses n and l, respectively, and ω_p, ζ_p, and $(A_{n,l})_p$ are the natural frequency, modal damping ratio, and modal constant of the p-th mode, respectively. Thus, if the three modal parameters can be determined for each mode, the FRF can then be approximated by a modal model. Provided the damping is light and the modes are well separated, these can be determined as follows:

1. Calculate the receptance from the acceleration FRF by dividing by $-\omega^2$.
2. Determine the natural frequencies ω_p by manually checking the frequencies at which the modulus of the receptance FRF is a maximum (or when the imaginary part of the receptance is a minimum or a maximum).
3. Use the half-power point method described in Chapter 2 to determine the modal damping ratios ζ_p. Note that it is easier to do this by checking the frequencies ω_1, ω_p, and ω_2 for mode p when the corresponding phase angle is $-45°$, $-90°$, and $-135°$, respectively. The modal damping ratio is then given by $\zeta_p \approx (\omega_2 - \omega_1)/(2\omega_p)$.
4. Determine the maximum value of the receptance by checking the maximum or minimum values of the imaginary part of the receptance. This gives $\text{Im}\{(\overline{X}_n/\overline{F}_l)_p\}_{\text{min/max}}$. Use this value to determine $(A_{n,l})_p$ by noting that this occurs when $\omega = \omega_p$, so that $(A_{n,l})_p = 2\zeta_p\omega_p^2 \, \text{Im}\{(\overline{X}_n/\overline{F}_l)_p\}_{\text{min/max}}$. Note that the sign of the imaginary part of the receptance at a natural frequency depends upon the position of the response measurement. For a point receptance the imaginary part is always negative, but for a transfer receptance it can be either positive or negative for any particular natural frequency.

Once the modal properties have been determined for each mode, the FRF can be plotted and compared with the measured FRF. This is carried out in MATLAB Example 9.2.

MATLAB Example 9.2

In this example, the modal properties of the system shown in Figure 9.2 are determined and a modal model of the FRF is constructed and compared with the measured FRF.

```
% clear all

% Exercise_9_1.m                         % Run MATLAB Example 9.1 and delete
                                          the figures

%% Calculation of receptance
Re=He./(-(2*pi*ff).^2)';      % [m/N]    % receptance

%% Modal properties
f1=9.84; f2=25.71;            % [Hz]     % estimation of natural
                                          frequencies from receptance FRF

x1=0.0148;x2=0.0003222;       % [m/N]    % estimation of receptance at the
                                          resonance frequencies from the
                                          Imag. part of the receptance
```

(Continued)

MATLAB Example 9.2   (Continued)

```
z1=(9.9-9.78)/(2*9.84);z2=(26.1-25.23)/(2*25.71);    % estimation of the modal damping
                                                        ratios using the half-power points

w=2*pi*ff;w1=2*pi*f1;w2=2*pi*f2;          % [rad/s]   % frequency and nat. freqs.
A1=x1.*(2*z1*w1^2); A2=x2.*(2*z2*w2^2);              % modal constants
Rm=A1./(w1^2-w.^2+j*2*z1*w*w1)+A2./(w2^2-...         % estimated receptance from modal
w.^2+j*2*z2*w*w2);                                      parameters

%% Figures
figure
plot(ff,20*log10(abs(Re)));hold on
plot(ff,20*log10(abs(Rm)),'--k')
axis square; axis([0,40,-120,-20]); grid
xlabel('frequency (Hz)');
ylabel('|receptance| (dB ref 1 m/N)');
figure
plot(ff,180/pi*unwrap(angle(Re)));hold on
plot(ff,180/pi*unwrap(angle(Rm)),'--k',...
'linewidth',2)
axis square; axis([0,40,-200,0]); grid
xlabel('frequency (Hz)');
ylabel('phase (degrees)');
figure
plot(real(Re),imag(Re));hold on
plot(real(Rm),imag(Rm),'--k');hold on
plot(real(R),imag(R),'k','linewidth',1.5)
axis square; axis([-0.01,0.01,-0.02,0]); grid
xlabel('real \{receptance\} (m/N)');
ylabel('imag \{receptance\} (m/N)');
```

Results

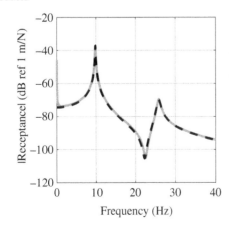

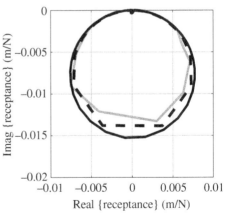

(Continued)

MATLAB Example 9.2   (Continued)

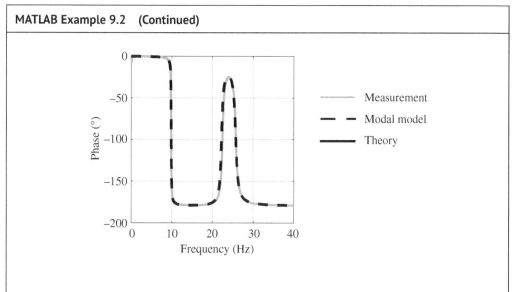

Comments:

1. Note that even though the frequency resolution is very good, the peak in the FRF at the first resonance frequency is still not completely resolved. This is evident from the shape of the Nyquist plot. This means that there is a small error in the modal model, which manifests itself as a slight shift in the estimated anti-resonance frequency. This can be resolved by using a finer frequency resolution.
2. Note the error in the measured receptance at very low frequencies. This is because the receptance was determined by dividing the receptance by $-\omega^2$. The calculated receptance at zero frequency is 0/0 in principle. However, this is contaminated by noise, which causes the error.
3. An exercise for the reader is to estimate modal models for the other point accelerance, and the transfer accelerance.
4. An exercise for the reader is to determine modal models using data from other measurements using chirp excitation with the shaker, and also excitation using an impact hammer. Compare the results.

9.3   Beam: FRF Estimation

In this section a virtual experiment is carried out on a beam, in a similar way to that carried out on the 2DOF lumped parameter system described in the previous section. The main difference between the two structures is that the beam has an infinite number of DOF, and hence an infinite number of natural frequencies, whereas the lumped parameter system has a finite number of DOF and hence a finite number of natural frequencies. In the virtual experiment, the beam can only be excited over a finite range of frequencies, and so only a finite number of modes will be excited. The experiment is shown in Figure 9.3. It consists of a 0.75 m long aluminium cantilever beam with a cross-section 2 cm × 1 cm. It is clamped at the left-hand end and excited by an electrodynamic shaker 10 cm from the clamped end. The force applied to the beam is measured by a force gauge

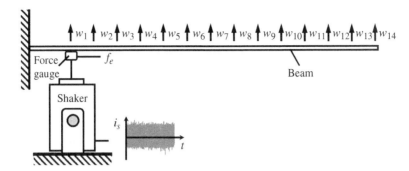

Figure 9.3 Measurement of receptance FRFs and modal properties of a cantilever beam.

as shown in the figure, and the displacement of the beam is measured at 14 positions that are 5 cm apart, also shown in the figure. Note that in an actual experiment it is likely that accelerometers or a laser velocity sensor would be used to measure the responses of the beam. However, if acceleration responses of the beam are simulated, as in the previous experiment with the 2DOF system, a significant amount of aliasing occurs at low frequencies for the reasons discussed in Chapter 4. The aliasing is more profound with a beam, however, because the mean level of the accelerance increases with frequency, unlike with the lumped parameter system, where it has a constant value at high frequencies. Accordingly, to avoid aliasing issues, displacement measurements are made in the virtual experiment, which could be achieved in practice by using a displacement sensing laser, for example. The large number of measurement positions shown on the beam has been chosen so that high fidelity estimates of the first three mode shapes can be made from the measured data (this is further discussed in Section 9.3.1).

Random noise, which has a constant amplitude up to half the sampling frequency and a standard deviation of unity, is supplied to the shaker. As with the previous experiment, the force generated by the shaker is proportional to the current supplied, and is divided into two parts, one part driving the shaker and the other part f_e, is applied to the beam and is measured by the force gauge. The resulting displacement signals are calculated using a model of the beam, by convolving the applied force f_e, with the displacement IRFs of beam corresponding to each measurement position, i.e. by

$$x_i(t) = f_e(t) * h_i(t),\tag{9.3}$$

where $h_i(t) = \mathcal{F}^{-1}(H_i)$ in which $\mathcal{F}^{-1}$ denotes the inverse Fourier transform, and H_i is the corresponding receptance FRF, which can be written in terms of a modal summation as described in Chapter 8, and is given by

$$\frac{\overline{W}(x_i)}{\overline{F}(x_j)} = \sum_{p=1}^{\infty} \frac{\phi_p(x_j)\phi_p(x_i)}{\tilde{m}_p\left(\omega_p^2 - \omega^2 + \mathrm{j}2\zeta_p\omega\omega_p\right)},\tag{9.4}$$

where $\phi_p(x_j)$ and $\phi_p(x_i)$ are the mode shapes of the p-th natural frequency evaluated at the position where the force is applied (x_j), and the positions where the displacement is measured (x_i); $\tilde{m}_p$, ω_p, and ζ_p are the modal mass, natural frequency, and modal damping ratio for the p-th mode, respectively. As mentioned above, a finite number of modes are included in the simulation, and for the specific beam studied in this chapter this involves 10 modes, which adequately covers the frequency range of 0–1500 Hz, which is up to half the sampling frequency.

Once the displacement time histories have been determined using Eq. (9.3), the force and the displacement time histories can be used to determine the receptance FRFs of the system. This is illustrated in MATLAB Example 9.3 for three measurement positions.

MATLAB Example 9.3

In this example, the receptance FRFs for three measurement positions on a cantilever beam is estimated from a system excited by a shaker with random noise.

```
clear all

%% Parameters
% cantilever beam
E=69e9;                                 % [N/m²]     % Young's modulus of aluminium
rho=2700;                               % [kg/m³]    % density of aluminium
l=0.75;b=0.02;d=0.01;S=b*d;I=b*d^3/12;              % geometrical parameters
z=0.01;n=2*z;                                        % damping ratio and loss factor
Ed=E*(1+j*n);                           % [N/m²]     % complex Young's modulus
m=rho*S*l;                              % [kg]       % mass of the beam

% frequency parameters
fs=3000;df=0.01;dt=1/fs;                             % frequency and time parameters
f=0.001:df:fs/2;                                     % frequency vector
w=2*pi*f;

%% Beam FRFs
n=0;
for x=0.1:0.05:l
 n=n+1;
 [ht,Htt] = calcFRF(E,I,rho,S,z,m,l,x,w,fs);        % calculate FRFs and IRFs
 H(n,:)=Htt;
 h(n,:)=ht;
end

%% Shaker
ms = 0.1;                               % [kg]       % mass
ws=2*pi*10;                             % [rad/s]    % natural frequency (rad/s)
ks=ws^2*ms;                             % [N/m]      % stiffness
zs=0.1;cs=2*zs*sqrt(ms*ks);             % [, Ns/m]   % damping
Ks=ks-w.^2*ms+j*w*cs;                   % [N/m]      % dynamic stiffness of shaker

%% Input and output
T=120;t=0:dt:T;                         % [s]        % signal duration; time vector
fe = randn(1,length(t));                % [N]        % random signal
fe=fe-mean(fe);                                      % set the mean to zero
K=1./H(1,:);                            % [N/m]      % dynamic stiffness of structure
Fe=K./(K+Ks);                                        % force ratio of applied force
G=[Fe,fliplr(conj(Fe(1:length(Fe)-1)))];             % double sided spectrum
g=fs*ifft(G);                                        % IRF to determine force applied
fes = conv(real(g),fe)/fs;              % [N]        % force applied to structure
fes = fes(1:length(fe));                % [N]        % force applied to structure
N=length(fes);                                       % Number of points

[dis] = calcResp(fes,fe,fs,h);                       % calculation of displ. responses

%% Frequency domain calculations
[Sffe,Sff,Sww,He,CoHe,ff]= ...                       % calculation of freq. domain
DisFRF(fes,fe,dis,N,fs,dt);                            quantitities
```

(Continued)

MATLAB Example 9.3   (Continued)

```
%% Figures
figure
plot(t,fes,'linewidth',2,'color',[0.7 0.7 0.7])      % force time history
hold on
plot(t,fe,'linewidth',2,'color',[0.3 0.3 0.3])
axis square; axis([0,120,-8,8]); grid
xlabel('time (s)');ylabel('force (N)');

p=1;                                                  % measurement position

figure                                                % disp. time history
plot(t,dis(p,:),'color',[0.7 0.7 0.7])
axis square; axis([0,120,-30e-5,30e-5]); grid
xlabel('time (s)');ylabel('displacement (m)');

figure                                                % force PSD
semilogx(ff,10*log10(abs(Sff)))
hold on
semilogx(ff,10*log10(abs(Sffe)))
axis square; axis([1,1000,-80,0]); grid
xlabel('frequency (Hz)');
ylabel('force PSD (dB ref 1 N^2/Hz)');

figure                                                % disp. PSD
semilogx(ff,10*log10(abs(Sww(p,:))))
axis square; axis([1,1000,-200,-60]); grid
xlabel('frequency (Hz)');
ylabel('displ. PSD (dB ref 1 m^2/Hz)');

figure                                                % FRF modulus
semilogx(ff,20*log10(abs(He(p,:))))
hold on
semilogx(f,20*log10(abs(H(p,:))))
hold on
axis square; axis([1,1000,-160,-40]); grid
xlabel('frequency (Hz)');
ylabel('|receptance| (dB ref 1 m/N)');

figure                                                % FRF phase
semilogx(ff,180/pi*unwrap(angle(He(p,:))))
hold on
plot(f,180/pi*unwrap(angle(H(p,:))))
axis square; axis([1,1000,-1000,0]); grid
xlabel('frequency (Hz)');
ylabel('phase (degrees)');

figure                                                % coherence
semilogx(ff,CoHe(p,:),'color',[0.7 0.7 0.7])
axis square; axis([1,1000,0,1]); grid
xlabel('frequency (Hz)');
ylabel('coherence');

%% Functions
function [ht,Htt] = ...                               % function to calculate FRF and IRF
```

(Continued)

MATLAB Example 9.3   (Continued)

```
calcFRF(E,I,rho,S,z,m,l,x,w,fs)
 nmax=10;
 % number of modes
 kl(1)=1.87510;kl(2)=4.69409;kl(3)=7.85476;        % kl values 1-3
 kl(4)=10.9956;kl(5)=14.1372;                      % kl values 4,5
 n=6:nmax;
 kl(n)=(2*n-1)*pi/2;                               % kl values > 5
 for n=1:nmax
  A=(sinh(kl(n))-...
  sin(kl(n)))./(cosh(kl(n))+cos(kl(n)));
  xf=0.1;                                          % position of applied force
  phi1=cosh(kl(n)*xf/l)-cos(kl(n)*xf/l)-...        % mode shape at force position
  A.*(sinh(kl(n)*xf/l)-sin(kl(n)*xf/l));
  phi2=cosh(kl(n)*x/l)-cos(kl(n)*x/l)-...          % mode shape at response position
  A.*(sinh(kl(n)*x/l)-sin(kl(n)*x/l));
  wn=sqrt((E*I)./(rho*S))*(kl(n)).^2;              % natural frequency

  Ht(n,:)=phi1*phi2./(m*(wn^2-w.^2+j*2*w*wn*z));   % receptance FRF of each mode
 end
 Htt=sum(Ht);                                      % overall receptance FRF

 %IRF
 Hd=[Htt fliplr(conj(Htt))];                       % form the double-sided spectrum
 Hdt=Hd(1:length(Hd)-1);                           % set the length of the FRF
 ht=fs*ifft(Hdt);                                  % calculation of the IRF
end

function [dis] = calcResp(fes,fe,fs,h)             % function to calculate
 for n=1:14                                           displacement responses
  dd = conv(real(h(n,:)),fes)/fs;                  % displacement response
  dis(n,:) = dd(1:length(fe));                     % displacement response
 end
end

function [Sffe,Sff,Sww,He,CoHe,ff] = ...           % function to calculate frequency
DisFRF(fes,fe,dis,N,fs,dt)                         % domain quantities
 Na = 8;                                           % number of averages
 nfft=round(N/Na);                                 % number of points in the DFT
 noverlap=round(nfft/2);                           % number of points in the overlap
 Sffe=cpsd(fes,fes,hann(nfft),noverlap,nfft,fs);   % PSD of force generated
 Sff=cpsd(fe,fe,hann(nfft),noverlap,nfft,fs);      % PSD of force applied

 for n=1:14                                        % number of measurement positions
  Sww(n,:)=cpsd(dis(n,:),dis(n,:),hann(nfft),...   % PSD of displacement response
  noverlap,nfft,fs);
  He(n,:)=tfestimate(fes,dis(n,:),hann(nfft),...   % FRF
  noverlap,nfft,fs);
  CoHe(n,:)=mscohere(fes,dis(n,:),hann(nfft),...   % coherence
  noverlap,nfft,fs);
 end

 dff=1/(nfft*dt);                                  % frequency resolution
 ff=0:dff:fs/2;                                    % frequency vector
end
```

(Continued)

MATLAB Example 9.3 (Continued)

Results

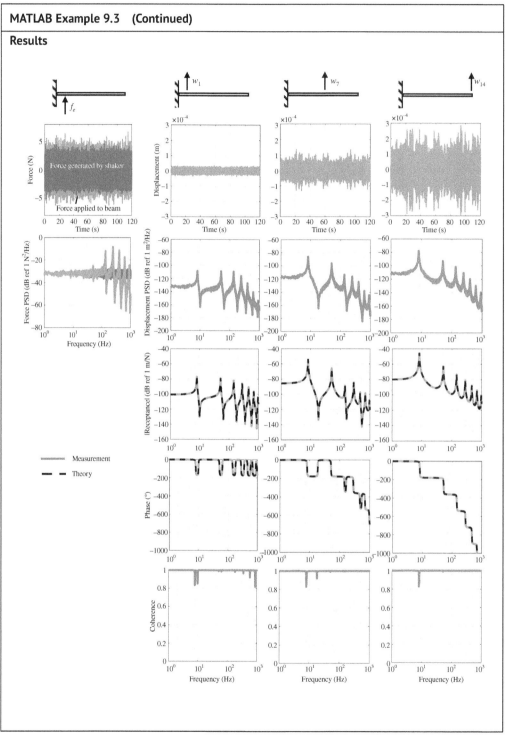

(Continued)

MATLAB Example 9.3   (Continued)

Comments:

1. This example gives the receptance FRFs for several positions on a cantilever beam using random noise excitation to excite the beam close to the clamped end. Fourteen FRFs are calculated, but only three are shown. Note the unwrapped phase is plotted so that the actual phase shift between the force and the displacement at the various points on the beam can be compared. The reader can plot the other FRFs.
2. An exercise for the reader is to use chirp excitation with the shaker and to use an impact hammer. Compare the results. Also change the position of the shaker and then calculate the receptance FRFs.
3. An exercise for the reader is to investigate what happens when the mass and natural frequency of the shaker are changed. Also investigate the effects of changing damping in the structure.
4. An exercise for the reader is to calculate the acceleration time histories, and then investigate the aliasing that occurs in the estimation of the accelerance FRFs.

9.3.1   Determination of a Modal Model

As with the virtual experiment for the lumped parameter system, the modal parameters can be estimated from measured receptance FRFs for the beam. However, because it is a distributed parameter system, more measurements need to be taken to provide a good estimate of the mode shapes. The approach is similar to that described for the 2DOF system, but with additional complexity. As the modes are well separated, an SDOF approach to modal estimation can be followed, but because there are an infinite number of modes, and only a few modes are included in the estimated FRFs, the contribution of the high-frequency unmodelled modes to the FRFs need to be included in the modal model. This contribution is generally called a residual R (Ewins, 2000). For the cantilever beam (in fact for any grounded structure), the residual is an additional stiffness term. It is a stiffness because the frequency range of interest is less than that containing the natural frequencies of the unmodelled higher frequency modes. Thus, these modes act as a stiffness at low frequencies, similar to that described in Chapter 2 for an SDOF system. The receptance FRF in terms of the measured modal properties is, therefore, given by

$$\frac{\overline{W}(x_i)}{\overline{F}(x_j)} = \sum_{p=1}^{P} \frac{(A_{ij})_p}{\omega_p^2 - \omega^2 + \mathrm{j}2\zeta_p\omega\omega_p} + R_{ij}. \tag{9.5}$$

The residual R_{ij} for each FRF is different and is generally larger for a point receptance. This is because the individual modal responses for a point receptance have the same sign, so the overall receptance is greater than that for an individual mode. The modal responses for a transfer receptance can be either positive or negative, however, so residuals for some modes tend to cancel each other, reducing the overall residual compared to that for the point receptance. Including the residual term is, therefore, most important for a point receptance, which will become evident in MATLAB Example 9.4.

The mode-shapes can be determined by making a note of the imaginary part of the receptance (including the sign) at each natural frequency of interest for the chosen points of interest on the beam. For example, for the beam shown in Figure 9.3, there are 14 points of interest, so 14 numbers represent the mode-shape. With an experimentally determined mode-shape, the numbers are often normalised by the maximum or minimum displacement per unit force. This is illustrated in MATLAB Example 9.4.

The procedure to estimate the modal parameters ω_p, ζ_p, and $(A_{ij})_p$ is the same as in steps 2–4 in Section 9.2.1. The residual term can be determined by adding/subtracting a number until Eq. (9.2) matches the measured receptance at frequencies well below the fundamental frequency of the structure. This procedure is also illustrated in MATLAB Example 9.4.

Once the modal properties and residual have been determined, the estimated FRF can be plotted and compared with the measured FRF. This is carried out in MATLAB Example 9.4 using the first three modes of the structure.

MATLAB Example 9.4

In this example, the modal properties of the system shown in Figure 9.3 are determined, and a modal model of the FRF including the first three natural frequencies is constructed and compared with the measured FRF.

```
% clear all

% Exercise_9_3.m                          % Run MATLAB Example 9.3 and
                                            delete the figures

%% Mode shapes
xx=[0.1 0.15 0.2 0.25 0.3 0.35 0.4 0.45 ... % positions on the beam where the
0.5 0.55 0.6 0.65 0.7 0.75];                 mode-shapes are estimated

p1=[0.13 0.29 0.49 0.74 1.03 1.35 1.7 ...   % obtain the displacement values
2.07 2.45 2.85 3.25 3.67 4.07 4.48];         from the imaginary part of the
p2=[0.11 0.22 0.34 0.44 0.50 0.53 0.51 ...   receptance at the first 3
0.44 0.31 0.15 -0.05 -0.27 -0.50 -0.74];     resonance frequencies
p3=[0.08 0.13 0.16 0.16 0.12 0.04 -0.04 ...
-0.1 -0.14 -0.14 -0.09 0 0.1 0.22];

%% Theoretical mode-shapes for the first 3 modes
kl=1.87510;
[xf,pt]=modeCalc(kl,1);                     % function for mode-shapes
p1t=pt;                                     % first mode-shape
kl=4.69409;
[xf,pt]=modeCalc(kl,1);                     % function for mode-shapes
p2t=pt;                                     % second mode-shape
kl=7.85476;
[xf,pt]=modeCalc(kl,1);                     % function for mode-shapes
p3t=pt;                                     % third mode-shape

%% Modal properties
% modal model for W1/Fe;
f1=8.13;f2=51.2;f3=143.3;                   % estimate the natural frequencies
                                              from the receptance FRF

x1=0.1315e-3;x2=0.1138e-3;x3=7.83e-5;       % estimate the responses at the
                                              resonance frequencies from the
                                              Imag. Part of the receptance

z1=(8.27-8.07)/(2*f1);z2=(51.73-50.73)/(2*f2);  % estimate the modal damping ratios
z3=(144.8-141.9)/(2*f3);                         using the half-power points
w=2*pi*ff;w1=2*pi*f1;w2=2*pi*f2;w3=2*pi*f3;  % frequency and natural frequencies
```

(Continued)

MATLAB Example 9.4   (Continued)

```
A1=x1.*(2*z1*w1^2);A2=x2.*(2*z2*w2^2);      % determine the modal constants
A3=x3.*(2*z3*w3^2);
Rm(1,:)=A1./(w1^2-w.^2+j*2*z1*w*w1)+A2./(w2^2-    % estimate receptance from modal
w.^2+j*2*z2*w*w2)+A3./(w3^2-w.^2+j*2*z3*w*w3);     parameters
AA=1/(He(1,1)-Rm(1));                        % estimate the residual
Rmm(1,:)= Rm(1,:) + w./w*1./AA;              % add the residual

% modal model W7/Fe;
f1=8.13;f2=51.2;f3=143.3;                    % estimate the natural frequencies
x1=1.699e-3;x2=0.5085e-3;x3=-3.55e-5;        % responses at the resonance freqs.
z1=(8.27-8.07)/(2*f1);z2=(51.73-50.73)/(2*f2);   % estimate the modal damping ratios
z3=(144.8-141.9)/(2*f3);                         using the half-power points
w=2*pi*ff;w1=2*pi*f1;w2=2*pi*f2;w3=2*pi*f3;  % frequency and natural frequencies
A1=x1.*(2*z1*w1^2);A2=x2.*(2*z2*w2^2);       % determine the modal constants
A3=x3.*(2*z3*w3^2);
Rm(7,:)=A1./(w1^2-w.^2+j*2*z1*w*w1)+A2./(w2^2-    % estimate receptance from modal
w.^2+j*2*z2*w*w2)+A3./(w3^2-w.^2+j*2*z3*w*w3);     parameters
AA=1/(He(7,1)-Rm(7,1));                      % estimate the residual
Rmm(7,:)= Rm(7,:) + w./w*1./AA;              % add the residual

% modal model W14/Fe;
f1=8.13;f2=51.2;f3=143.3;                    % estimate the natural frequencies
x1=4.483e-3;x2=-0.7371e-3;x3=0.2183e-3;      % responses at the resonance freqs.
z1=(8.27-8.07)/(2*f1); z2=(51.73-50.73)/(2*f2);  % estimate the modal damping ratios
z3=(144.8-141.9)/(2*f3);                         using the half-power points
w=2*pi*ff;w1=2*pi*f1;w2=2*pi*f2;w3=2*pi*f3;  % frequency and natural frequencies
A1=x1.*(2*z1*w1^2);A2=x2.*(2*z2*w2^2);       % determine the modal constants
A3=x3.*(2*z3*w3^2);
Rm(14,:)=A1./(w1^2-w.^2+j*2*z1*w*w1)+...     % estimate receptance from modal
w.^2+j*2*z2*w*w2)+A3./(w3^2-...                   parameters
w.^2+j*2*z3*w*w3);
AA=1/(He(14,1)-Rm(14,1));                    % estimate the residual
Rmm(14,:)= Rm(14,:) + w./w*1./AA;            % add the residual

%% Plot the results
figure
plot(xf,p1t/max(p1t),'color',[0.7 0.7 0.7])  % first mode-shape
hold on
plot(xx,p1/4.48,'ok','Markersize',10)
axis square; axis([0,0.75,-0.1,1]); grid
xlabel('beam position (m)');
ylabel('mode shape');

figure
plot(xf,-p2t/min(p2t),'color',[0.7 0.7 0.7])  % second mode shape
hold on
plot(xx,p2/0.74,'ok','Markersize',10)
axis square; axis([0,0.75,-1,1]); grid
xlabel('beam position (m)');
ylabel('mode shape');

figure
plot(xf,p3t/max(p3t),'color',[0.7 0.7 0.7])   % third mode shape
```

(Continued)

MATLAB Example 9.4   (Continued)

```
hold on
plot(xx,p3/0.22,'ok','Markersize',10)
axis square; axis([0,0.75,-1,1]); grid
xlabel('beam position (m)');
ylabel('mode shape');

p=1;                                        % choose position on beam, 1,7 or 14
figure
semilogx(ff,20*log10(abs(He(p,:))))         % modulus of FRF
hold on
semilogx(ff,20*log10(abs(Rm(p,:))))
hold on
semilogx(ff,20*log10(abs(Rmm(p,:))))
axis square; axis([1,1000,-160,-40]); grid
xlabel('frequency (Hz)');
ylabel('|receptance| (dB ref 1 m/N)');

figure
semilogx(ff,180/pi*(angle(He(p,:))))        % phase
hold on
semilogx(ff,180/pi*(angle(Rm(p,:))))
hold on
semilogx(ff,180/pi*(angle(Rmm(p,:))))
axis square; axis([1,1000,-200,+200]); grid
xlabel('frequency (Hz)');
ylabel('phase (degrees)');

%% Function
function [xf,pt]=modeCalc(kl,l);            % function to calculate mode-shape
 A=(sinh(kl)-sin(kl))./(cosh(kl)+cos(kl));
 xf=0:0.01:l;
 pt=cosh(kl*xf/l)-cos(kl*xf/l)-...
 A.*(sinh(kl*xf/l)-sin(kl*xf/l));
end
```

Results

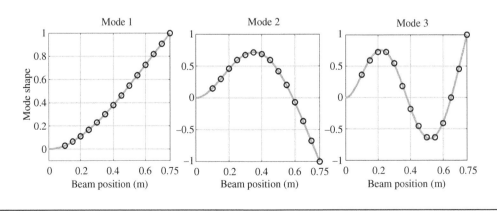

(Continued)

MATLAB Example 9.4 (Continued)

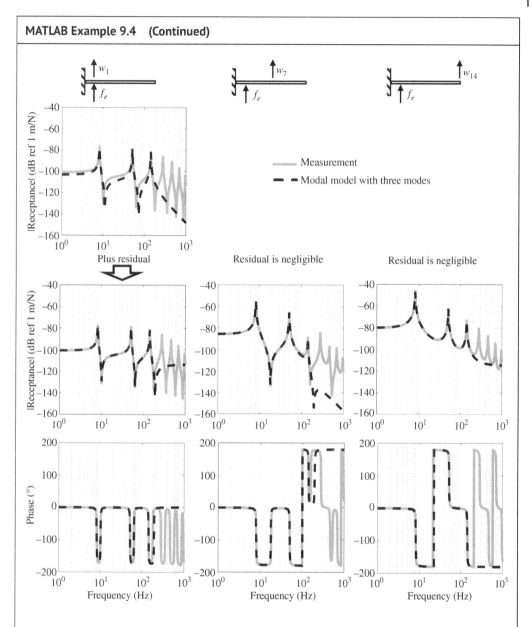

Comments:

1. The mode-shapes are well estimated using the imaginary part of the receptance at the corresponding natural frequencies. To do this, however, a large number of measurements need to be made.
2. Note that inclusion of the residual in the modal model is important with the point FRF to obtain a good match between the measurements and the model.
3. An exercise for the reader is to determine the modal model for other measurement positions.
4. An exercise for the reader is to determine the modal model using data from other measurements using chirp excitation with the shaker and an impact hammer. Compare the results.

9.4   The Vibration Absorber as a Vibration Control Device

Since its inception (Ormondroyd and Den Hartog, 1928), the vibration absorber has been employed as a vibration control device for almost 100 years. It is used in one of two distinct ways. Either it is tuned to a troublesome resonance frequency, where it is configured so that it adds damping to the host structure, or it is tuned to a troublesome forcing frequency, to provide a high impedance so that the response of the host structure is minimised at this frequency. In the former case it is called a 'vibration absorber', and in the latter case it is called a 'vibration neutraliser'. Although the device has many different forms, some of which are described in Den Hartog (1956) and Reed (2002), it can be analysed as a base-excited mass-spring-damper system. Of particular interest in this chapter is the vibration absorber. The theory underpinning its operation and performance is described, and then two examples are given, in which a vibration absorber is used to suppress vibration on a lumped parameter and a distributed parameter system.

9.4.1   Theory

To study the way in which a vibration absorber affects the vibration of a host structure, a frequency domain approach is used, in which the component parts of the system are described in terms of their FRFs. A generic example is illustrated in Figure 9.4, which shows the vibrating host structure and the vibration absorber as disconnected parts. Three points are of interest on the host structure. The structure is excited at point 1. Point 2 is where the vibration absorber is attached and point 3 is any point on the structure. The host structure alone can be described in terms of a matrix of FRFs as

$$\begin{Bmatrix} \overline{X}_1 \\ \overline{X}_2 \\ \overline{X}_3 \end{Bmatrix} = \begin{bmatrix} H_{11} & H_{12} & H_{13} \\ H_{21} & H_{22} & H_{23} \\ H_{31} & H_{32} & H_{33} \end{bmatrix} \begin{Bmatrix} \overline{F}_1 \\ \overline{F}_2 \\ \overline{F}_3 \end{Bmatrix}, \tag{9.6}$$

where $\overline{X}_i$ and $\overline{F}_j$ are the complex amplitudes of the displacement at point i and forces at point j, respectively, and H_{ij} are the receptance FRFs between these points. The vibration absorber can be described in terms of receptance at its base, which is the reciprocal of its dynamic stiffness K_a, and is given by Brennan (1997)

$$\frac{\overline{X}_a}{\overline{F}_a} = H_a = \frac{1}{K_a} = \frac{k_a - \omega^2 m_a + j\omega c_a}{-\omega^2 m_a (k_a + j\omega c_a)}. \tag{9.7}$$

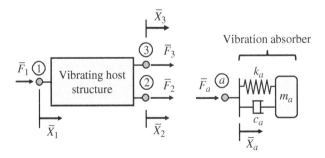

Figure 9.4   Connecting a vibration absorber to a general vibrating system.

When the absorber is connected to the vibrating structure, the following conditions apply:

$\overline{F}_3 = 0$ (no force applied),

$\overline{F}_2 = -\overline{F}_a$ (force equilibrium),

$\overline{X}_2 = \overline{X}_3$ (continuity of displacement)

Applying these conditions, Eqs. (9.6) and (9.7) combine to give

$$\frac{\overline{X}_3}{\overline{F}_1} = H_{31} - \frac{H_{21}H_{32}}{H_{22} + H_a} \tag{9.8a}$$

This is a general formulation from which the vibration at other points can be determined. For example, if points 1 and 2 are of interest, then in the formulation of Eq. (9.8a), 3 can be replaced with 1 and 2, respectively, to give

$$\frac{\overline{X}_1}{\overline{F}_1} = H_{11} - \frac{H_{21}H_{12}}{H_{22} + H_a}, \tag{9.8b}$$

$$\frac{\overline{X}_2}{\overline{F}_1} = \frac{H_{21}H_a}{H_{22} + H_a}, \tag{9.8c}$$

and if the absorber is moved to the source position, then Eq. (9.8a) becomes

$$\frac{\overline{X}_3}{\overline{F}_1} = \frac{H_{31}H_a}{H_{11} + H_a}. \tag{9.8d}$$

Thus, the vibration absorber affects the vibration of different points on the structure in different ways. It is most effective, globally, when it is placed close to the source. From Eq. (9.7), note that when the forcing frequency is equal to the natural frequency of the vibration absorber, i.e. when $\omega = \sqrt{k_a/m_a}$, and if damping is small, the absorber behaves as a damper with an equivalent damping coefficient of $m_a k_a/c_a$. Thus, the vibration absorber can add a large amount of damping to a host structure, but over a limited bandwidth.

9.4.2 Effect of a Vibration Absorber on an SDOF System

A convenient way to illustrate the way in which a vibration absorber suppresses vibration is to investigate the way it affects the vibration of an SDOF system. Such a system with a vibration absorber attached is shown in Figure 9.5a, which can be considered in terms of its component parts as shown in Figure 9.5b. Note that the excitation force and the displacement responses are shown in terms of frequency domain quantities, where $\overline{F}$ is the amplitude of the harmonic excitation force, and $\overline{X}, \overline{X}_a$ are the complex displacement amplitudes of the masses of the host structure and absorber, respectively. The displacement of the host structure mass per unit input force is given by

$$\frac{\overline{X}}{\overline{F}} = \frac{1}{K + K_a}, \tag{9.9}$$

where $K_a = 1/H_a$ is the dynamic stiffness of the absorber, which is given by the reciprocal of Eq. (9.7), and K is the dynamic stiffness of the host structure given by $K = k - \omega^2 m + j\omega c$. To simplify the analysis, the damping in the host structure is set to zero. Substituting for K and K_a in Eq. (9.9)

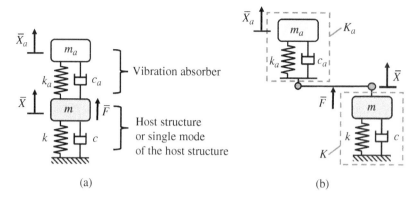

(a)   (b)

Figure 9.5 A vibration absorber attached to an SDOF host structure or a single mode of the host structure: (a) physical diagram, (b) dynamic stiffness model of the system.

and noting that $\gamma = \omega_n/\omega_a$, $\mu = m_a/m$, $\zeta_a = c_a/(2\sqrt{m_a k_a})$, and $\Omega = \omega/\omega_n$ in which $\omega_n = \sqrt{k/m}$, results in

$$\frac{k\overline{X}}{\overline{F}} = \frac{1 - \gamma^2\Omega^2 + j2\zeta_a\gamma\Omega}{\gamma^2\Omega^4 - (1 + \gamma^2 + \mu)\Omega^2 + 1 + j2\zeta_a\gamma\Omega(1 - (1 + \mu)\Omega^2)}. \tag{9.10}$$

Noting that $\dfrac{k\overline{X}_a}{\overline{F}} = \dfrac{k\overline{X}}{\overline{F}}\dfrac{\overline{X}_a}{\overline{X}}$, and $\dfrac{\overline{X}_a}{\overline{X}} = \dfrac{k_a + j\omega c_a}{k_a - \omega^2 m_a + j\omega c_a}$, which is the displacement transmissibility between the mass of the host structure and the absorber mass, results in

$$\frac{k\overline{X}_a}{\overline{F}} = \frac{1 + j2\zeta_a\gamma\Omega}{\gamma^2\Omega^4 - (1 + \gamma^2 + \mu)\Omega^2 + 1 + j2\zeta_a\gamma\Omega(1 - (1 + \mu)\Omega^2)}. \tag{9.11}$$

It can be seen that the factors which determine the dynamics of the host structure with a vibration absorber attached are the mass ratio μ, the natural frequency of the absorber compared with the natural frequency of the host structure γ, and the absorber damping ratio ζ_a. Various optimum parameters have been proposed (Den Hartog, 1956), but the following parameters result in an effective practical absorber

$$\mu = 0.05, \tag{9.12a}$$

$$\gamma = 1 + \mu, \tag{9.12b}$$

$$\zeta_{a(opt)} = \sqrt{\frac{3\mu}{8(1 + \mu)^3}}. \tag{9.12c}$$

Equation (9.10) is plotted in the lower part of Figure 9.6 for $\mu = 0.05$ and $\gamma = 1 + \mu$ for various values of damping. By setting the natural frequency of the absorber as in Eq. (9.12b), the amplitudes of vibration at points A and B in Figure 9.6 become the same. It can be seen that if the damping is very small, then although the vibration at the natural frequency of the absorber is small, there is a large response around the two resonance frequencies of the composite system. If the damping is very large, then the relative motion of the absorber mass compared with that of the host structure mass is very small and so the absorber is ineffective. The optimum damping given by Eq. (9.12c) ensures that the maximum amplitude of vibration is approximately the same as that at points A and B. Thus, the absorber has the effect of achieving a relatively 'flat' FRF at frequencies close to the resonance frequency of the host structure ($\Omega = 1$), which is due to the targeted high (optimum) damping characteristic of the absorber in this frequency region.

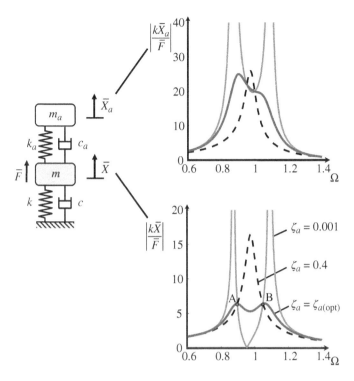

Figure 9.6 Normalised displacement FRFs of the absorber mass and the host structure mass due to force excitation of the host structure.

The corresponding vibration of the absorber mass is plotted in the top part of Figure 9.6 using the same parameters. It can be seen that when $\zeta_a \leq \zeta_{a(opt)}$, then at the resonance frequency of the host structure alone ($\Omega = 1$), the amplitudes of the FRFs have a similar value, which is independent of damping. This value can be approximately determined by setting $\Omega = 1$ in Eq. (9.11) and noting that $2\zeta_a \gamma \ll 1$, which results in $|k\overline{X}_a/\overline{F}|_{\Omega=1} \approx 1/\mu$, i.e. the displacement of the absorber mass at this frequency is governed by the mass ratio. If the absorber mass is required to vibrate less due to design constraints, then the absorber mass should be increased.

9.4.3 Vibration Absorber Attached to an SDOF System – Virtual Experiment

In many practical situations a vibration absorber is fitted retrospectively to a structure where the vibration is too large. There are many case studies which demonstrate this, but a particularly high-profile case is the vibration of the millennium bridge in London (Newland, 2003; Belykh et al., 2021). To illustrate the design process for a simple structure, an SDOF system is considered. An experiment is often carried out to determine the mass and natural frequency of this system. Such an experiment is illustrated in Figure 9.7. The host structure is excited using an electrodynamic shaker and the accelerance is measured, the modulus of which is plotted in the lower part of Figure 9.7. It can be seen that the natural frequency can be determined approximately from the frequency of the resonance peak (for a lightly damped structure), and the high-frequency asymptote gives the mass. The process of carrying out a virtual experiment on the host structure alone is described in MATLAB Example 9.5a below.

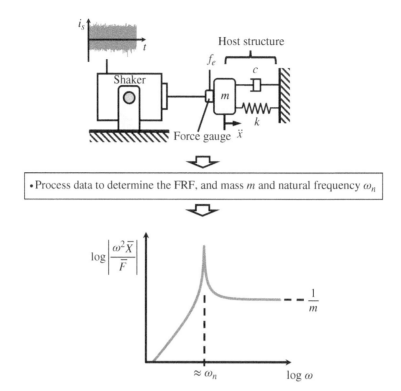

Figure 9.7 Experiment to measure the accelerance of the host structure, and from this to determine the mass and natural frequency.

MATLAB Example 9.5a

In this example, the accelerance FRF for the host SDOF is measured in a virtual experiment in which the shaker is driven with random noise as shown in Figure 9.7.

```
clear all

%% Parameters
% host structure
m=1;                           % [kg]          % mass
k=1e4;                         % [N/m]         % stiffness
wn=sqrt(k/m);                  % [rad/s]       % natural frequency
z=0.01;c=2*z*sqrt(k*m);        % [ ,Ns/m]      % damping ratio and coefficient

% shaker
ms = 0.1;                      % [kg]          % mass
ws=2*pi*10;                    % [rad/s]       % natural frequency (rad/s)
ks=ws^2*ms;                    % [N/m]         % stiffness
zs=0.1; cs=2*zs*sqrt(ms*ks);   % [ ,Ns/m]      % damping

%% FRF calculation
fs=1000;df=0.01;dt=1/fs;       % [Hz,s]        % frequency/time parameters
f=0:df:fs/2; w=2*pi*f;         % [Hz,rad/s]    % frequency vector
H=1./(k-w.^2*m+j*w*c);         % [m/N]         % receptance
Ha=-w.^2.*H;                   % [m/Ns²]       % accelerance
```

(Continued)

MATLAB Example 9.5a   (Continued)

```
%% IRF
Hd=[Ha fliplr(conj(Ha))];              % [m/N]        % form the double-sided spectrum
Ht=Hd(1:length(Hd)-1);                 % [m/N]        % set the length of the FRF
h=fs*ifft(Ht);                         % [m/Ns]       % calculation of the IRF

%% Input and output
T=250;t=0:dt:T;                        % [s]          % signal duration; time vector
fe = randn(1,length(t));               % [N]          % random signal
fe=fe-mean(fe);                                        % set the mean to zero
K=1./H;                                % [N/m]        % dynamic stiffness of structure
Ks=ks-w.^2*ms+j*w*cs;                  % [N/m]        % dynamic stiffness of shaker
Fe=K./(K+Ks);                          % [N]          % force ratio of applied force
G=[Fe,fliplr(conj(Fe(1:length(Fe)-1)))];              % double sided spectrum
g=fs*ifft(G);                          % [N/s]        % IRF to determine force applied
fes = conv(real(g),fe)/fs;             % [N]          % force applied to structure
fes = fes(1:length(fe));               % [N]          % force applied to structure
N=length(fes);                                         % number of points
a = conv(h,fes)/fs;                    % [m/s^2]      % acceleration response
a = a(1:length(fe));                   % [m/s^2]      % acceleration response

%% Frequency domain calculations
Na = 8;                                                % number of averages
nfft=round(N/Na);                                      % number of points in the DFT
noverlap=round(nfft/2);                                % number of points in the overlap
Sff=cpsd(fes,fes,hann(nfft),noverlap,nfft,fs);         % PSD of force applied
Saa=cpsd(a,a,hann(nfft),noverlap,nfft,fs);             % PSD of acceleration response
He=tfestimate(fes,a,hann(nfft),noverlap,nfft,fs);      % FRF
CoHe=mscohere(fes,a,hann(nfft),noverlap,nfft,fs);      % coherence
dff=1/(nfft*dt); ff=0:dff:fs/2;                        % frequency resolution/vector

%% Plot of the results
figure
plot(t,fes,'linewidth',2)                              % force time history
axis square; axis([0,250,-6,6]); grid
xlabel('time (s)'); ylabel('force (N)');

figure
plot(t,a,'linewidth',2)                                % acceleration time history
axis square; axis([0,250,-8,8]); grid
xlabel('time (s)');ylabel('acceleration (m/s^2)');

figure
semilogx(ff,10*log10(abs(Sff)),'linewidth',4)          % force PSD
axis square; axis([1,1000,-50,-10]); grid
xlabel('frequency (Hz)');
ylabel('force PSD (dB ref 1 N^2/Hz)');

figure
semilogx(ff,10*log10(abs(Saa)),'linewidth',4)          % acceleration PSD
axis square; axis([1,1000,-80,20]); grid
xlabel('frequency (Hz)');
ylabel('acc. PSD (dB ref 1 m^2/s^4Hz)');
```

(Continued)

MATLAB Example 9.5a   (Continued)

```
figure
semilogx(ff,20*log10(abs(He)),'linewidth',4)          % accelerance FRF modulus
hold on
plot(f,20*log10(abs(Ha)),'--k','linewidth',2)
axis square; axis([1,1000,-40,40]); grid
xlabel('frequency (Hz)');
ylabel('|accelerance| (dB ref 1 m/Ns^2)');

figure
semilogx(ff,180/pi*unwrap(angle(He)),'linewidth',4)   % phase
hold on
semilogx(f,180/pi*unwrap(angle(Ha)),'--k')
axis square; axis([1,1000,0,200]); grid
xlabel('frequency (Hz)');ylabel('phase (degrees)');

figure
semilogx(ff,CoHe,'linewidth',4)                       % coherence
axis square; axis([1,1000,0,1]); grid
xlabel('frequency (Hz)');ylabel('coherence');

figure
plot(real(He),imag(He),'linewidth',4)                 % Nyquist plot
hold on
plot(real(Ha),imag(Ha),'--k','linewidth',2)
axis square; axis([-30,30,0,60]); grid
xlabel('real \{accelerance\} (m/Ns^2)');
ylabel('imag \{accelerance\} (m/Ns^2)');
```

Results

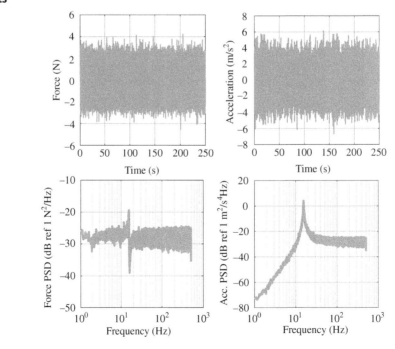

(Continued)

MATLAB Example 9.5a   (Continued)

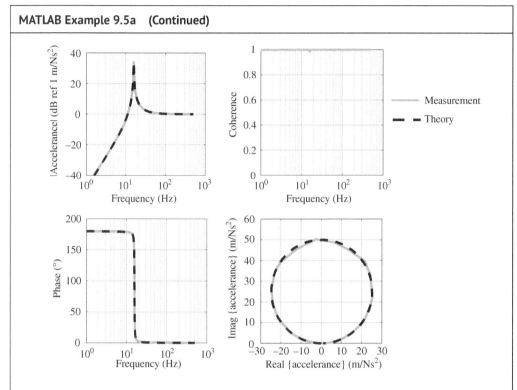

Comments:

1. This example gives the accelerance for the SDOF host structure when it is excited using random noise with a shaker. An exercise for the reader is to plot the mobility and receptance FRFs.

2. From this measurement, determine the mass of the structure and the natural frequency as shown in Figure 9.7. The parameters are needed to calculate the optimum parameters of the vibration absorber.

Once the mass and the natural frequency of the host structure have been determined, the vibration absorber can be designed using Eqs. (9.12a)–(9.12c). The mass of the absorber is chosen first, and in the example shown here, it is 5% of the mass of the host structure. If a larger vibration absorber mass is used, then it vibrates proportionately less and the frequency range between the two peaks in the FRF widens. Once the mass of the vibration absorber has been set, then its stiffness and damping values may be chosen. The vibration absorber can then be designed, built, and tested. Perhaps the most difficult parameter value to achieve in practice is the damping, as this is often realised in combination with the stiffness element.

A vibration experiment to test the vibration absorber is shown in Figure 9.8. However, a problem occurs in the virtual experiment described in MATLAB Example 9.5b that does not manifest itself in the real experiment. In the virtual experiment, as in the real experiment, an input is needed to excite the system. In the virtual experiment, however, the output is calculated using a model, unlike in the real experiment, where the output is measured. The problem occurs due to the dynamics of the model of the absorber, when excited at its base. Referring to Figure 9.4, this is because the receptance FRF $\overline{X}_a/\overline{F}_a$ is infinite at low frequency, the mobility FRF $j\omega\overline{X}_a/\overline{F}_a$ is infinite at

- Choose mass of the absorber, e.g. $m_a = 0.05\,m$, so that $\mu = 0.05$
- Choose the stiffness by tuning the absorber so that $\omega_a = \omega_n/(1 + \mu)$
- Choose the damping so that $\zeta_a = \sqrt{3\mu/\left[8(1 + \mu)^3\right]}$
- Carry out an experiment to verify the parameters

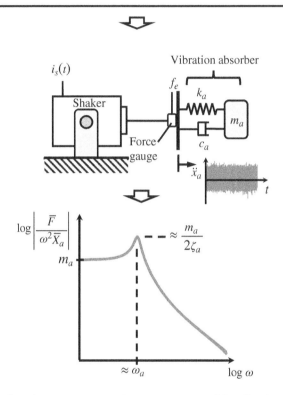

Figure 9.8 Experiment to measure the apparent mass of the vibration absorber.

low and high frequency, and the accelerance $-\omega^2 \overline{X}_a/\overline{F}_a$ is infinite at high frequency. This causes a problem in determining the time domain output when excited by a force that has a constant PSD as a function of frequency. A solution to the problem is to use the apparent mass, given by $M_a = \overline{F}_a/(-\omega^2\overline{X}_a)$, to determine the force for a prescribed acceleration of the vibration absorber base. The apparent mass is shown in the lower part of Figure 9.8, in which it can be seen that it is constant at low frequencies, has a peak at the resonance frequency, and then decreases at higher frequencies, tending to zero.

Thus, the apparent mass of the vibration absorber is the best model to use in the estimation of the time histories to be used in the virtual experiment to measure the dynamic properties of the vibration absorber in MATLAB Exercise 9.5b.

MATLAB Example 9.5b

In this example, the apparent mass of the vibration absorber to be attached to the SDOF host structure is measured in a virtual experiment in which the shaker is excited with random noise as shown in Figure 9.8 The base acceleration of the vibration absorber is set so that its PSD

(Continued)

MATLAB Example 9.5b   (Continued)

is constant with frequency. This could be achieved in a practical test set up by controlling the current supplied to the shaker.

```
% clear all

%% Parameters
% structure
m=1;                          % [kg]          % mass
k=1e4;                        % [N/m]         % stiffness
wn=sqrt(k/m);                 % [rad/s]       % natural frequency

% absorber
ma=0.05;                      % [kg]          % mass of absorber
mu=ma/m;                                      % mass ratio
wa=wn/(1+mu);                 % [rad/s]       % absorber natural frequency
ka=wa^2*ma;                   % [N/m]         % stiffness of the absorber
za=sqrt(3/8*mu/(1+mu)^3);                     % absorber damping ratio
ca=2*za*sqrt(ma*ka);          % [Ns/m]        % absorber damping coefficient

% shaker
ms=0.1;                       % [kg]          % mass
ws=2*pi*10;                   % [rad/s]       % natural frequency (rad/s)
ks=ws^2*ms;                   % [N/m]         % stiffness
zs=0.1; cs=2*zs*sqrt(ms*ks);  % [ ,Ns/m]      % damping

%% FRF calculation
fs=1000;df=0.01;dt=1/fs;      % [Hz,s]        % frequency/time parameters
f=0:df:fs/2; w=2*pi*f;        % [Hz,rad/s]    % frequency vector
Ka=ma*(ka+j*w*ca)./(ka-w.^2*ma+j*w*ca); % [kg]  % apparent mass

%% IRF
Hd=[Ka fliplr(conj(Ka))];     % [kg]          % form the double-sided spectrum
Ht=Hd(1:length(Hd)-1);        % [kg]          % set the length of the FRF
h=fs*real(ifft(Ht));          % [kg/s]        % calculation of the IRF

%% Input and output
T=250;t=0:dt:T;               % [s]           % signal duration; time vector
a=randn(1,length(t));         % [m/s^2]       % random acceleration signal
a=a-mean(a);                  % [m/s^2]       % set the mean to zero
N=length(a);
fe = conv(h,a)/fs;            % [N]
fe = fe(1:length(a));         % [N]           % force signal

%% Frequency domain calculations
Na = 8;                                       % number of averages
nfft=round(N/Na);                             % number of points in the DFT
noverlap=round(nfft/2);                       % number of points in the overlap
Sff=cpsd(fe,fe,hann(nfft),noverlap,nfft,fs);  % PSD of force signal
Saa=cpsd(a,a,hann(nfft),noverlap,nfft,fs);    % PSD of acceleration signal
Kae=tfestimate(a,fe,hann(nfft),noverlap,nfft,fs); % apparent mass FRF
CoHe=mscohere(a,fe,hann(nfft),noverlap,nfft,fs);  % coherence
dff=1/(nfft*dt);ff=0:dff:fs/2;                % frequency resolution/vector
```

(Continued)

MATLAB Example 9.5b   (Continued)

```
%% Plot the results
figure
plot(t,fe,'linewidth',2)                          % force time history
axis square; axis([0,250,-0.2,0.2]); grid
xlabel('time (s)'); ylabel('force (N)');

figure
plot(t,a,'linewidth',2)                           % acceleration time history
axis square; axis([0,250,-6,6]); grid
xlabel('time (s)');ylabel('acceleration (m/s^2)');

figure
semilogx(ff,10*log10(abs(Sff)),'linewidth',4)     % force PSD
axis square; axis([1,1000,-100,-30]); grid
xlabel('frequency (Hz)');
ylabel('force PSD (dB ref 1 N^2/Hz)');

figure
semilogx(ff,10*log10(abs(Saa)),'linewidth',4)     % acceleration PSD
axis square; axis([1,1000,-50,-10]); grid
xlabel('frequency (Hz)');
ylabel('acc. PSD (dB ref 1 m^2/s^4Hz)');

figure
semilogx(ff,20*log10(abs(Kae)),'linewidth',4)     % apparent mass modulus
hold on
semilogx(f,20*log10(abs(Ka)),'--k','linewidth',2)
axis square; axis([1,1000,-80,0]); grid
xlabel('frequency (Hz)');
ylabel('|apparent mass| (dB ref 1 kg)');

figure
semilogx(ff,180/pi*unwrap(angle(Kae)))            % apparent mass phase
hold on
semilogx(f,180/pi*unwrap(angle(Ka)),'--k')
axis square; axis([1,1000,-200,0]); grid
xlabel('frequency (Hz)');ylabel('phase (degrees)');

figure
semilogx(ff,CoHe,'linewidth',4)                   % coherence
axis square; axis([1,1000,0,1]); grid
xlabel('frequency (Hz)');ylabel('coherence');

figure
plot(real(Kae),imag(Kae),'linewidth',4)           % Nyquist plot
hold on
plot(real(Ka),imag(Ka),'--k','linewidth',2)
axis square; axis([-0.15,0.15,-0.3,0]); grid
xlabel('real \{apparent mass\} (kg)');
ylabel('imag \{apparent mass\} (kg)');
```

(Continued)

MATLAB Example 9.5b   (Continued)

Results

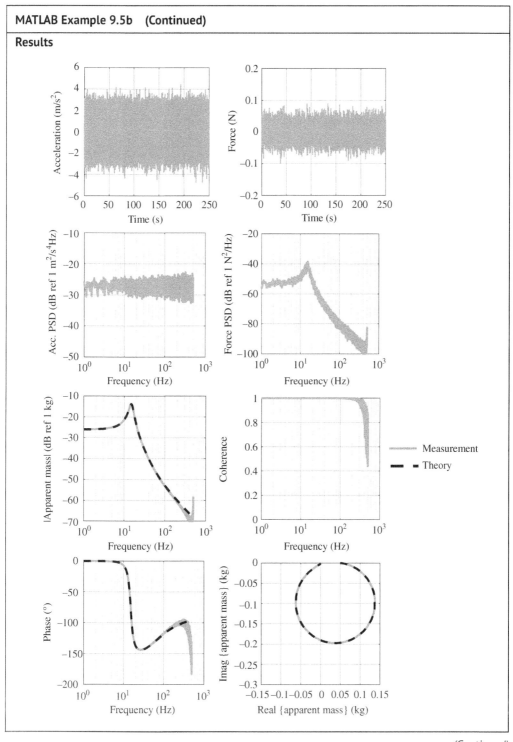

(Continued)

MATLAB Example 9.5b   (Continued)

Comments:

1. In this virtual experiment, the vibration absorber, which is designed using the procedure shown in Figure 9.8, is tested using an electrodynamic shaker.
2. The output from this virtual experiment is the apparent mass of the vibration absorber at its attachment point. The apparent mass is estimated in this case because other frequency domain quantities either tend to infinity at low or high frequencies. This means that the IRF, which is needed to determine the output response, cannot be easily calculated (note that this is not necessarily an issue in a real experiment). Thus, for the virtual experiment with the vibration absorber, the PSD of the acceleration of the base, which is attached to the shaker, is maintained at a constant level as a function of frequency. In a practical system this could be achieved by using a control system. The PSD of the resulting force will then vary with frequency, which can be seen in the figure.
3. An exercise for the reader is to plot the accelerance, mobility, and receptance FRFs and the impedance and the dynamic stiffness of the base-excited absorber.

Having designed and tested the vibration absorber, its effect on the host structure can be investigated by combining the measurements of the host structure alone and the vibration absorber in the computer before assembling the physical system and testing it. Equation (9.9) can be used for this purpose, in which the dynamic stiffness of the host structure K can be determined from the measured acceleration, and the dynamic stiffness of the vibration absorber K_a can be determined from the measured apparent mass. It can be seen that it is a simple operation, and the result can also be compared with model predictions. Predictions of the vibration absorber effectiveness are illustrated in MATLAB Example 9.5c using data measured during the virtual experiments conducted in MATLAB Examples 9.5a and b, and this is compared with the output from a model of the coupled system, both with and without the absorber attached. It is clear from these results that the main effect of the vibration absorber is to add damping to the host structure.

MATLAB Example 9.5c

In this example, the measurements of the apparent mass of the host structure (calculated from the accelerance) and the vibration absorber are combined to predict the receptance of the host structure with the vibration absorber attached. This is then checked using a theoretical approach in which the coupled system is considered.

```
% clear all

Exercise_9_5a                              % delete the figures
Exercise_9_5b                              % delete the figures and remove
                                             the clear all command at the
                                             beginning of the program

%% Measured quantities
Ha=(He'./(-(2*pi*ff).^2))';             % [m/N]   % receptance of SDOF host
                                                    structure without absorber

Htae=(1./(Kae'+1./He')./(-(2*pi*ff).^2))'; % [m/N]  % receptance of SDOF host
                                                    structure with absorber
```

(Continued)

MATLAB Example 9.5c   (Continued)

```
%% Check on the sum of the measured quantities
mu=0.05;                                              % mass ratio
ma=mu*m;                           % [kg]             % mass of the vibration absorber
wa=wn/(1+mu);                      % [rad/s]          % tuned frequency of the absorber
ka=wa^2*ma;                        % [N/m]            % stiffness of the absorber
za=sqrt(3/8*mu/(1+mu)^3);                             % optimum absorber damping ratio
ca=2*za*sqrt(ma*ka);               % [Ns/m]           % optimum absorber damping coeff.
n=0;df=0.01;                                          % frequency resolution
for fc=0:df:100                    % [Hz]             % excitation frequency
  n=n+1;
  wc=2*pi*fc;                      % [rad/s]
  M=[m 0; 0 ma];K=[k+ka -ka; -ka ka];                 % mass, stiff. and damp. matrices
  C=[c+ca -ca; -ca ca];
  F=[1;0];                         % [N]              % excitation force vector
  D=K-wc.^2*M+j*wc*C;              % [N/m]            % dyn. stiff. of complete system
  HH=inv(D)*F;                     % [m/N]            % receptance vector
  Hc(n)=HH(1);                     % [m/N]            % receptance of host-structure
end
fc=0:df:100;                       % [Hz]             % frequency vector

%% IRF
% calculate IRF of host structure alone
Ha(1)=Ha(2);                                          % remove infinity at zero Hz
Hd=[Ha' fliplr(conj(Ha'))]';                          % form the double-sided spectrum
Ht=Hd(1:length(Hd)-1);                                % set the length of the FRF
ha=fs*ifft(Ht);                                       % calculation of the IRF

% calculate IRF of host structure with absorber
Htae(1)=Htae(2);                                      % remove infinity at zero Hz
Hde=[Htae' fliplr(conj(Htae'))]';                     % form the double-sided spectrum
Hta=Hde(1:length(Hde)-1);                             % set the length of the FRF
hta=fs*ifft(Hta);                                     % calculation of the IRF

TT=1/ff(2);tt=0:dt:TT;                                % time vector

%% Figures
figure
semilogx(ff,20*log10(abs(Ha)))                        % modulus of the receptance FRF
hold on
semilogx(f,20*log10(abs(H)),':k')
hold on
semilogx(ff,20*log10(abs(Htae)))
hold on
semilogx(fc,20*log10(abs(Hc)),'--k')
axis square; axis([1,100,-120,-40]); grid
xlabel('frequency (Hz)');
ylabel('|receptance| (dB ref 1 m/N)');
figure
semilogx(ff,180/pi*unwrap(angle(Ha)))
hold on
semilogx(f,180/pi*unwrap(angle(H)),':k')
hold on
```

(Continued)

MATLAB Example 9.5c   (Continued)

```
semilogx(ff,180/pi*unwrap(angle(Htae)))              % phase of the receptance FRF
hold on
semilogx(fc,180/pi*unwrap(angle(Hc)),'--k')
axis square; axis([1,100,-200,0]); grid
xlabel('frequency (Hz)');ylabel('phase (degrees)');
figure
plot(tt,ha,'linewidth',2,'color',[0.6 0.6 0.6])      % displacement IRF
hold on
plot(tt,hta,'k','linewidth',2)
axis square; axis([0,3,-12e-3,12e-3]); grid
xlabel('time (s)');ylabel('IRF (m/Ns)');
```

Results

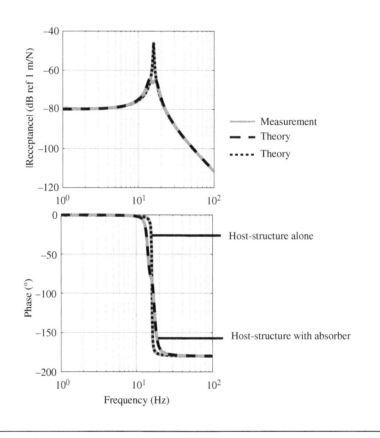

(Continued)

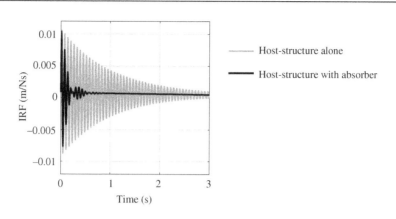

Comments:

1. In this exercise the receptance FRF of the SDOF host structure with the vibration absorber is compared with corresponding FRF of the host structure alone. The FRF is estimated two ways. One is by combining the measurements of the host structure and the vibration absorber that were carried out separately, and the other is by a theoretical approach in which the combined host structure and vibration absorber are described in terms of coupled differential equations of motion.

2. The displacement IRF of the host structure with the vibration absorber is also calculated and compared with the displacement IRF of the host structure alone. The effect of pre-dominantly adding damping by attaching an optimally tuned vibration absorber is evident as the vibration decays away much more quickly when the absorber is attached. However, the additional dynamics due to the attachment of the vibration absorber is also evident as the envelope of the IRF does not decay away monotonically as it does when the vibration absorber is not attached.

3. The approach, in which the measured apparent masses of the host structure and the vibra-tion absorber are combined, is one that can be used in practice to predict the effectiveness of the vibration absorber.

4. An exercise for the reader is to modify the program in MATLAB Example 9.1 to carry out the virtual experiment shown in Figure 9.9 to measure the receptance FRF of the host structure with and without the vibration absorber attached. Check that the FRFs are the same as shown in this example.

5. Exercises for the reader are to:
   (a) change the mass of the vibration absorber and investigate the effect that this has on the vibration of the host structure,
   (b) detune the vibration absorber and investigate the effect that this has on the receptance FRF of the host structure,
   (c) investigate the effect on the receptance FRF of the host structure when the damping in the vibration absorber is changed from its optimum value.

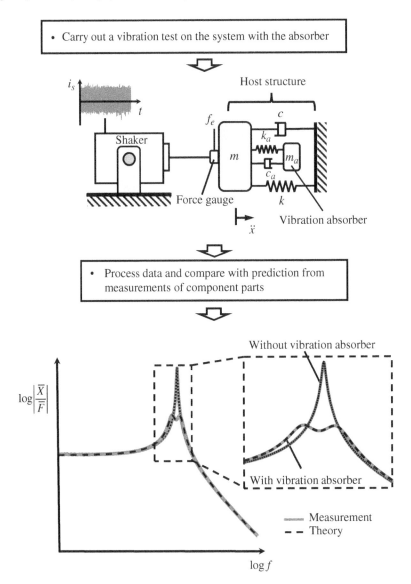

Figure 9.9 Experiment to measure the effectiveness of the vibration absorber.

A final check on the dynamic behaviour of the host structure with the absorber attached can be carried out as shown in Figure 9.9. The results should be the same as in the lower part of Figure 9.9, and in the case when the measurements of the host structure and the vibration absorber are combined in the computer as shown in MATLAB Example 9.5c. In a practical situation there may be some differences from the ideal behaviour, which could be due to the following:

- When the absorber is attached to the host structure, the mass of the vibration absorber (which has been assumed to be zero) is added to the mass of the host structure. This changes the mass ratio, which in turn affects the optimum stiffness and damping of the vibration absorber.
- The connection between the vibration absorber and the host structure may not have zero damping and infinite stiffness. This will also change the natural frequency of the vibration absorber,

and hence the optimum frequency to which it should be tuned, and the damping added to the host structure. This problem can be partially overcome by allowing the vibration absorber stiffness and damping to be adjusted in situ.

9.4.4 Vibration Absorber Attached to a Cantilever Beam – Virtual Experiment

In the previous section, the effect of attaching a vibration absorber, tuned according to Den Hartog's method (Den Hartog, 1956), was investigated for an SDOF host structure. In practical situations, however, most structures are MDOF systems. In this section, the effect on the vibration of an MDOF system due to the attachment of a vibration absorber is investigated. To illustrate the effect, a cantilever-beam host structure, such as that discussed in Section 9.3, is considered. The specific set-up is shown in Figure 9.10. The beam is excited by a shaker close to the root of the beam, and the vibration absorber is attached at the tip of the beam. The numbers 1–3 in the circles correspond to those in Figure 9.4, and the modelling and measurements of the system are based on this generic framework, which is described in Section 9.4.1.

The vibration absorber is designed according to Den Hartog's method, such that it predominantly adds damping to particular mode of a structure, where the natural frequencies of the structure are well separated, such as for a lightly damped beam. In this case the vibration close to the targeted natural frequency is dominated by that mode. For example, for frequencies close to the p-th mode of vibration, the point receptance FRF at position x_i, where the vibration absorber is attached, can be approximated by,

$$\frac{\overline{W}(x_i)}{\overline{F}(x_i)} \approx \frac{(A_{ii})_p}{\omega_p^2 - \omega^2 + j2\zeta_p\omega\omega_p}, \tag{9.13}$$

The modal mass for the beam at this position is thus given by $m_p = 1/(A_{ii})_p$. The vibration absorber with mass m_a has a mass ratio of $\mu = m_a/m_p$, and its natural frequency and damping can be chosen according to Eq. (9.12).

Schematic diagrams of the measurements required to determine the optimum parameters for the vibration absorber, and to predict the vibration of the beam when the vibration absorber is attached, are shown in Figure 9.11. The shaker is first positioned at the free end of the beam, which corresponds to point 14 in Figure 9.3. Note that the shaker is not shown in the figure for clarity, but the amplitude of the force it applies to the beam is $\overline{F}_{14}$. The point receptance of the beam is then

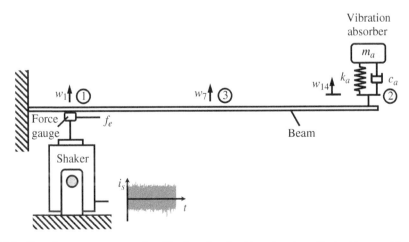

Figure 9.10 Experiment to measure the effective of a vibration absorber attached to a cantilever beam.

measured. At frequencies close to the natural frequency of interest, this can be approximated using Eq. (9.13). The two other FRFs are measured with the shaker at this position, as shown in the top part of Figure 9.11. Note also that the beam FRFs could be measured using an impact hammer, as discussed in Chapter 5. Once the point receptance FRF has been measured at point 14, the vibration absorber parameters can be determined, so that the vibration absorber may be designed and built. To complete the measurements on the beam, the shaker is moved to position 1 as shown in the centre part of Figure 9.11. Once the beam and the vibration absorber FRFs have been measured, the vibration of the beam with the absorber attached as shown in Figure 9.10, can be predicted using Eq. (9.8a), which can be written as

$$\frac{\overline{W}_i}{\overline{F}_1} = H_{i1} - \frac{H_{21}H_{i2}}{H_{22} + H_a}, \tag{9.14}$$

where the subscripts correspond to the numbers in the circles and $i = 1, 2, 3$ correspond to the measurement position of interest. The predicted FRFs corresponding to the three measurement points, i, due to excitation at point 1, are shown in MATLAB Example 9.6. The associated IRFs are also calculated.

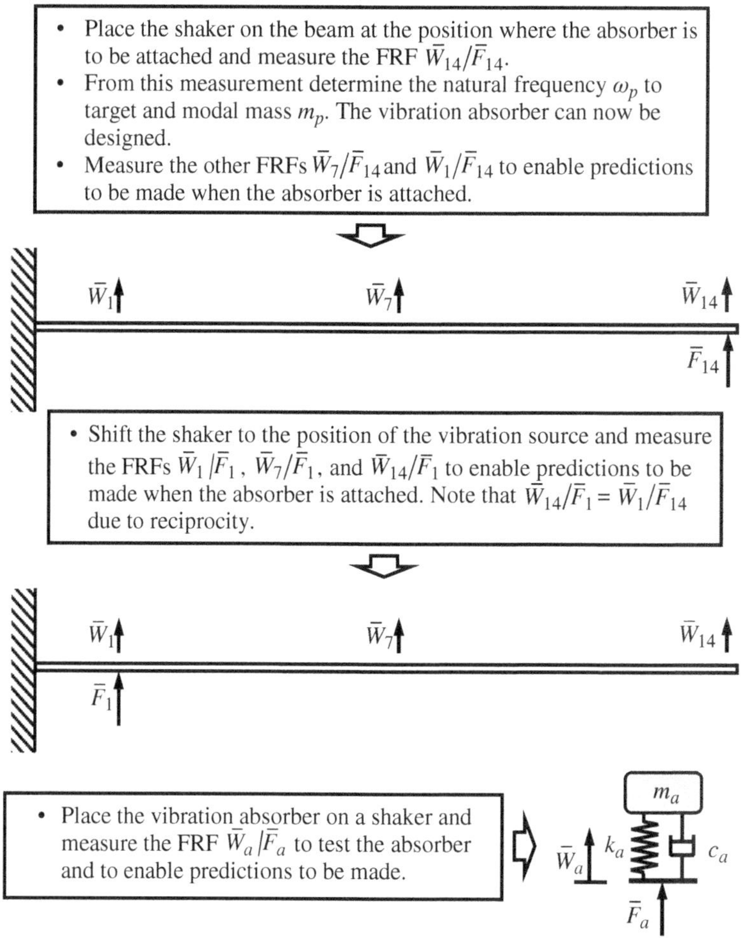

Figure 9.11 Experiments required to design the vibration absorber for the beam host-structure, excited as shown in Figure 9.10.

MATLAB Example 9.6

In this example, the effect of attaching a vibration absorber, which is tuned according to Den Hartog's method, to the tip of a cantilever beam is investigated.

```
clear all
%% Parameters
% cantilever beam
E=69e9;                               % [N/m²]        % Young's modulus of aluminium
rho=2700;                             % [kg/m³]       % density of aluminium
l=0.75;b=0.02;d=0.01;S=b*d;I=b*d^3/12; %             % geometrical parameters
z=0.01;n=2*z;                         %               % damping ratio and loss factor
Ed=E*(1+j*n);                         % [N/m²]        % complex Young's modulus
m=rho*S*l;                            % [kg]          % mass of the beam
% frequency parameters
fs=3000;df=0.01;dt=1/fs;              % [Hz,s]        % frequency parameters
f=0:df:fs/2;w=2*pi*f;                 % [Hz,rad/s]    % frequency vector

%% Positions along the beam
xp=0.1:0.05:l;                        % [m]           % positions along the beam
fp=1;                                 %               % force position in displ. vector
ap=14;                                %               % absorber position
op=14;                                %               % measurement position

%% Beam FRFs
n=0;
for x=0.1:0.05:l;                     % [m]
 n=n+1;
 xf=xp(fp);                           % [m]           % position of excitation force
 [Htt,wnn] = calcFRF(E,I,rho,S,z,m,l,x,xf,w,fs);  % calculate beam FRFs and IRFs
 H1(n,:)=Htt;                         % [m/N]         % beam FRFs wrt excitation force
 xf=xp(ap);                           % [m]           % position of absorber
 [Htt,wnn] = calcFRF(E,I,rho,S,z,m,l,x,xf,w,fs);  % calculate beam FRFs and IRFs
 H2(n,:)=Htt;                         % [m/N]         % beam FRFs wrt absorber position
end

%% Modal properties (first 3 modes)
x=0.1:0.05:l;                         % [m]           % position along the beam
fn=wnn/(2*pi);                        % [Hz]          % natural freqs. in Hz
z1=(8.27-8.07)/(2*fn(1));             %               % 1st modal damping ratio
z2=(51.73-50.73)/(2*fn(2));           %               % 2nd modal damping ratio
z3=(144.8-141.9)/(2*fn(3));           %               % 3rd modal damping ratio
xx1=max(abs(H2(ap,700:1000)));        % [m/N]         % max value of 1st modal response
xx2=max(abs(H2(ap,5000:5500)));       % [m/N]         % max value of 2nd modal response
xx3=max(abs(H2(ap,14000:15000)));     % [m/N]         % max value of 3rd modal response
A1=xx1.*(2*z1*wnn(1)^2);              % [1/kg]        % 1st modal constant
A2=xx2.*(2*z2*wnn(2)^2);              % [1/kg]        % 2nd modal constant
A3=xx3.*(2*z3*wnn(3)^2);              % [1/kg]        % 3rd modal constant

%% Vibration absorber design
mu=0.05;                              %               % mass ratio
ma=0.05/A1;                           % [m]           % mass of absorber for 1st mode
```

(Continued)

MATLAB Example 9.6   (Continued)

```
wa=wnn(1)/(1+mu);                    % [rad/s]      % absorber nat. freq. (1st mode)
ka=wa^2*ma;                          % [N/m]        % stiffness of the absorber
za=sqrt(3/8*mu/(1+mu)^3);                           % absorber damping ratio
ca=2*za*sqrt(ma*ka);                 % [Ns/m]       % absorber damping coefficient
Ka=-w.^2.*ma.*(ka+j*w*ca)./(ka-w.^2*ma+j*w*ca);     % absorber dynamic stiffness
Ha=1./Ka;                            % [m/N]        % absorber receptance

%% FRF of beam with absorber attached
H1c=H1(op,:)-H1(ap,:).*H2(op,:)./(H2(ap,:)+Ha);     % Beam FRF with absorber attached

%% IRFs
% host structure alone
Hd=[H1(op,:) fliplr(conj((H1(op,:))))];             % form the double-sided spectrum
Ht=Hd(1:length(Hd)-1);                              % set the length of the FRF
ha=fs*ifft(Ht);                      % [m/Ns]       % calculation of the IRF

% host structure with absorber
Hde=[H1c fliplr(conj(H1c))];                        % form the double-sided spectrum
Hta=Hde(1:length(Hde)-1);                           % set the length of the FRF
hta=fs*ifft(Hta);                    % [m/Ns]       % calculation of the IRF

TT=1/f(2);tt=0:dt:TT;                % [s]          % time vector

%% Figures
figure
semilogx(f,20*log10(abs(H1(op,:))),'linewidth',4)   % modulus of the receptance FRF
hold on
semilogx(f,20*log10(abs(H1c)),'k','linewidth',2);
axis square; axis([1,1000,-160,-40]); grid
xlabel('frequency (Hz)');
ylabel('|receptance| (dB ref 1 m/N)');
figure
plot(tt,ha,'linewidth',2,'color',[0.6 0.6 0.6])     % IRF
hold on
plot(tt,hta,'linewidth',2,'color',[0.3 0.3 0.3])
axis square; axis([0,3,-15e-3,15e-3]); grid
xlabel('time (s)');ylabel('IRF (m/Ns)');

%% Function
function [Htt,wnn]=calcFRF(E,I,rho,S,z,m,l,x,xf,w,fs) % function to calculate FRF
 nmax=10;                                           % number of modes

 kl(1)=1.87510;kl(2)=4.69409;kl(3)=7.85476;         % kl values 1-3
 kl(4)=10.9956;kl(5)=14.1372;                       % kl values 4,5
 n=6:nmax;
 kl(n)=(2*n-1)*pi/2;                                % kl values > 5
 for n=1:nmax
  A=(sinh(kl(n)))-...
  sin(kl(n)))./(cosh(kl(n))+cos(kl(n)));
  phi1=cosh(kl(n)*xf/l)-cos(kl(n)*xf/l)-...
  A.*(sinh(kl(n)*xf/l)-sin(kl(n)*xf/l));
  phi2=cosh(kl(n)*x/l)-cos(kl(n)*x/l)-...           % response position
```

(Continued)

MATLAB Example 9.6   (Continued)

```
 A.*(sinh(kl(n)*x/l)-sin(kl(n)*x/l));
 wn=sqrt((E*I)./(rho*S))*(kl(n)).^2;              % natural frequency
 wnn(n)=wn;
 Ht(n,:)=phi1*phi2./(m*(wn^2-w.^2+j*2*w*wn*z));   % FRF of each mode
 end
 Htt=sum(Ht);                                     % overall receptance FRF
end
```

Results

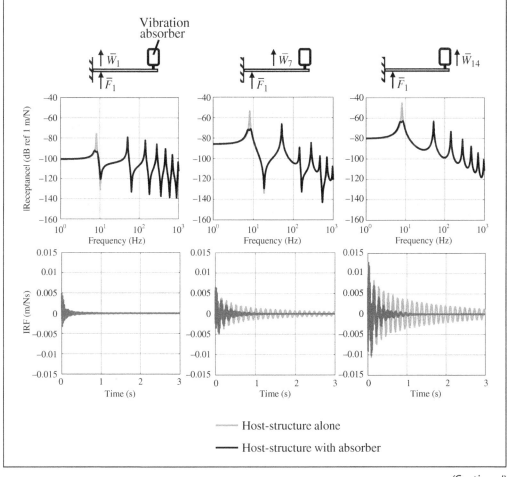

Host-structure alone

Host-structure with absorber

(Continued)

MATLAB Example 9.6   (Continued)

Comments:

1. In this exercise the receptance formulation was used to investigate the effectiveness of a vibration absorber on an MDOF system such as a cantilever beam. The optimum parameters of the absorber were calculated using the point receptance of the beam at the vibration absorber attachment point.
2. The effectiveness of the vibration absorber was illustrated using both FRFs and IRFs at several measurement positions on the beam.
3. An exercise for the reader is to modify the programs used in MATLAB Examples 9.5a–c to carry out a virtual experiment to:
   (a) measure the point receptance FRF of the beam at the vibration absorber attachment point on the beam so that the vibration absorber properties can be determined,
   (b) measure various FRFs on the beam with and without the vibration absorber attached.
4. Further exercises for the reader are to:
   (a) place the vibration absorber at different points on the beam and determine the effect on the vibration at other measured points,
   (b) tune the vibration absorber to the second and third natural frequencies and investigate the effect that this has on the resulting beam vibration in both the frequency and time domains,
   (c) change the mass of the vibration absorber and investigate the effect that this has on the vibration of the beam,
   (d) detune the vibration absorber and investigate the effect that this has on the vibration of the beam,
   (e) investigate the effect of changing the damping in the vibration absorber from its optimum value.

9.5   Summary

In this chapter, many of the topics discussed in previous chapters have been used in the conduct of some virtual experiments for multi-degree-of-freedom (MDOF) systems. Two structures have been studied. One was a 2DOF lumped parameter system that has both resonance and anti-resonance frequencies, and the other was a cantilever beam that has distributed mass and stiffness. The beam has an infinite number of DOF, but it was modelled as a finite DOF system using the modal approach described in Chapter 8 to capture the dynamics within a finite frequency range. It was shown how to estimate the mode shapes of a structure using the measured FRFs and to construct a modal model based on measured data.

The addition of a vibration absorber to two structures has been considered. One of these structures was an SDOF system, and the other was a beam system in which many DOF were considered. The choice of the optimum parameters for the vibration absorber was made by considering the dynamic properties of the host structure, in particular the point receptance at the point where the absorber was attached. The receptance approach to predict the dynamic behaviour of the host structure with and without the absorber attached has been illustrated. The formulation of this approach is summarised in Figure 9.12. This was described in Section 9.4.1 and involves the estimation of

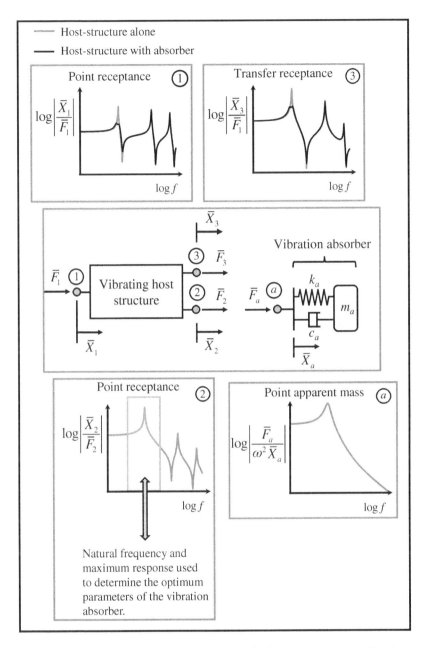

Figure 9.12 Schematic diagram showing the dynamics of a host structure and a vibration absorber.

receptances, and combining these to predict the dynamic behaviour of the host structure with the absorber attached. The displacement of point i on the host structure dues to a force applied at point 1 with an absorber attached at point 2 is given by

$$\frac{\overline{W}_i}{\overline{F}_1} = H_{i1} - \frac{H_{21}H_{i2}}{H_{22} + H_a} \tag{9.15}$$

where H_{ij} is the receptance FRF between points i and j, and H_a is the receptance FRF of the vibration absorber, which is related to the apparent mass M_a, by $M_a = 1/(-\omega^2 H_a)$. The optimum stiffness of the absorber for a lightly damped host structure, which has well-separated natural frequencies, is given by $k_a = \omega_a^2 m_a$, where m_a is the chosen mass of the vibration absorber and $\omega_a = \omega_p(1 + \mu)$, where ω_p is the natural frequency of the p-th mode (the targeted mode) and $\mu = m_a/m_p$, in which $m_p = 1/(A_{22})_p$, where $(A_{22})_p$ is the amplitude of the point receptance of the host structure at the targeted natural frequency where the absorber is to be attached. Following the design methodology given by (Den Hartog, 1956), the absorber damping is given by $\zeta_{a(\text{opt})} = \sqrt{3\mu/[8(1 + \mu)^3]}$.

References

Avitabile, P. (2017). *Modal Testing: A Practitioner's Guide*. Wiley.

Belykh, I., Bocian, M., Champneys, A.R., et al. (2021). Emergence of the London Millennium Bridge instability without synchronisation. *Nature Communications*, 12, 7223. https://doi.org/10.1038/s41467-021-27568-y.

Brennan, M.J. (1997) Vibration control using a tunable vibration neutraliser. *Proceedings of the Institution of Mechanical Engineers, Part C: Journal of Mechanical Engineering Science*, 211(2), 91–108. https://doi.org/10.1243/0954406971521683.

Den Hartog, J.P. (1956). *Mechanical Vibrations*, 4th Edition. McGraw-Hill.

Reed, E.F. (2002). *Dynamic Vibration Absorbers and Auxiliary Mass Dampers, Chapter 6 in Shock and Vibration Handbook*, 5th Edition, (eds. C.M. Harris and A.G. Piersol). McGraw-Hill.

Ewins, D.J. (2000). *Modal Testing: Theory, Practice and Application*, 2nd Edition. Research Studies Press.

Newland, D.E. (2003). Vibration of the London Millennium Bridge: Cause and cure. *International Journal of Acoustics and Vibration*, 8(1), 9–14. https://doi.org/10.20855/ijav.2003.8.1124.

Ormondroyd, J. and Den Hartog, J. P. (1928). The theory of the dynamic vibration absorber. *Transactions of the American Society of Mechanical Engineers, Applied Mechanics*, 50, 9–22.

Appendix A

Numerical Differentiation and Integration

In vibration engineering, time domain measurements are most often made using displacement, velocity, or acceleration sensors. Moreover, the data are often stored as sampled time histories in a computer. There is often a requirement to view the data in another form to that measured. For example, the displacement of a system may be of interest, but acceleration is measured due to the availability of transducers. It is thus necessary to be able to differentiate and integrate signals numerically, which are operations that can be carried out in either the time or frequency domain.

A.1 Differentiation in the Time Domain

Consider part of a velocity signal as a function of time $\dot{x}(t)$, as shown in Figure A.1. If they are sampled data, the simplest way to differentiate the signal to determine the acceleration is to use a finite difference approximation. To illustrate this approach, a straight line is drawn between two points, $\dot{x}(t_1)$, which is the velocity at time t_1, and $\dot{x}(t_1 + \Delta t)$, which is the velocity Δt seconds later. The gradient of this line, which is an approximation of the acceleration at a time $(t_1 + \Delta t/2)$, is given by

$$\ddot{x}\left(t_1 + \frac{\Delta t}{2}\right) \approx \frac{\dot{x}(t_1 + \Delta t) - \dot{x}(t_1)}{\Delta t}. \tag{A.1}$$

Equation (A.1) is easily calculated in MATLAB (Lindfield and Penny, 2012) and is illustrated using a sine wave in MATLAB Example A.1. Note that the derivative is readily checked in this case because if $\dot{x}(t) \approx \sin \omega t$, then $\ddot{x}(t) \approx \omega \cos \omega t$. Note also that if there are n points in the velocity vector, then

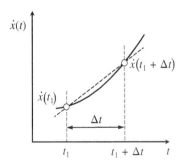

Figure A.1 Part of a velocity signal as a function of time.

Virtual Experiments in Mechanical Vibrations: Structural Dynamics and Signal Processing,
First Edition. Michael J. Brennan and Bin Tang.
© 2023 John Wiley & Sons Ltd. Published 2023 by John Wiley & Sons Ltd.
Companion website: www.wiley.com/go/brennan/virtualexperimentsinmechanicalvibrations

there are $n-1$ points in the acceleration vector, which is a result of using the finite difference method.

A.2   Integration in the Time Domain

Consider again, part of a velocity signal as a function of time $\dot{x}(t)$, as shown in Figure A.1. The simplest way to integrate the signal to determine the displacement is by cumulative integration using the trapezoidal rule. As with differentiation, a straight line is drawn between two points, $\dot{x}(t_1)$, which is the velocity at time, and $\dot{x}(t_1 + \Delta t)$, which is the velocity Δt seconds later. The displacement of the signal at time $(t_1 + \Delta t)$ is the displacement at time t_1 plus the area under the velocity curve between time $(t_1 + \Delta t)$ and time t_1, which is approximated by the area of the trapezium, i.e.

$$x\left(t_1 + \frac{\Delta t}{2}\right) \approx x\left(t_1\right) + \frac{\Delta t}{2}\left(\dot{x}\left(t_1 + \Delta t\right) + \dot{x}\left(t_1\right)\right). \tag{A.2}$$

Equation (A.2) is easily calculated in MATLAB and is illustrated using a sine wave in MATLAB Example A.1. Note that the integral is readily checked in this case because if $\dot{x}(t) \approx \sin \omega t$, then $x(t) \approx \frac{-1}{\omega}\cos \omega t + x(0)$. Note also that if there are n points in the velocity vector, then there are n points in the displacement vector, and the initial displacement needs to be known (which is the constant of integration).

MATLAB Example A.1

In this example the acceleration and displacement are calculated from a sinusoidal velocity signal.

```
clear all

%% Time vectors
dt=0.05; dt1=0.001;          % [s]          % time resolution in seconds
T=1;                         % [s]          % duration of time signal
t=0:dt:T; t1=0:dt1:T;        % [s]          % time vectors
f=1;                         % [Hz]         % frequency in Hz
w=2*pi*f;                    % [rad/s]      % frequency in rad/s

%% Velocity
v=sin(w*t1);                 % [m/s]        % velocity - high time resolution
vm=sin(w*t);                 % [m/s]        % velocity - low time resolution

%% Differentiation
a=w.*cos(w*t1);              % [m/s^2]      % acceleration - high time resolution
am=diff(vm)/dt;              % [m/s^2]      % numerical differentiation
tt=dt/2:dt:T-dt/2;           % [s]          % define time vector (one point less)

%% Integration
x=-1/w*cos(w*t1);            % [m]          % displacement - high time resolution
xt=cumtrapz(vm)*dt + x(1);   % [m]          % numerical integration
```

(Continued)

MATLAB Example A.1   (Continued)

```
%% Plot the results
figure
plot(t1,x,t,xt,'o')                          % displacement as a function of time
grid;axis square
xlabel('time (s)');
ylabel('displacement (m)');

figure
plot(t1,v,t,vm,'o')                          % velocity as a function of time
grid;axis square
xlabel('time (s)');
ylabel('velocity (m/s)');

figure
plot(t1,a,tt,am,'o')                         % acceleration as a function of time
grid;axis square
xlabel('time (s)');
ylabel('acceleration (m/s^2)');
```

Results

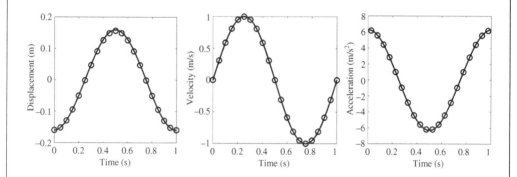

Comments:

1. The solid lines in the graphs above are the actual signals and the circles are the sampled data which are processed.
2. The displacement from n discrete points of velocity by numerical integration also has n discrete points. As well as calculating the cumulative integral using the trapezoidal rule, the initial value of the displacement is needed.
3. The acceleration calculated from the discrete points of velocity by numerical differentiation is also at discrete points. These points are in between the points for velocity, and as mentioned above, there is one less point in the acceleration vector. Accordingly, a new time vector must be calculated.

(Continued)

MATLAB Example A.1   (Continued)

4. The displacement and acceleration estimates improve as the time resolution increases, i.e. as Δt reduces.

5. If the signal contains noise, the high-frequency components of the noise are amplified by the operation of differentiation, but are attenuated by integration. This is because the amplitude of acceleration is the product of the velocity amplitude and the angular frequency, and the amplitude of the displacement is the product of the velocity amplitude and the reciprocal of the angular frequency.

6. If integration is to be carried out on experimental data as described in this example, it is preferable to remove the DC value and the linear trend of the data before carrying out the integration. This is done using the `detrend` command in MATLAB.

7. Exercises for the reader are to:
   (a) Add noise to the velocity sine wave using the `randn` function and investigate what happens when the signal is differentiated and integrated.
   (b) Add two sine functions together with different frequencies and phases, and investigate the results of numerical differentiation and integration. Compare the results with the theoretical solutions.

A.3   Differentiation and Integration in the Frequency Domain

The operations of differentiation and integration are much simpler in the frequency domain, and these are shown in Table A.1.

Table A.1   Differentiation and integration in the time and frequency domains.

	Time domain	Frequency domain
Differentiation	$\dfrac{d}{dt}$	$\times j\omega$
Integration	$\int dt$	$\div j\omega$

Reference

Lindfield, G.R. and Penny, J.E.T. (2012). *Numerical Methods Using MATLAB®*, 3rd Edition. Elsevier.

Appendix B

The Hilbert Transform

The Hilbert transform is named after David Hilbert, a German mathematician (Hilbert, 1902), who introduced the transform to solve a special case of the Riemann–Hilbert problem for analytic functions. It has subsequently found many uses in electrical and mechanical engineering (Feldman, 2011). The Hilbert transform of $x(t)$ is the convolution of $x(t)$ with $h(t) = 1/\pi t$ (see Appendix G for a discussion on convolution). Note that this has a singularity at $t = 0$. The frequency domain representation of the Hilbert transform is

$$H(j\omega) = -j\text{sgn}\omega = \begin{cases} j & \text{for} & \omega < 0 \\ 0 & \text{for} & \omega = 0 \\ -j & \text{for} & \omega > 0 \end{cases}, \tag{B.1}$$

which shows that the Hilbert transform effectively shifts the original time history $x(t)$ by 90°. The Hilbert transform can be used to form the analytic signal from a time domain signal. An analytic signal $a(t)$ is a complex time domain signal, which has a real part that corresponds to the original signal $x(t)$, and the imaginary part is the Hilbert transform of $x(t)$, i.e. $\hat{x}(t)$, such that

$$a(t) = x(t) + j\hat{x}(t). \tag{B.2}$$

The envelope of the original time history is given by $|a(t)| = \sqrt{x^2(t) + \hat{x}^2(t)}$ and the instantaneous phase is given by $\phi(t) = \tan^{-1}(\hat{x}(t)/x(t))$.

To illustrate the application of the Hilbert transform, consider an amplitude- and phase-modulated displacement signal given by

$$x(t) = A(t)\sin(\omega t + \phi(t)), \tag{B.3}$$

where $A(t)$ and $\phi(t)$ are time-varying amplitude and time-varying phase, respectively, given by $A(t) = 1 + \alpha \sin \omega_a t$, and $\phi(t) = \beta \cos \omega_b t$, in which α and β are modulation amplitudes and ω_a and ω_b are amplitude and phase modulation frequencies. A typical envelope together with the original displacement signal, and the time-varying (or instantaneous) phase are shown in Figure B.1.

MATLAB Example B.1

In this example the Hilbert transform is used to calculate the envelope of a signal composed of two sine waves that have slightly different amplitudes and closely spaced frequencies, such that the beating phenomenon is observed.

(Continued)

Virtual Experiments in Mechanical Vibrations: Structural Dynamics and Signal Processing,
First Edition. Michael J. Brennan and Bin Tang.
© 2023 John Wiley & Sons Ltd. Published 2023 by John Wiley & Sons Ltd.
Companion website: www.wiley.com/go/brennan/virtualexperimentsinmechanicalvibrations

MATLAB Example B.1   (Continued)

```
clear all

%% Time and frequency data
dt=0.0001;                % [s]        % time resolution
T=3;t=0:dt:T;             % [s]        % duration of time signal and time vector
f=10;                     % [Hz]       % frequency in Hz
w=2*pi*f;                 % [rad/s]    % frequency in rad/s

%% Signals and the envelope
x=1*sin(w*t);             % [m]        % signal x
y=0.9*sin(0.95*w*t);      % [m]        % signal y
z=x+y;                    % [m]        % sum of the signals x and y
zh=hilbert(z);            % [m]        % Hilbert transform of z

%% Plot the results
plot(t,z,'k','linewidth',2)                    % plot z and its envelope
hold on
plot(t,abs(zh),t,-abs(zh),'linewidth',2)
axis([0,3,-3,3])
axis square; grid
set(gca,'fontsize',16)
xlabel('time (s)');
ylabel('displacement (m)');
```

Results

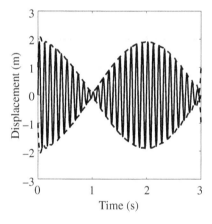

Comments:

1. The dashed line shows the envelope of the signal. Note that the envelope has some ripples at the beginning and end of the time series, which is due to the windowing effect as discussed in Chapter 2.
2. An exercise for the reader is to generate other signals and to calculate the envelope of these signals. Think of other ways of calculating the envelope of a signal.

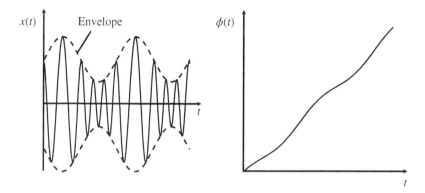

Figure B.1 Amplitude- and phase-modulated signal together with its envelope and instantaneous phase calculated using the Hilbert transform.

References

Feldman, M. (2011). *Hilbert Transform Applications in Mechanical Vibration*. Wiley.

Hilbert, D. (1902). *Grundlagen der Geometrie* (The Foundations of Geometry). Translated by Townsend, E.J. (1950). Open Court Publishing.

Appendix C

The Decibel: A Brief Description

The decibel (dB) is one-tenth of a bel (B) and is a *relative* unit of measurement. It originated in the early twentieth century in Bell Telephone Laboratories who were quantifying signal loss in telegraph and telephone circuits (Martin, 1929). It is named after telecommunications pioneer Alexander Graham Bell (1847–1922). The decibel is a logarithmic quantity and is based on a ratio of powers (or squared quantities). For example, suppose there is a measurement in terms of voltage V, then the level V^2 with respect to a reference voltage V_0^2, in dB is given by

$$\text{Level in dB} = 10\log_{10}\left(\frac{V^2}{V_0^2}\right), \tag{C.1}$$

where $\log_{10}$ denotes logarithm to the base 10. The reference level should also be stated. For example, if the reference level V_0^2 is 1 V^2, and V^2 is 100 V^2, the correct way to state the result in dB would be 20 dB (ref 1 V^2). This would be appropriate if a power or energy quantity is of interest. An example in terms of mechanical systems could be if the kinetic energy of a vibrating mass is of interest. Suppose the energy level is 0.5 J, this would be written in dB as -3 dB (ref 1 J).

An alternative way to represent Eq. (C.1) if non-squared quantities are of interest is

$$\text{Level in dB} = 20\log_{10}\left(\frac{V}{V_0}\right). \tag{C.2}$$

An example in this case could be a displacement FRF which has a value of 1×10^{-3} m/N. The value in dB would be -60 dB (ref 1 m/N).

As the dynamic range (ratio of the largest to the smallest value) tends to be very large in FRFs of vibrating systems, the amplitude is generally plotted on a logarithmic scale. An alternative to this is to plot the amplitude in terms of dB, and this is often used by vibration engineers, because (after some experience) it is easy to interpret.

The differences between the linear scale, the $\log_{10}$ scale, and dB scale are illustrated in Figure C.1. The way in which the $\log_{10}$ and dB scales compress and stretch parts of the axis can be seen. For example, in the linear scale, the range between 0 and 10 occupies 10% of the axis, and the range 10–100 occupies 90% of the axis, whereas on the $\log_{10}$ and dB axes, the range 0.01–1, and 1–100 each occupy 50% of their respective axis.

Virtual Experiments in Mechanical Vibrations: Structural Dynamics and Signal Processing.
First Edition. Michael J. Brennan and Bin Tang.
© 2023 John Wiley & Sons Ltd. Published 2023 by John Wiley & Sons Ltd.
Companion website: www.wiley.com/go/brennan/virtualexperimentsinmechanicalvibrations

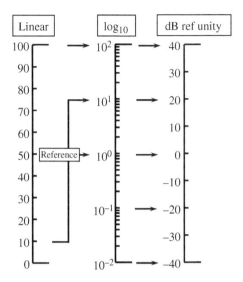

Figure C.1 An illustration of the differences between the linear, $\log_{10}$, and dB scales.

Reference

Martin, W. H. (1929). Decibel – The name for the transmission unit. *Bell System Technical Journal*, 8(1), 1–2. https://doi.org/10.1002/j.1538-7305.1929.tb02302.x.

Appendix D

Numerical Integration of Equations of Motion

In some parts of this book, there is a need to generate a time series relating to the dynamic motion of a system due to an input force. If the equation of motion for the system is known, then one convenient way to do this is to solve this equation numerically. As an equation of motion describes the *dynamic* behaviour of a system, it involves inertia forces which are the products of masses and their acceleration. Thus, for lumped parameter systems, the equations of motion are ordinary differential equations (ODEs) of second order. The equations can be solved numerically in MATLAB using a specific function (`ode45`) as this is convenient for the systems considered in this book. In this appendix, the way in which a differential equation can be solved numerically is outlined. It is not the intention to treat this subject in detail as there are many specific textbooks devoted to this topic, for example Lambert (1991), Hairer et al. (1993), Press et al. (2007), and Kreyszig (2011). Note that the foundation for numerical analysis was laid down by Taylor in 1715 who first described a series approximation to a function (Taylor, 1715, 21–23 [Prop. VII, Thm. III, Cor. II]). The intention is to give the reader an appreciation of how a differential equation is solved numerically, and to provide an example of how this may be done using MATLAB.

The method used is the classical *Runge–Kutta* method of *fourth order* with variable time increment (step size). This algorithm is named after the German mathematicians Carle Runge (1856–1927) and Martin Kutta (1867–1944). It uses four function values per time increment, and the step size can be varied automatically to ensure a predetermined level of accuracy. The Runge–Kutta method is discussed later, but first *Euler's* method is discussed as it helps to explain the principle of the method based on the Taylor series expansion of a function. As most of the ordinary differential equations of interest in this book are functions of time, this is taken as the independent variable.

D.1 Euler's Method

Euler's method is a crude method which can be used to solve *first-order* differential equations numerically. It is named after the Swiss mathematician and physicist Leonhard Euler (1707–1783). As mentioned above, second-order differential equations of time are of primary interest in this book, so these equations must first be written in terms of first-order equations.

Consider the equation of motion of an SDOF system given by

$$m\ddot{x} + c\dot{x} + kx = f_e, \tag{D.1}$$

Virtual Experiments in Mechanical Vibrations: Structural Dynamics and Signal Processing,
First Edition. Michael J. Brennan and Bin Tang.
© 2023 John Wiley & Sons Ltd. Published 2023 by John Wiley & Sons Ltd.
Companion website: www.wiley.com/go/brennan/virtualexperimentsinmechanicalvibrations

where m, k, and c are mass, stiffness, and damping, respectively, and f_e is the force applied to the mass and is a function of time t. Equation (D.1) can be written as

$$\ddot{x} = \frac{1}{m}(f_e - c\dot{x} - kx).\tag{D.2}$$

To apply Euler's method, Eq. (D.2) is written in terms of two first-order differential equations, as

$$\dot{x} = y,\tag{D.3a}$$

and

$$\dot{y} = \frac{1}{m}(f_e - c\dot{x} - kx),\tag{D.3b}$$

which can be combined and written in vector matrix form as

$$\dot{\mathbf{x}} = \mathbf{A}\mathbf{x} + \mathbf{b},\tag{D.4}$$

where $\mathbf{A} = \begin{bmatrix} 0 & 1 \\ -\frac{k}{m} & -\frac{c}{m} \end{bmatrix}$, $\mathbf{b} = \left\{ \begin{array}{c} 0 \\ \frac{f_e}{m} \end{array} \right\}$, $\mathbf{x} = \left\{ \begin{array}{c} x \\ y \end{array} \right\}$. Equation (D.4) is solved using a step-by-step method, starting from the initial conditions $\mathbf{x}_0$ at time t_0. *Approximate* values of the vector $\mathbf{x}$, which consists of the displacement and velocity of the mass, are determined at each step. At time $t_i + \Delta t$, the vector $\mathbf{x}(t_i + \Delta t)$ can be determined by adding an increment $\Delta\mathbf{x}$ to the vector $\mathbf{x}(t_i)$ at time t_i, i.e.,

$$\mathbf{x}(t_i + \Delta t) = \mathbf{x}(t_i) + \Delta\mathbf{x}.\tag{D.5}$$

As the initial conditions are known, the problem becomes one of determining $\Delta\mathbf{x}$ for each time increment Δt. Now, consider the Taylor series (Kreyszig, 2011)

$$\mathbf{x}(t_i + \Delta t) = \mathbf{x}(t_i) + \Delta t \dot{\mathbf{x}}(t_i) + \frac{(\Delta t)^2}{2}\ddot{\mathbf{x}}(t_i) + \dots.\tag{D.6}$$

For small values of Δt the higher powers $(\Delta t)^2$, $(\Delta t)^3$, etc. are small and can be neglected. Thus Eq. (D.6) can be approximated as

$$\mathbf{x}(t_i + \Delta t) \approx \mathbf{x}(t_i) + \Delta t \dot{\mathbf{x}}(t_i),\tag{D.7}$$

By comparing Eqs. (D.5) and (D.7), it can be seen that $\Delta\mathbf{x} \approx \Delta t \dot{\mathbf{x}}(t_i)$. To illustrate the approach, the approximation to a curve for a time increment Δt is shown in Figure D.1 for the variable x in the vector $\mathbf{x}$. The curve is approximated by the straight line which is determined using the slope of the curve $\dot{x}$ at t_i. It is clear that if x changes rapidly within the time increment then there will be a large error in the approximation. There have been many methods developed to improve the accuracy of solution of this type, and the interested reader is referred to Press et al. (2007) for further details.

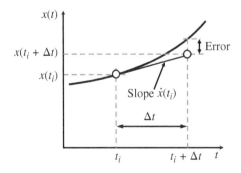

Figure D.1 An illustration of Euler's method in the numerical integration of a function.

D.2   The Runge–Kutta Method

The Runge–Kutta method offers much improved accuracy compared to Euler's method and is thus commonly used to solve differential equations. Rather than use only the slope at the beginning of the time step at t_i, it uses an estimate of the slope at the end of the time step at $t_i + \Delta t$ and two estimates of the slope at the mid-point in the time step at $t_i + \Delta t/2$. It then calculates a weighted average of the estimates of $\Delta \mathbf{x}$ to give

$$\Delta \mathbf{x} = \frac{\Delta \mathbf{x}_A}{6} + \frac{\Delta \mathbf{x}_B}{3} + \frac{\Delta \mathbf{x}_C}{3} + \frac{\Delta \mathbf{x}_D}{6}, \tag{D.8}$$

where $\Delta \mathbf{x}_A$ is related to the time derivative $\dot{\mathbf{x}}(t_i)$ using $\mathbf{x}(t_i)$, $\Delta \mathbf{x}_B$ is related to an estimate of the time derivative $\dot{\mathbf{x}}(t_i + \Delta t/2)$ using $\mathbf{x}(t_i) + \Delta \mathbf{x}_A/2$, $\Delta \mathbf{x}_C$ is also related to the time derivative $\dot{\mathbf{x}}(t_i + \Delta t/2)$ but using $\mathbf{x}(t_i) + \Delta \mathbf{x}_B/2$, and $\Delta \mathbf{x}_D$ is related to the time derivative of $\dot{\mathbf{x}}(t_i + \Delta t)$ using $\mathbf{x}(t_i) + \Delta \mathbf{x}_C$. Thus

$$\Delta \mathbf{x}_A = \Delta t \dot{\mathbf{x}}(t_i) = \Delta t [\mathbf{A} \mathbf{x}(t_i) + \mathbf{b}(t_i)] \tag{D.9a}$$

$$\Delta \mathbf{x}_B = \Delta t \left[\mathbf{A} \left(\mathbf{x}(t_i) + \frac{\Delta \mathbf{x}_A}{2} \right) + \mathbf{b}(t_i + \Delta t/2) \right] \tag{D.9b}$$

$$\Delta \mathbf{x}_C = \Delta t \left[\mathbf{A} \left(\mathbf{x}(t_i) + \frac{\Delta \mathbf{x}_B}{2} \right) + \mathbf{b}(t_i + \Delta t/2) \right] \tag{D.9c}$$

$$\Delta \mathbf{x}_D = \Delta t [\mathbf{A}(\mathbf{x}(t_i) + \Delta \mathbf{x}_C) + \mathbf{b}(t_i + \Delta t)] \tag{D.9d}$$

This is illustrated in Figure D.2 for the displacement x, which shows the four estimates of the slopes used to calculate Δx_A, Δx_B, Δx_C, and Δx_D, which are then used in Eq. (D.8) to calculate Δx.

It can be seen from Figure D.1 that the accuracy of the first-order solution, in which the function is approximated by a straight line within the time step, is dependent on the size of the time step. Comparing Eqs. (D.5) and (D.6), the step-by-step method can be written as

$$\mathbf{x}(t_i + \Delta t) \approx \mathbf{x}(t_i) + \Delta \mathbf{x} + \mathbf{e} \tag{D.10}$$

where the vector of errors $\mathbf{e}$, is given approximately by $\mathbf{e} = \frac{(\Delta t)^2}{2} \ddot{\mathbf{x}}(t_i)$. Thus, if $\ddot{\mathbf{x}}(t_i)$ is calculated for each time step, then the step size can be adjusted for each time step so that the maximum error does not exceed a certain value, which is called the tolerance. In the MATLAB ode45 function the default maximum relative error is set to 1×10^{-5}.

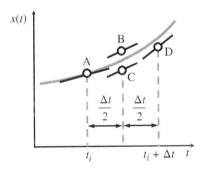

Figure D.2   An illustration of the Runge–Kutta method.

MATLAB Example D.1

Example of how to solve a second-order differential equation using ode45.m. In this example, a single degree-of-freedom system is excited by an arbitrary force, and the resulting response in the time domain is calculated by numerical integration of the equation of motion.

```
clear all

%% Variables
m = 1;                          % [kg]           % mass
k = 1000;                       % [N/m]          % stiffness
z = 0.05;                                        % damping ratio
c = 2*z*sqrt(m*k);              % [Ns/m]         % damping factor

A = [0 1; -k/m -c/m];                            % system matrices
B = [0; 1/m];

dt = 0.005;                     % [s]            % time resolution in seconds
fs = 1/dt;                      % [Hz]           % sampling frequency
T = 12;                         % [s]            % duration of time signal
df=1/T;                         % [Hz]           % frequency resolution
t = 0:dt:T;                     % [s]            % time vector
f = 0:df:fs;                    % [Hz]           % frequency vector
L = length(t);                                   % number of samples

%% Definition of an input force time series
f1 = zeros(size(t));
Tn = T*5/12;                    % [s]            % duration of the force pulse
Ln = round(Tn/dt);                               % number of samples in the pulse
f = [zeros(1,100), ones(1,Ln), ...              % force time history
zeros(1,(length(t)-(Ln+100)))];

%% Runge-Kutta Method
n=t;                                             % dummy variable for the look up
                                                   table used interp1.m
[t,x] = ode45(@(t,x) pulse(t,x,A,B,f,n),t,[0 0]);% calculating the response due to
                                                   the force pulse
%% Plot the results
figure
plot(t,f);axis([0,12,0,1.2])                     % force time history
xlabel('Time (s)')
ylabel('Input Force (N)')
figure
plot(t,x(:,1))                                   % displacement time history
xlabel('Time (s)')
ylabel('Displacement (m)')
figure
plot(t,x(:,2))                                   % velocity time history
xlabel('Time (s)')
ylabel('Velocity (m/s)')

function dx = pulse(t,x,A,B,f,n)                 % gives force f at time t
  force = interp1(n,f,t);                        % defines system of differential
  dx=A*x+B*force;                                  equations
end
```

(Continued)

MATLAB Example D.1   (Continued)

Results

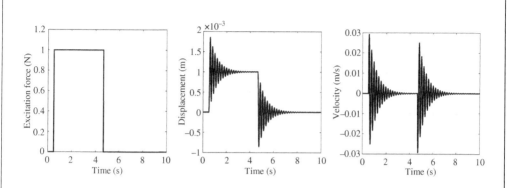

Comments

1. An exercise for the reader is to change the type of force excitation to sinusoidal excitation and to check that the result agrees with the steady-state theoretical prediction for an excitation frequency, which is
   (a) much less than the resonance frequency,
   (b) equal to the resonance frequency,
   (c) well above the resonance frequency.
2. An exercise for the reader is to reduce the value of damping to $\zeta = 0.001$ and repeat 1 above. What happens to the displacement and velocity in each case? Try to determine the physical reason why this changes.
3. An exercise for the reader is to calculate the acceleration for the examples above using the numerical method discussed in Appendix A.

References

Hairer, E., Nørsett, S.P., and Wanner, G. (1993). *Solving Ordinary Differential Equations I: Nonstiff Problems*, 2nd Edition. Springer-Verlag.

Kreyszig, E. (2011). *Advanced Engineering Mathematics*, 10th Edition. Wiley.

Lambert, J.D. (1991). *Numerical Methods for Ordinary Differential Systems: The Initial Value Problem*. Wiley.

Press, H.W., Teukolsky, S.A., Vetterling, W.T., and Flannery, B.P. (2007). *Numerical Recipes: The Art of Scientific Computing*, 3rd Edition. Cambridge University Press.

Taylor, B. (1715). *Methodus Incrementorum Directa et Inversa [Direct and Indirect Methods of Incrementation]*. Londini: Typis Pearsonianis prostant apud Gul. Innys ad Insignia Principis in Coemeterio Paulino (in Latin).

Appendix E

The Delta Function

The delta function is a mathematical concept that is extremely useful in vibration engineering, as it facilitates the derivation of several important quantities such as the impulse response function (IRF) and the discrete Fourier transform (DFT). The delta function was introduced by physicist Paul Dirac as a tool for the normalisation of state vectors (Dirac, 1930), and so it is sometimes called the Dirac delta function. From an engineering perspective it is probably best thought of as a unit impulse with the condition that the duration of the impulse tends to zero. This is illustrated in Figure E.1, which shows an impulse that has unit area, acting for a short time duration ε seconds. If $\varepsilon \to 0$ such that the impulse acts over an infinitesimally small time period, then the result is the delta function $\delta(t)$ defined by

$$\delta(t) = 0 \quad \text{for} \quad t \neq 0 \quad \text{and} \quad \int_{-\infty}^{\infty} \delta(t)\mathrm{d}t = 1. \tag{E.1}$$

Note that the unit of $\delta(t)$ in this case is 1/s as the unit of t is the second. An alternative representation of the delta function is shown in Figure E.2, as the limiting case of a sinc function in which

$$\delta(t) = \lim_{2f \to \infty} \left(2f \times \frac{\sin(2\pi ft)}{2\pi ft} \right) = \lim_{2f \to \infty} (2f \times \mathrm{sinc}(2ft)). \tag{E.2}$$

Examining Figure E.1, it can be seen that the impulse can be thought of as being composed of two time-shifted Heaviside or unit step functions $u(t)$, such that the impulse I is given by $I = (1/\varepsilon)[u(t + \varepsilon/2) - u(t - \varepsilon/2)]$. Letting $\varepsilon = \Delta t$ and then letting $\Delta t \to 0$ means that the delta function is the same as the time derivative of $u(t)$ such that

$$\delta(t) = \frac{\mathrm{d}}{\mathrm{d}t} u(t). \tag{E.3}$$

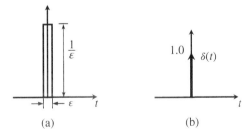

Figure E.1 The unit impulse and the delta function. (a) A unit impulse. (b) Graphical representation of the delta function.

Virtual Experiments in Mechanical Vibrations: Structural Dynamics and Signal Processing,
First Edition. Michael J. Brennan and Bin Tang.
© 2023 John Wiley & Sons Ltd. Published 2023 by John Wiley & Sons Ltd.
Companion website: www.wiley.com/go/brennan/virtualexperimentsinmechanicalvibrations

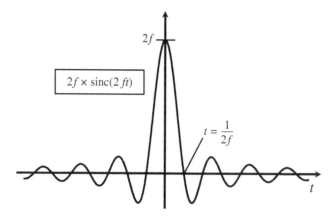

Figure E.2 Sinc function approximation of a delta function.

E.1 Properties of the Delta Function

There are several properties of the delta function that makes it useful in the context of vibration engineering. They are as follows:

1. The delta function is an even function, so that $\delta(t) = \delta(-t)$.
2. A time-shifted delta function located at $t = a$ is given by $\delta(t - a)$.
3. The delta function can 'sift out' the value of a function at the value at which the delta function is located. This is given by the integral of the product of the function and the delta function so that

$$x(a) = \int_{-\infty}^{\infty} x(t)\delta(t - a)\mathrm{d}t. \tag{E.4}$$

4. The sum of all frequency components in the spectrum of white noise with unit amplitude is equal to a delta function in the time domain. This can be shown by starting with the relationship

$$\int_{-\infty}^{\infty} e^{\pm j2\pi ft}\mathrm{d}f = \lim_{f_{\max} \to \infty} \int_{-f_{\max}}^{f_{\max}} \cos 2\pi ft \mathrm{d}f, \tag{E.5}$$

which holds because $e^{\pm j2\pi ft} = \cos 2\pi ft \pm j\sin 2\pi ft$ and cos and sin are even and odd functions, respectively. Thus $\lim_{f_{\max} \to \infty} \int_{-f_{\max}}^{f_{\max}} \sin 2\pi ft \mathrm{d}f \to 0$. Evaluating the integral in Eq. (E.5) results in $\lim_{f_{\max} \to \infty} \int_{-f_{\max}}^{f_{\max}} \cos 2\pi ft \mathrm{d}f = \lim_{f_{\max} \to \infty} (2f_{\max} \times \mathrm{sinc}(2f_{\max} t))$, which has the same form as Eq. (E.2), therefore $\int_{-\infty}^{\infty} e^{\pm j2\pi ft}\mathrm{d}f = \delta(t)$. Also, note that $\int_{-\infty}^{\infty} e^{\pm j2\pi ft}\mathrm{d}t = \delta(f)$, which means that if a time domain signal has a value of unity for all time, then this is equal to a delta function in the frequency domain. These properties are illustrated in Figure E.3.

5. The scaling property of the delta function is such that $\delta(at) = \delta(t)/|a|$ for $a \neq 0$. This can be seen by starting with the integral in Eq. (E.1) but replacing $\delta(t)$ with $\delta(at)$. The integral can be rewritten using a change in variable $\bar{t} = at$ (for $a > 0$) so that $\mathrm{d}t = (1/a)\mathrm{d}\bar{t}$ and

$$\int_{-\infty}^{\infty} \delta(at)\mathrm{d}t = \frac{1}{a}\int_{-\infty}^{\infty} \delta(\bar{t})\mathrm{d}\bar{t} = \frac{1}{a}. \tag{E.6a}$$

If $a < 0$, then the limits of integration change so that

$$\int_{-\infty}^{\infty} \delta(at)\mathrm{d}t = \frac{1}{a}\int_{\infty}^{-\infty} \delta(\bar{t})\mathrm{d}\bar{t} = \frac{1}{-a}\int_{-\infty}^{\infty} \delta(\bar{t})\mathrm{d}\bar{t} = \frac{1}{-a}. \tag{E.6b}$$

Therefore $\delta(at) = \delta(t)/|a|$.

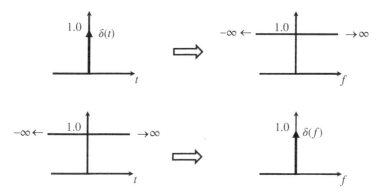

Figure E.3   Relationship between a delta function in one domain and its equivalent quantity in the other domain.

E.2   Fourier Series Representation of a Train of Delta Functions

A train of delta functions is shown in Figure E.4, which is described by

$$i(t) = \sum_{n=-\infty}^{\infty} \delta(t - n\Delta t),$$ (E.7)

Following Eq. (3.9), this has a Fourier series representation of

$$i(t) = \sum_{n=-\infty}^{\infty} I_n e^{j2\pi f_n t},$$ (E.8)

where $f_n = n/\Delta t$ and $I_n = \frac{1}{\Delta t}\int_{-\Delta t/2}^{\Delta t/2} i(t)e^{-j2\pi nt/\Delta t}dt$. In the interval $-\Delta t/2 \leq t \leq \Delta t/2$, $i(t) = \delta(t)$, so using the sifting property of the delta function given above, $I_n = 1/\Delta t$, which means that the train of delta functions given by Eq. (E.7) can be written in terms of its Fourier series as

$$i(t) = \frac{1}{\Delta t} \sum_{n=-\infty}^{\infty} e^{j2\pi nt/\Delta t}.$$ (E.9)

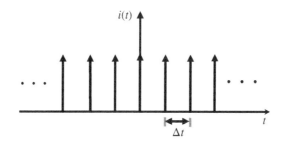

Figure E.4   A train of delta functions.

Reference

Dirac, P. (1930). *The Principles of Quantum Mechanics*, 1st Edition. Oxford University Press.

Appendix F

Aliasing

When a continuous time series is sampled and the discrete Fourier transform (DFT) of the sampled time history is calculated, the changes made by sampling can have profound effects on the resulting spectrum. One of these effects occurs because of under sampling, which means that high-frequency components of the original time series are not accurately captured and manifest themselves at low frequencies. This is called *aliasing* and is a result of the Nyquist–Shannon sampling theorem, which serves as a bridge between continuous-time signals and discrete-time signals. It establishes a sufficient condition for a sample rate such that a discrete sequence of samples can capture all the information from a continuous-time signal of finite bandwidth. The theorem is named after Harry Nyquist and Claude Shannon. In 1924 Harry Nyquist determined that the number of independent pulses that could be transmitted by a telegraph channel per unit time is limited to twice the bandwidth of the channel (Nyquist, 1924), and in 1948, Claude Shannon articulated his sampling theorem, which states that a signal can be completely determined if it is sampled at a frequency which is at least twice the bandwidth of the signal (Shannon, 1948).

In vibration engineering, aliasing is most often associated with measurements, where it is important to ensure that aliasing does not occur by setting the sampling frequency to be at least twice that of the highest frequency in the measured signals. It is, however, equally important to be aware that aliasing can occur when carrying out numerical simulations of vibrating systems using a computer in which sampled time histories are processed. In this case aliasing *always* occurs even when differential equations are solved by numerical integration as described in Appendix D.

Aliasing can be illustrated by considering the rotating vector shown in Figure F.1, which has an arbitrary amplitude and rotates at an angular velocity of ω rad/s. The rotating vector is sampled every Δt seconds, i.e. the sampling frequency is $f_s = 1/\Delta t$ Hz (or the angular sampling frequency $\omega_s = 2\pi f_s$ rad/s). If the angular sampling frequency is much larger than the angular velocity, as illustrated in the left-hand part of Figure F.1, then there is no ambiguity in the estimated frequency and the direction of rotation of the vector – all is well. However, if the sampling frequency is reduced so that the rotating vector is only sampled twice per revolution as shown in the central part of Figure F.1, then it would not be possible to determine the direction of rotation. This frequency is when $\omega_s = 2\omega$ (or $f_s = 2f$) and is called the Nyquist frequency, after Harry Nyquist. If the sampling frequency is less than the Nyquist frequency, then aliasing occurs. This is illustrated in the right-hand part of Figure F.1, which shows the case when $\omega_s/2 < \omega < \omega_s$. It can be seen that the rotating vector appears to rotate in the opposite direction to the actual rotation, and the apparent angular velocity of the rotating vector is much slower. Generally, the rotating vector $|\overline{X}|e^{\pm j\omega t}$ is indistinguishable from $|\overline{X}|e^{\pm j(\omega+n\omega_s)t}$, where n is a positive integer.

Virtual Experiments in Mechanical Vibrations: Structural Dynamics and Signal Processing,
First Edition. Michael J. Brennan and Bin Tang.
© 2023 John Wiley & Sons Ltd. Published 2023 by John Wiley & Sons Ltd.
Companion website: www.wiley.com/go/brennan/virtualexperimentsinmechanicalvibrations

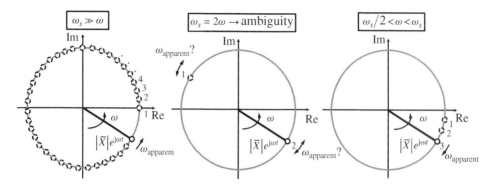

Figure F.1 Illustration of the sampling (and aliasing) of a rotating vector.

MATLAB Example F.1

In this example, the concept of aliasing is illustrated using an animated rotating vector.

```
clear all

N=100;                                  % No. of samples per cycle
                                        (choose 100, 2, 1.1, 0.9)
df=2*pi/N;                % [rad/s]     % angular resolution
for phi=0:df:20*pi        % [rad]       % angle
 X=exp(j*phi);                          % rotating vector
 x=[0 real(X)]; y=[0 imag(X)];          % x and y co-ordinates
 pause (1/(N))                          % creates a pause between plots

 th=0:0.01:2*pi;          % [rad]       % create the circle
 Z=exp(j*th);
 zx=real(Z);zy=imag(Z);

%% Plot the results
 plot(zx,zy,'linewidth',2,'Color',[.7 .7 .7])
 hold on
 plot(x,y,'k','linewidth',4)
 hold on
 axis([-1.2,1.2,-1.2,1.2])
 axis square
 hold off
end
```

(Continued)

MATLAB Example F.1   (Continued)

Results

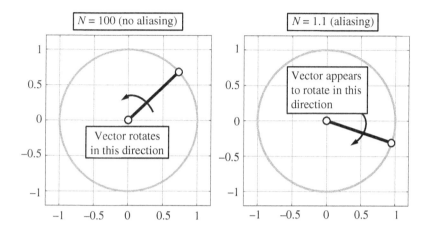

Comments

1. An exercise for the reader is to explore what happens to the rotating vector when N changes. The two examples shown are for $N = 100$ samples per cycle, when no aliasing occurs, and $N = 1.1$ when aliasing occurs. Also try changing the sign of the exponent so that the vector rotates in the opposite direction.

An example of aliasing on the FRF calculated using the DFT is further studied using the receptance function shown in Figure 3.11, and MATLAB Example 3.1. This involves the DFT of the displacement IRF given by $h(t) = 1/(m\omega_d)e^{-\zeta\omega_n t}\sin(\omega_d t)$ for $t \geq 0$. The theoretical receptance FRF determined using the FT is given by Eq. (3.15b), i.e. $H(j\omega) = 1/(k - \omega^2 m + j\omega c)$, for $-\infty < \omega < \infty$. Note that the receptance has frequency components for all positive and negative frequencies, i.e. its frequency content is of infinite extent. However, because of sampling in the time domain, and the subsequent DFT, the computed FRF is only valid in the frequency range $0 < \omega < \omega_s/2$ ($\omega_s = 2\pi f_s$). Thus, there are aliased components for all frequencies greater than $f_s/2$ and less than $-f_s/2$. This is illustrated in Figure F.2, which shows the DFT of the displacement IRF, and the theoretical receptance FRF. It is clear that at frequency f_a, which is slightly less than $f_s/2$, the DFT of the IRF shown by the solid black circle contains tangible aliased components, because its value is greater than the theoretical value of $|H(f_a)|$. However, because the amplitude of the FRF reduces with the square of frequency, the aliased high-frequency components become vanishingly small at very high frequencies. At frequency f_b, which is much less than $f_s/2$, aliasing still occurs, but has much less effect, because the amplitude of the aliases is much less than the true value of $|H(f_b)|$.

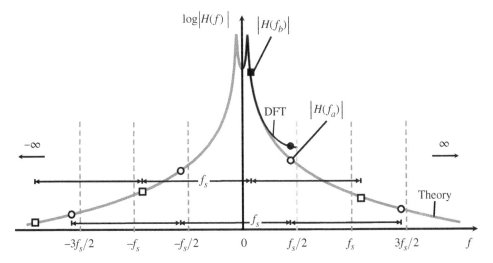

Figure F.2 Illustration of the effect of aliasing on the receptance FRF of an SDOF system. The DFT of the displacement IRF at frequencies f_a and f_b shown by the black circle and black square consists of the values of $|H(f_a)|$ and $|H(f_b)|$, respectively, and infinite sums of all their aliases at positive and negative frequencies, shown by the hollow circles and squares, respectively.

The aliased FRF for positive and negative frequencies is given by

$$H_+^{(\text{alias})}(f) = \sum_{n=0}^{\infty} H(f + nf_s) \quad \text{for } 0 \le f \le f_s \tag{F.1}$$

$$H_-^{(\text{alias})}(f) = \sum_{n=0}^{\infty} H(f - nf_s) \quad \text{for } -f_s \le f \le 0 \tag{F.2}$$

which are plotted in the top graph Figure F.3. As noted previously $\left(H_+^{(\text{alias})}(f)\right)^* = H_-^{(\text{alias})}(f)$ because the modulus is an even function and the phase is an odd function. Also plotted in this graph is the FT of the displacement IRF from zero frequency to $\pm f_s/2$. The aliasing effect is clear at frequencies close to $\pm f_s/2$. As shown in Figure F.1 and in MATLAB Example F.1, rotating vectors in the frequency range $-f_s \le f \le -f_s/2$ appear as rotating vectors in the frequency range $0 \le f \le f_s/2$. This also happens to vectors rotating in the other direction. Thus, the complete aliased FRF can be calculated by combining the aliased components with the corresponding actual components in the manner shown in the top graph of Figure F.3. The result is shown in the centre graph in Figure F.3, in which the combined aliased FRF for positive and negative frequencies are $\overline{H}_+^{(\text{alias})}(f)$ and $\overline{H}_-^{(\text{alias})}(f)$, respectively.

As discussed in Chapter 3, the FRF of a sampled time history is calculated using the DFT. The FRF shown can be compared directly with the FRF calculated using the DFT, by shifting the negative frequency components to the frequency range $f_s/2 < f \le f_s$ as shown in the bottom graph in Figure F.3. Of course, the FRF calculated using the DFT must be scaled by dividing by the sampling frequency for this comparison.

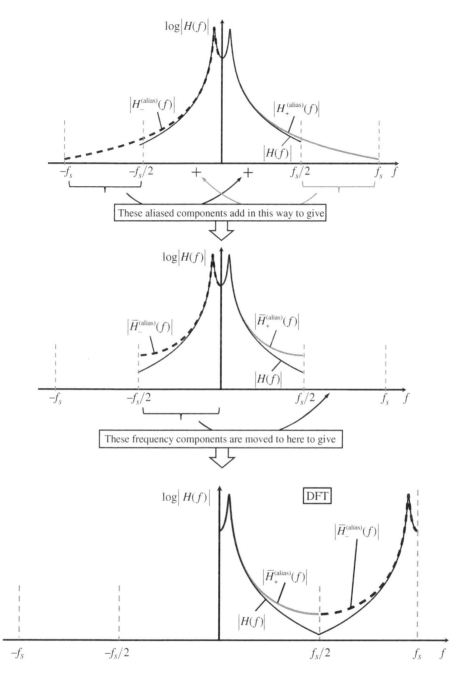

Figure F.3 Figure showing how the theoretical receptance FRF and the infinite sum of its aliases at positive and negative frequencies given by $H_+^{(\text{alias})}(f)$ and $H_-^{(\text{alias})}(f)$, respectively, are related to the DFT of the displacement IRF.

MATLAB Example F.2

In this example, the effects due to aliasing on the receptance FRF of an SDOF system determined from the theoretical model and that calculated using the DFT of the displacement IRF
are compared.

```
clear all

%% parameters
m = 10;                          % [kg]      % mass
k = 1000;                        % [N/m]     % stiffness
z = 0.5;                                     % damping ratio
c = 2*z*sqrt(m*k);               % [Ns/m]    % damping coefficient
wn=sqrt(k/m); wd=sqrt(1-z^2)*wn; % [rad/s]   % undamped and damped natural
                                             frequencies

%% Time and frequency parameters
T=10;                            % [s]       % time window
fs=10;                           % [Hz]      % sampling frequency
dt=1/fs; t=0:dt:T;               % [s]       % time resolution and time vector
N=length(t);                                 % Number of points
df=1/T; f=0:df:fs;               % [Hz]      % frequency resolution and frequency
                                             vector

%% IRF
h=1/(m*wd)*exp(-z*wn*t).*sin(wd*t); % [m/Ns]  % impulse response function

%% Calculation of DFT
H=dt*fft(h);

%% Calculation of aliased FRF
for p=1:1000                              % calculate the frequency range that
 f1=(p-1)*fs:df:p*fs;                     is included in the sum of aliases
 w1=2*pi*f1;                              for the theoretical FRF.
 HP(p,:)=1./(k-w1.^2*m+j*w1*c);           % aliased FRF for +ve freq.
 f2=-p*fs:df:-(p-1)*fs;
 w2=2*pi*f2;
 HM(p,:)=1./(k-w2.^2*m+j*w2*c);           % aliased FRF for -ve freq.
end

MP=sum(HP); MS=(sum(HM));                 % summing all aliases
HT1=MP+MS;                                % total aliased FRF
HT=HT1(1:(N+1)/2);
HA=[HT fliplr(conj(HT))];                 % double sided spectrum
HA1=HA(1:length(f));

%% Plot the results
figure
plot(f,abs(HA1))                          % modulus
hold on
plot(f,abs(H),'o')
xlabel('frequency (Hz)');
ylabel('|receptance| (m/N)');

figure
plot(f,180/pi*angle(HA1))                 % phase
hold on
plot(f,180/pi*angle(H),'o')
xlabel('frequency (Hz)');
ylabel('phase (degrees)');
```

(Continued)

MATLAB Example F.2   (Continued)

Results

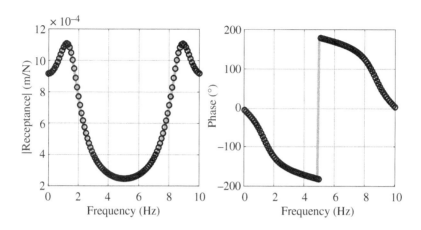

Comments

1. An exercise for the reader is to explore what happens to the FRFs when the sampling frequency and/or the frequency resolution is changed.
2. An exercise for the reader is to investigate the number of aliased frequency components needed to be included in the theoretical FRF to give a good approximation to that calculated using the DFT.

References

Nyquist, H. (1924). Certain factors affecting telegraph speed. *The Bell System Technical Journal*, 3(2), 324–346. https://doi.org/10.1002/j.1538-7305.1924.tb01361.x.

Shannon, C.E., (1948). A mathematical theory of communication. *The Bell System Technical Journal*, 27(3), 379–423. https://doi.org/10.1002/j.1538-7305.1948.tb01338.x.

Appendix G

Convolution

In the literature, convolution is generally defined as the operation between the time domain input $f(t)$ of a linear, time-invariant (LTI) system, and its impulse response function (IRF) $h(t)$ to give the time domain output $x(t)$. It involves an integral and was first articulated by Leonhard Euler (1768, pp. 230–255). However, it was not called convolution until many years later, as discussed by Domínguez (2015). In 1832 Jean-Marie Duhamel (1834) derived a similar expression to that derived by Euler, but to calculate the response of a vibrating system to an arbitrary force. For this reason, the convolution integral is often called the Duhamel integral in the vibration literature. For a vibrating system such as that shown in Figure G.1, convolution of the impulse response function and a force time history is written down mathematically as

$$x(t) = h(t) * f_e(t), \tag{G.1}$$

where * denotes the convolution operation. It is shown in Chapters 2 and 3 that the relationship between the input and output in the frequency domain is given by

$$X(j\omega) = H(j\omega)F(j\omega), \tag{G.2}$$

where $X(j\omega) = \mathcal{F}\{x(t)\}$, $H(j\omega) = \mathcal{F}\{h(t)\}$, and $F(j\omega) = \mathcal{F}\{f_e(t)\}$, in which $\mathcal{F}$ denotes the Fourier transform. Comparing Eqs. (G.1) and (G.2) it can be seen that convolution in the time domain is equivalent to multiplication in the frequency domain, and this is shown later in this appendix. First, it is necessary to determine the mathematical details of convolution.

Consider the force time series $f_e(t)$ shown in the top part of Figure G.2. It can be considered as a series of impulses of very short time duration $\Delta\tau$. One of the impulses, shown as a shaded rectangle, occurs at time τ such that the force at this time instant is $f_e(\tau)$. In the middle and lower

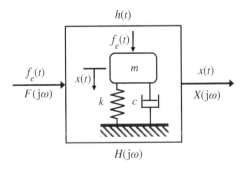

Figure G.1 A schematic diagram of an LTI vibrating system.

Virtual Experiments in Mechanical Vibrations: Structural Dynamics and Signal Processing,
First Edition. Michael J. Brennan and Bin Tang.
© 2023 John Wiley & Sons Ltd. Published 2023 by John Wiley & Sons Ltd.
Companion website: www.wiley.com/go/brennan/virtualexperimentsinmechanicalvibrations

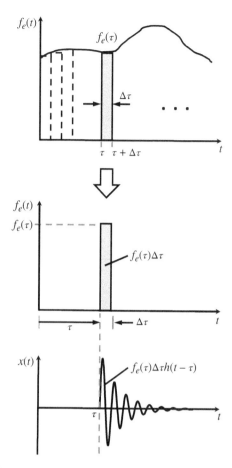

Figure G.2 Illustration of a force input described as a series of impulses, and the response of an SDOF system to one of the impulses (duration of $\Delta\tau$ is exaggerated for clarity).

parts of Figure G.2 are the impulse and the response of the system due to the impulse, respectively. The magnitude of the impulse is the area of the shaded rectangle, which is $f_e(\tau)\Delta\tau$. The response at time τ due to the impulse is the product of the magnitude of the impulse and the IRF, i.e. $f_e(\tau)\Delta\tau h(t-\tau)$. The total response at time t is the sum of all the responses, so that

$$x(t) \approx \sum f_e(\tau)h(t-\tau)\Delta\tau. \tag{G.3}$$

By setting $\Delta\tau \to 0$, the summation in Eq. (G.3) becomes

$$x(t) = \int_0^t f_e(\tau)h(t-\tau)d\tau. \tag{G.4a}$$

If the force is non-zero for $\tau < 0$, then the lower integration limit of 0 can extend to $-\infty$. Note the convolution integral in Eq. (G.4a) includes only impulses for $\tau < t$, and so it describes the operation for a causal system, i.e. the output is dependent only on past inputs. An example of the convolution of a force input to an SDOF system and the IRF of the system is shown in Figure G.3.

The convolution integral has the commutative property such that $h(t)*f_e(t) = f_e(t)*h(t)$ so that Eq. (G.4a) with a lower limit of integration of $-\infty$ can be written as

$$x(t) = \int_0^\infty f_e(t-\tau)h(\tau)d\tau. \tag{G.4b}$$

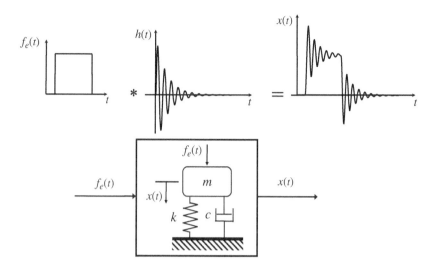

Figure G.3   An example of convolution for an SDOF system.

If the input is a unit impulse such that it can be represented by a delta function, then by using the sifting property of the delta function described in Appendix E shows that in this case $x(t) = h(t)$.

MATLAB Example G.1

In this example, the convolution of a force input and the IRF of an SDOF system is illustrated by way of an animation.

```
clear all

%% Parameters
m = 1;                          % [kg]          % see MATLAB example 3.1
k = 10000;                      % [N/m]
z = 0.1; c = 2*z*sqrt(m*k);     % [Ns/m]
wn=sqrt(k/m); wd=sqrt(1-z^2)*wn; % [rad/s]

%% Time and frequency parameters
fs=500;                         % [Hz]          % see MATLAB example 3.1
T=0.5; dt=1/fs; t=0:dt:T;       % [s]

%% IRF
ho=1/(m*wd)*exp(-z*wn*t).*sin(wd*t); % [m/Ns]   % displacement IRF
hn=ho/max(ho);                                  % normalised IRF
tmin = min(t)-abs(max(t)-min(t))-0.2;           % time range for plot
tmax = max(t)+abs(max(t)-min(t))+0.2;

%% Force input
N=1; N=200;                                     % use N=1 for a scaled delta func-
N1=round(length(t)/10); N2=length(t)-N1-N;      % tion, and N=200 for step changes in
                                                % force input

fo=[zeros(1,N1) ones(1,N) zeros(1,N2)];         % force input
```

(Continued)

MATLAB Example G.1   (Continued)

```
%% Convolution
c = dt*conv(ho,fo);

%% Animation
h = fliplr(hn);                             % flip the IRF
tf = fliplr(-t);
tf = tf + ( min(t)-max(tf) );               % time window

tc = [ tf t(2:end)];                        % time range of output
tc = tc+max(t);
set(figure,'Position', [40, 40, 1450, 700]);   % figure position
gr=[.6 .6 .6];                              % grey shade

%% Plot the results
ax = subplot(2,1,1);                        % first plot
p = plot(tf, h,'k','linewidth',2); hold on
q = plot(t, fo,'linewidth',2,'Color',gr);
axis([tmin,tmax,1.2*min(hn),1.2*max(hn)])
ym = get(ax, 'ylim');
xlabel('time (s)');
ylabel('normalised IRF and force');

sl=line([min(t) min(t)],[ym(1) ym(2)]);     % vertical lines for shaded region
el=line([min(t) min(t)], [ym(1) ym(2)]);    % shaded region
hold on; grid on;
sg = rectangle('Position', [min(t) ym(1) 0
ym(2)-ym(1)],'FaceColor', [.9 .9 .9]);

ax2=subplot(2,1,2);                         % second plot
r=plot(tc,c,'k','linewidth',2);
grid on; hold on;
s=plot(tc,c,'linewidth',2,'Color',gr);
grid on; hold on;
uistack(s,'bottom');
axis([tmin,tmax,1.2*min(c),1.2*max(c)])
xlabel('time (s)');
ylabel('displacement (m)');

%% Animation
for n=1:length(tc)
  pause(0.01);                              % controls animation speed
  tf=tf+dt;
  set(p,'XData',tf,'YData',h);
  sx=min(max(tf(1),min(t)), max(t));        % left-hand boundary of overlap
  sxa=[sx sx]; set(sl,'XData',sxa);         % right-hand boundary of overlap
  ex=min(tf(end),max(t));                   % shading of overlap region
  exa = [ex ex];set(el,'XData',exa);
  rpos=[sx ym(1) max(1e-6,ex-sx) ym(2)-ym(1)];
  set(sg,'Position',rpos);
  uistack(sg,'bottom');
  set(r,'XData',tc(1:n),'YData',c(1:n));    % plot of convolved function
end
```

(Continued)

MATLAB Example G.1   (Continued)

Results

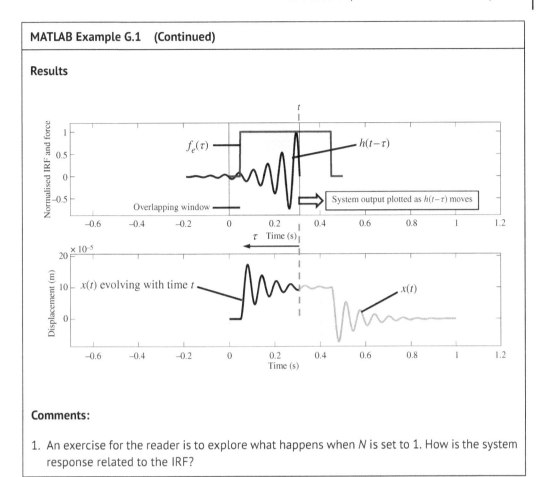

Comments:

1. An exercise for the reader is to explore what happens when N is set to 1. How is the system response related to the IRF?

G.1   Relationship Between Convolution and Multiplication

To determine the relationship between convolution in the time domain and multiplication in the frequency domain, both sides of Eq. (G.4b) are Fourier transformed to give

$$X(\mathrm{j}\omega) = \mathcal{F}\{x(t)\} = \int_{-\infty}^{\infty}\int_{0}^{\infty} f_e(t-\tau)h(\tau)e^{\mathrm{j}\omega t}\,\mathrm{d}\tau\,\mathrm{d}t. \tag{G.5}$$

Letting $t - \tau = u$, Eq. (G.5) can be written as

$$X(\mathrm{j}\omega) = \int_{-\infty}^{\infty}\int_{0}^{\infty} f_e(u)h(\tau)e^{\mathrm{j}\omega(u+\tau)}\,\mathrm{d}\tau\,\mathrm{d}u, \tag{G.6a}$$

which can be written in terms of the product of two integrals to give

$$X(\mathrm{j}\omega) = \int_{-\infty}^{\infty} f_e(u)e^{\mathrm{j}\omega u}\,\mathrm{d}u \int_{0}^{\infty} h(\tau)e^{\mathrm{j}\omega\tau}\,\mathrm{d}\tau. \tag{G.6b}$$

Now $F(j\omega) = \int_{-\infty}^{\infty} f_e(u)e^{j\omega u}du$ and $H(j\omega) = \int_0^{\infty} h(\tau)e^{j\omega\tau}d\tau$, which means that $X(j\omega) = F(j\omega)H(j\omega)$. This result shows that convolution in the time domain becomes multiplication in the frequency domain, i.e.

$$x(t) \ = \ f_e(t) \ * \ h(t)$$
$$\text{FT} \qquad \text{FT} \qquad \text{FT}$$
$$\downarrow \qquad \downarrow \qquad \downarrow$$
$$X(j\omega) \ = \ F(j\omega) \ \times \ H(j\omega)$$

There is also the relationship that multiplication in the time domain becomes convolution in the frequency domains, such that

$$x(t) \ = \ f_e(t) \ \times \ h(t)$$
$$\text{FT} \qquad \text{FT} \qquad \text{FT}$$
$$\downarrow \qquad \downarrow \qquad \downarrow$$
$$X(j\omega) \ = \ F(j\omega) \ * \ H(j\omega)$$

Example G.1   Consider the impulse response of an SDOF system given by

$$h(t) = \frac{1}{m\omega_d}e^{-\zeta\omega_n t}\sin(\omega_d t) \quad \text{for} \quad t \geq 0. \tag{G.7}$$

This can be written as the product of the exponential envelope $a(t) = u(t)e^{-\zeta\omega_n t}$ in which $u(t)$ is the Heaviside function, and the oscillatory component $b(t) = \frac{1}{m\omega_d}\sin(\omega_d t)$, so that

$$h(t) = a(t)b(t). \tag{G.8}$$

Calculating the FT of each component in Eq. (G.8) gives

$$H(j\omega) = A(j\omega) * B(j\omega), \tag{G.9}$$

in which it can be seen that the FT of the envelope is convolved with the FT of the oscillatory component to give the FRF of the system. For an SDOF system, there is analytical expression for the FRF of the envelope, which is

$$A(j\omega) = \frac{1}{\zeta\omega_n + j\omega}, \quad \text{for} \quad -\infty < \omega < \infty, \tag{G.10}$$

where $\omega_n = \sqrt{k/m}$ is the undamped natural frequency and $\zeta = c/(2m\omega_n)$ is the damping ratio. There is also an analytical expression for the FT of the oscillatory component, which is

$$B(j(\omega)) = \frac{1}{j2m\omega_d}2\pi[\delta(\omega - \omega_d) - \delta(\omega + \omega_d)], \quad \text{for} \quad -\infty < \omega < \infty, \tag{G.11}$$

where $\omega_d = \omega_n\sqrt{1 - \zeta^2}$ is the damped natural frequency. The time and frequency domain descriptions of the system are illustrated in Figure G.4. It can be seen that the FRF of the system is the convolution of the envelope with two frequency- and phase-shifted delta functions. The FRF can be determined analytically using the convolution in the frequency domain, by first noting that Eq. (G.9) can be written as

$$H(j\omega) = \frac{1}{2\pi}\int_{-\infty}^{\infty} A(jv)\, B\,(j(\omega - v))dv, \tag{G.12a}$$

where $A(jv) = \frac{1}{\zeta\omega_n + jv}$ and $B(j(\omega - v)) = \frac{\pi}{jm\omega_d}[\delta(\omega - \omega_d - v) - \delta(\omega + \omega_d - v)]$.

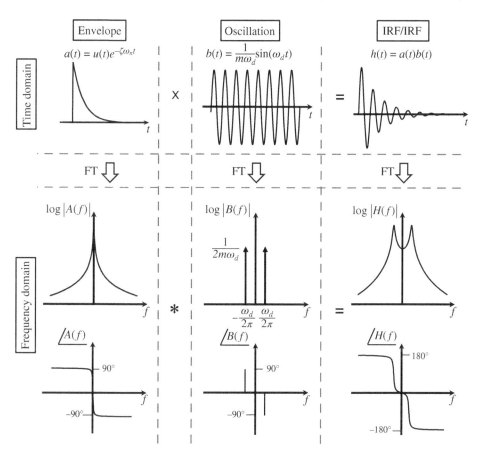

Figure G.4 Illustration of convolution in the frequency domain for an SDOF system.

Substituting for $A(jv)$ and $B(j(\omega - v))$ into Eq. (G.12a) gives

$$H(j\omega) = \frac{1}{j2m\omega_d} \int_{-\infty}^{\infty} \frac{1}{\zeta\omega_n + jv} [\delta(\omega - \omega_d - v) - \delta(\omega + \omega_d - v)]dv. \tag{G.12b}$$

Using the sifting properties of the delta function shown in Appendix E results in

$$H(j\omega) = \frac{1}{j2m\omega_d} \left(\frac{1}{\zeta\omega_n + j(\omega - \omega_d)} - \frac{1}{\zeta\omega_n + j(\omega + \omega_d)} \right), \quad \text{for } -\infty < \omega < \infty, \tag{G.13a}$$

which simplifies to

$$H(j\omega) = \frac{1}{m\left(\omega_n^2 - \omega^2 + j2\zeta\omega\omega_n\right)}, \quad \text{for } -\infty < \omega < \infty. \tag{G.13b}$$

It is interesting to compare the form of the FRF of the system given in Eq. (G.13a), with the FRF of the envelope given in Eq. (G.10). It can be seen that the FRF of the system is simply the weighted combination of two frequency-shifted FRFs of the envelope, $A(j\omega)$, where the frequency shift is $\pm\omega_d$, i.e.

$$H(j\omega) = \frac{1}{j2m\omega_d}[A(j(\omega - \omega_d)) - A(j(\omega + \omega_d))]. \tag{G.13c}$$

MATLAB Example G.2

In this example, the convolution of the FT of the envelope and the FT of the oscillatory term of the IRF of an SDOF system is illustrated by way of an animation.

```
clear all

%% Parameters
m = 1;                              % [kg]              % see MATLAB example 3.1
k = 10000;                          % [N/m]
z = 0.1; c = 2*z*sqrt(m*k);         % [Ns/m]
wn=sqrt(k/m); wd=sqrt(1-z^2)*wn;    % [rad/s]
fd=wd/(2*pi);                       % [Hz]

%% FRFs
df=0.4;f=-100:df:100;               % [Hz]              % frequency vector
w=2*pi*f;                           % [rad/s]
H=1./(z*wn+j*w);                    % [1/Hz]            % FRF of envelope
HH=1./(k-w.^2*m+j*w*c);             % [m/N]             % FRF of SDOF system
fr=-0.2*100:df:0.2*100;                                % frequency vector
W=zeros(length(fr),1);A=1/(2*j)*1/(m*wd);              % W is the FRF of the oscilla-
W((length(fr)+1)/2-round(fd/df))=1/df*A;               tory term. It has units of (m/N)
W((length(fr)+1)/2+round(fd/df))=-1/df*A;

%% Convolvution
C=conv(W,H)*df;

% Plots
fmin = 1.5*min(f); fmax = 1.5*max(f);                  % set freq. range for graph
Wf=fliplr(W);                                          % flip the envelope
ff=fliplr(-fr);
ff=ff+(min(f)-max(ff));                                % slide range of W
fc=[ff f(2:end)]; fc=fc+max(fr);                       % range of convolved function
set(figure,'Position', [40, 40, 1450, 700]);           % set the position of animation

subplot(2,1,1);                                        % plot of separate functions
HN=abs(H)/max(abs(H));WN=abs(W)/max(abs(W));           % normalized functions
p=plot(f,HN,'k','linewidth',4);hold on                 % normalized FT of envelope
gr=[.6 .6 .6];                                         % define grey colour
q=plot(fr,WN,'k','linewidth',4,'Color',gr);            % normalized W
axis([fmin,fmax,0,1.1])
xlabel('frequency (Hz)');
ylabel('normalised modulus');
sl=line([min(f) min(f)],[1.1 1.1],'color','k');        % vertical line for overlap
hold on; grid on;
sg=rectangle('Position', [min(f) 1 0 0], ...           % shaded region
   'FaceColor', [.9 .9 .9]);

subplot(2,1,2);                                        % plot of convolved values in dB
CdB=20*log10(abs(C));HHdB=20*log10(abs(HH));
r=plot(fc,CdB,'k','linewidth',2);hold on
p=plot(f,HHdB,'k','linewidth',2,'Color',gr);
hold on;grid on;
axis([fmin,fmax,-120,max(CdB)+10])
xlabel('frequency (Hz)');
ylabel('modulus (dB ref 1N/m)');
```

(Continued)

MATLAB Example G.2   (Continued)

```
%% Animation block
for n=1:length(fc)
  pause(0);                                 % controls animation speed
  ff=ff+df;
  set(q,'XData',ff,'YData',WN);
  sx=min(max(ff(1),min(f)),max(f));         % left-hand boundary of overlap
  sxa=[sx sx];
  set(sl,'XData',sxa);
  ex=min(ff(end),max(f));                   % right-hand boundary of overlap
  exa=[ex ex];
  set(sl,'XData',exa);
  rpos=[sx 0 max(0.0001,ex-sx) 1.1];        % shading of overlap region
  set(sg,'Position',rpos);
  uistack(sg,'bottom');

  set(p,'XData',f,'YData',HHdB);            % plot of FRF
  uistack(p,'bottom');
  set(r,'XData',fc(1:n),'YData',CdB(1:n));  % plot of convolved function
end
```

Results

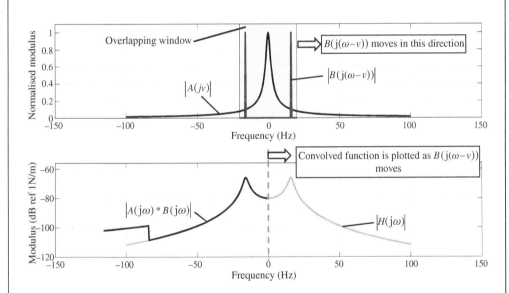

Comments:

1. An exercise for the reader is to explore what happens when mass, stiffness, and damping are changed.

G.2 Circular Convolution

Although the time response of a vibrating system can be calculated directly using convolution in the time domain as described by Eq. (G.1), it can also be calculated via the frequency domain by noting that $x(t) = \mathcal{F}^{-1}\{X(j\omega)\}$, where $X(j\omega) = F(j\omega)H(j\omega)$, in which $F(j\omega) = \mathcal{F}\{f_e(t)\}$ and $H(j\omega) = \mathcal{F}\{h(t)\}$. This process is illustrated in Figure G.5, but for discrete time data which have been sampled at a rate of $1/(\Delta t)$. In this case it is called *circular convolution* because sampling results in a periodic vibration response when it is calculated via the frequency domain. It is described mathematically by

$$x(n\Delta t) = f_e(n\Delta t) \circledast h(n\Delta t) \tag{G.14}$$

If the response is determined in the time domain by convolution, the duration of the force time history and the duration of the IRF may be different. However, if the response is calculated using circular convolution, then the time durations of the force time history and the IRF must be the same. The difference between convolution and circular convolution is illustrated using the example shown in Figure G.6. In this example the force involves a positive step change from zero to a DC level and after some time a step change returns the force to zero. The force time history has a duration of T seconds, as shown in Figure G.6a. This is convolved with the IRF shown in Figure G.6a, which also has a duration of T seconds. The result is the response $x(n\Delta t)$, which has a duration of $2T$ seconds, and is shown in Figure G.6a. Also shown in this figure is the response $x(n\Delta t)$ calculated using circular convolution. Note that the time duration of $x(n\Delta t)$ calculated in this way is only T seconds rather than $2T$ seconds when calculated using convolution. Any response of the system beyond T seconds cannot, of course, be captured using circular convolution, because the IDFT imposes a periodic structure on the IRF, as discussed in Chapter 3. With circular convolution, the response between T seconds and $2T$ seconds 'wrap around' and is added to the actual IRF as shown in Figure G.6a. To avoid wrap around, so that convolution and circular convolution give the same response for the first T seconds, it is necessary to double the time duration of the force input and the IRF by adding vectors of zeros for a time duration of T seconds. This effect is illustrated in Figure G.6b, where it can be seen that convolution and circular convolution now give the same

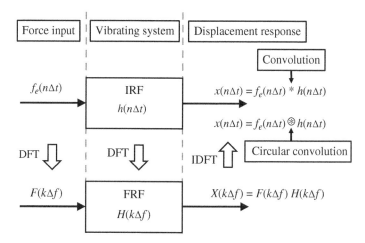

Figure G.5 Determination of the response of a vibrating system by convolution in the time domain, and by circular convolution via the frequency domain.

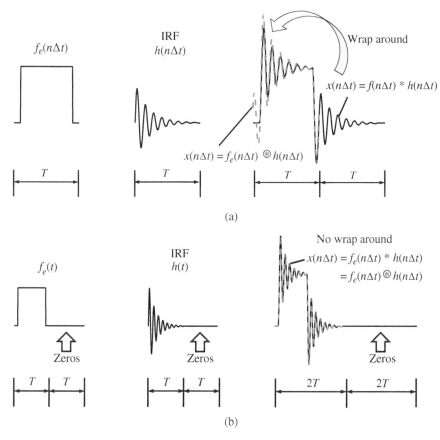

Figure G.6 Illustration of the difference between convolution and circular convolution and how to avoid wrap around. (a) Wrap around corners and (b) no wrap around corners.

output for the first $2T$ seconds, which is achieved by forcing the response to be zero between T seconds and $2T$ seconds.

MATLAB Example G.3

In this example, the response of an SDOF system calculated using convolution and circular convolution is illustrated.

```
clear all

%% Parameters
m = 1;                          % [kg]        % see MATLAB example 3.1
k = 10000;                      % [N/m]
z = 0.1; c = 2*z*sqrt(m*k);     % [Ns/m]
wn=sqrt(k/m); wd=sqrt(1-z^2)*wn; % [rad/s]
```

(Continued)

MATLAB Example G.3   (Continued)

```
%% Time and frequency parameters
T=0.5;                           % [s]         % T needs to be changed to
fs=500;                                        1 second to remove wrap around
dt=1/fs; t=0:dt:T;               % [s]

%% IRF
ho=1/(m*wd)*exp(-z*wn*t).*sin(wd*t); % [m/Ns]   % IRF
% ho((length(t)-1)/2:length(t))=0;             % this needs to be uncommented
                                                 when T=1
%% Force
N=1;N=200;
N1=25; N2=length(t)-N1-N;
fo=[zeros(1,N1) ones(1,N) zeros(1,N2)];        % time history of force

%% Convolution
y = dt*conv(ho,fo);                            % response by convolution
tt=0:dt:2*T;                                   % time vector for the response

%% Circular convolution
H=fft(ho)*dt;                                  % FRF
F=fft(fo)*dt;                                  % force spectrum
Y=H.*F;                                        % response spectrum
yc=ifft(Y)*fs;                                 % circular convolution

%% Plot the results
figure
plot(t,fo,'k','linewidth',2)                   % force
ylim([0,2])
axis square;grid;
xlabel('time (s)');
ylabel('force (N)');

figure
plot(t,ho,'k','linewidth',2)                   % displacement IRF
axis square;grid;
xlabel('time (s)');
ylabel('displacement IRF (m/Ns)');

figure
gr=[.6 .6 .6];
plot(tt,y,'k','linewidth',2);grid on; hold on; % response by convolution
plot(t,yc,'-k','linewidth',2,'color',gr);      % response by circular convolution
axis square;
xlabel('time (s)');
ylabel('displacement (m)');
```

(Continued)

MATLAB Example G.3	**(Continued)**

Results

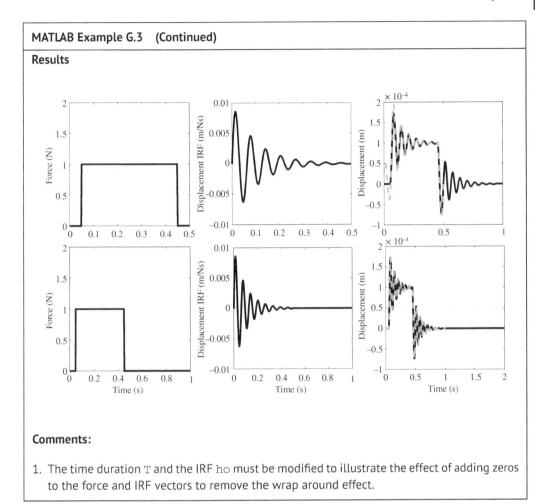

Comments:

1. The time duration T and the IRF ho must be modified to illustrate the effect of adding zeros to the force and IRF vectors to remove the wrap around effect.

References

Domínguez, A. (2015). A history of the convolution operation [Retrospectroscope], *IEEE Pulse*, 6(1), 38–49. https://doi.org/10.1109/MPUL.2014.2366903.

Duhamel, J.M.C. (1834). Sur les vibrations d'un système quelconque de points matériels, *Journal de l'École Polytechnique*, 14, 1–36.

Euler, L. (1768). *Institutiones Calculi Integralis*, vol. 2. Impensis Academiae Imperialis Scientiarum.

Appendix H

Some Influential Scientists in Topics Related to This Book

1550–1617

1635–1703

1643–1727

John Napier of Merchiston was a Scottish mathematician, physicist, and astronomer. He is best known as the discoverer of logarithms. Source: Wikimedia Commons/Public Domain.

Robert Hooke was an English polymath, active as a physicist, scientist, and architect. He is known for Hooke's law. Source: Mary Beale/Wikimedia Commons/Public Domain.

Sir Isaac Newton was an English mathematician, physicist, and astronomer. He is widely recognised as one of the greatest mathematicians and physicists. Source: Godfrey Kneller/Wikimedia Commons/Public Domain.

Virtual Experiments in Mechanical Vibrations: Structural Dynamics and Signal Processing,
First Edition. Michael J. Brennan and Bin Tang.
© 2023 John Wiley & Sons Ltd. Published 2023 by John Wiley & Sons Ltd.
Companion website: www.wiley.com/go/brennan/virtualexperimentsinmechanicalvibrations

1685–1731

Brook Taylor was an English mathematician. He is best known for creating the Taylor series. Source: Dr. Manuel/Wikimedia Commons/Public Domain.

1700–1782

Daniel Bernoulli was a Swiss mathematician and physicist. He is noted for his applications of mathematics to mechanics. Source: Wikimedia Commons/Public Domain.

1707–1783

Leonhard Euler was a Swiss mathematician and physicist. He is thought to be one of the greatest mathematicians in history. Source: Jakob Emanuel Handmann/Wikimedia Commons / Public Domain.

1755–1836

Marc-Antoine Parseval was a French mathematician, most famous for what is now known as Parseval's theorem. Source: Wikimedia Commons/Creative Commons CC0 License.

1768–1830

Jean-Baptiste Joseph Fourier was a French mathematician and physicist. He is best known for the Fourier series, which developed into Fourier analysis. Source: Louis-Léopold Boilly/Wikimedia Commons/Public Domain

1777–1855

Johann Carl Friedrich Gauss was a German mathematician and physicist who made significant contributions to many fields in mathematics and science. Source: Christian Albrecht Jensen/Wikimedia Commons/Public Domain.

1797–1872

1839–1921

1842–1919

Jean-Marie Constant Duhamel was a French mathematician and physicist. In vibrations he is known for the Duhamel integral.

Julius Ferdinand von Hann was an Austrian meteorologist. The Hanning window used in signal processing is named after him.

John William Strutt, 3rd Baron Rayleigh, was a British scientist who made extensive contributions to physics. He wrote the classic book *The Theory of Sound.*

1847–1922

1850–1925

1856–1927

Alexander Graham Bell was a Scottish scientist and engineer who invented the first practical telephone. The decibel (dB) was named in his honour.

Oliver Heaviside was an English mathematician and physicist. The unit step function was named as the Heaviside function after him.

Carl David Tolmé Runge was a German mathematician and physicist. He was co-developer of the Runge–Kutta method.

1857–1894

1862–1943

1867–1944

Heinrich Rudolf Hertz was a German physicist. The unit of frequency, cycle per second, or hertz (Hz), was named after him. Source: Robert Krewaldt/Wikimedia Commons/Public Domain.

David Hilbert was a very influential German mathematician. The Hilbert transform was named after him.

Martin Wilhelm Kutta was a German mathematician. He was co-developer of the Runge–Kutta method. Source: Wikimedia Commons/Public Domain.

1889–1976

1902–1984

1901–1989

Harry Nyquist was a Swedish physicist and engineer who made important contributions to theory in communications. The Nyquist frequency is named after him. Source: Harry Nyquist/Wikimedia Commons/American Institute of Physics.

Paul Adrien Maurice Dirac was an English theoretical physicist who was very influential in the twentieth century. The Dirac delta function was named after him.

Jacob Pieter Den Hartog was a Dutch–American mechanical engineer and academic. He was one of the great vibration engineers of the twentieth century.

1905–1982

1916–2001

Hendrik Wade Bode was an American engineer, researcher, and inventor. The frequency response function used in control theory is named the Bode plot after him.

Claude Elwood Shannon was an American mathematician and electrical engineer. He is known as 'the father of information theory'. Source: Jacobs, Konrad/Wikimedia Commons/CC BY-SA 2.0 DE.

Peter D. Welch is an American scientist in the area of computer simulation. He is known for Welch's method to reduce signal noise.

Index

Virtual Experiments in Mechanical Vibrations: Structural Dynamics and Signal Processing,
First Edition. Michael J. Brennan and Bin Tang.
© 2023 John Wiley & Sons Ltd. Published 2023 by John Wiley & Sons Ltd.
Companion website: www.wiley.com/go/brennan/virtualexperimentsinmechanicalvibrations